Django Web 开发从入门到实战

孟令菊 编著

清华大学出版社

北京

内 容 简 介

本书循序渐进地讲解了使用 Python 语言开发 Django Web 程序的知识，并通过具体实例的实现过程演示了使用 Django 框架的方法和流程。全书共 17 章，分别讲解了初步认识 Django，分析 Django 项目的结构，视图层，Django 数据库操作，使用模块，表单，站点配置和管理，站点的安全性，站点管理，系统优化、调试和部署，邮件发送模块，用户登录验证模块，计数器模块，Ajax 模块，分页模块，富文本编辑器模块，综合实战：民宿信息可视化。全书文字简洁而不失其技术深度，内容丰富全面，历史资料翔实齐全。并且本书易于阅读，以极简的文字介绍了复杂的案例，同时涵盖了其他同类图书中很少涉及的历史参考资料，是学习 Django Web 开发的完美教程。

本书适合已经了解了 Python 语言基础语法并且希望进一步提高 Python 开发水平的读者阅读，还可以作为大中专院校相关专业的师生用书和培训学校的专业性教材。

本书封面贴有清华大学出版社防伪标签，无标签者不得销售。
版权所有，侵权必究。举报：010-62782989，beiqinquan@tup.tsinghua.edu.cn。

图书在版编目(CIP)数据

Django Web 开发从入门到实战/孟令菊编著. —北京：清华大学出版社，2021.5
ISBN 978-7-302-57529-0

Ⅰ. ①D… Ⅱ. ① 孟… Ⅲ. ①软件工具—程序设计 Ⅳ. ①TP311.561

中国版本图书馆 CIP 数据核字(2021)第 025256 号

责任编辑：魏　莹
装帧设计：李　坤
责任校对：王明明
责任印制：沈　露

出版发行：清华大学出版社
网　　址：http://www.tup.com.cn, http://www.wqbook.com
地　　址：北京清华大学学研大厦 A 座　　邮　编：100084
社 总 机：010-62770175　　邮　购：010-62786544
投稿与读者服务：010-62776969, c-service@tup.tsinghua.edu.cn
质量反馈：010-62772015, zhiliang@tup.tsinghua.edu.cn

印 装 者：三河市金元印装有限公司
经　　销：全国新华书店
开　　本：185mm×260mm　　印　张：24.25　　字　数：590 千字
版　　次：2021 年 6 月第 1 版　　印　次：2021 年 6 月第 1 次印刷
定　　价：89.00 元

产品编号：087733-01

前言

从你开始学习编程的那一刻起，就注定了以后所要走的路：从编程学习者开始，依次经历实习生、程序员、软件工程师、架构师、CTO 等职位的磨砺；当你站在职位顶峰蓦然回首时，会发现自己的成功并不是偶然的，在程序员的成长之路会有不断修改代码、寻找并解决 Bug、不停测试程序和修改项目的经历；不可否认的是，只要你在自己的程序开发生涯中稳扎稳打，并且善于总结和学习，最终将会得到可喜的收获。

选择一本合适的书

对于一名程序开发者来说，究竟应该如何学习并提高自己的开发技术呢？答案就是买一本合适的程序开发书籍进行学习。但是，市面上许多编程书籍主要都是讲解基础知识，多偏向于理论，读者读了以后面对实战项目还是无从下手。

本书面向有一定 Python 基础的读者，传授使用 Python 语言开发 Django Web 程序的知识。本书主要讲解实现 Django Web 开发所必须具备的知识和技巧，这些知识能够帮助开发者迅速开发出需要的 Web 项目功能，提高开发效率。

本书的特色

1．内容全面

本书详细讲解 Django Web 开发所需要的开发技术，循序渐进地讲解了这些技术的使用方法和技巧，帮助读者快速步入 Python Web 开发的高手行列。

2．实例驱动教学

本书采用理论加实例的教学方式，通过这些实例实现了对知识点的横向切入和纵向比较，让读者有更多的实践演练机会，并且可以从不同的方位展现一个知识点的用法，真正实现了拔高的教学效果。

3．贴心提示和注意事项提醒

本书根据需要在各章安排了"注意""说明"和"技巧"等小板块，让读者可以在学习过程中更轻松地理解相关知识点及概念，更快地掌握相关技术的应用技巧。

本书的内容

本书循序渐进地讲解了使用 Python 语言开发 Django Web 程序的知识，并通过具体实例的实现过程演示了使用 Django 框架的方法和流程。全书共 17 章，分别讲解了初步认识

Django，分析 Django 项目的结构，视图层，Django 数据库操作，使用模块，表单，站点配置和管理，站点的安全性，站点管理，系统优化、调试和部署，邮件发送模块，用户登录验证模块，计数器模块，Ajax 模块，分页模块，富文本编辑器模块，综合实战：民宿信息可视化。全书文字简洁而不失其技术深度，内容丰富全面，历史资料翔实齐全。并且本书易于阅读，以极简的文字介绍了复杂的案例，同时涵盖了其他同类图书中很少涉及的历史参考资料，是学习 Django Web 开发的完美教程。

本书适用于已经了解了 Python 语言基础语法的读者，并且适用于希望进一步提高自己 Python 开发水平的读者，还可以作为大专院校相关专业的师生用书和培训学校的专业性教材。

本书的读者对象

- 软件工程师；
- Django 学习者和开发者；
- Python Web 学习者和开发者；
- 专业数据分析人员；
- 数据库工程师和管理员；
- 大学及中学教育工作者。

致谢

本书在编写过程中，得到了清华大学出版社编辑们的大力支持，正是各位编辑的求实、耐心和效率，才使得本书能够在较短的时间内出版。另外，也十分感谢我的家人给予的巨大支持。本人水平毕竟有限，书中纰漏之处在所难免，诚请读者提出宝贵的意见或建议，以便修订使之更臻完善。

最后感谢您购买本书，希望本书能成为您编程路上的领航者，祝您阅读快乐！

编 者

目录

第 1 章 初步认识 Django 1
1.1 Django Web 开发基础 2
1.1.1 Web 开发和 Web 框架介绍 2
1.1.2 Django 框架介绍 3
1.1.3 Django 框架的特点 4
1.2 Django 的设计模式 4
1.2.1 MVC 设计模式介绍 4
1.2.2 MTV 设计模式介绍 5
1.3 搭建 Django 开发环境 6
1.3.1 搭建 Python 环境 6
1.3.2 搭建 Django 环境 8
1.3.3 常用的 Django 命令 9
1.4 实现第一个 Django Web 项目 11
1.4.1 实战演练：使用 Django 命令创建 Django Web 项目 11
1.4.2 实战演练：使用 PyCharm 创建 Django Web 项目 14

第 2 章 分析 Django 项目的结构 19
2.1 实战演练：在线投票系统 20
2.2 编写第一个视图 20
2.3 实现数据库 22
2.3.1 配置数据库 22
2.3.2 创建数据库模型 23
2.3.3 启用模型 24
2.3.4 使用模型的 API 26
2.4 使用 admin 后台管理 Web 29
2.4.1 创建管理员用户 30
2.4.2 启动 Web 项目 30
2.4.3 进入 admin 站点 31
2.4.4 在 admin 中注册投票应用 31
2.4.5 体验便捷的管理功能 31
2.5 视图和模板 33
2.5.1 编写视图 33
2.5.2 编写一个真正有用的视图 34
2.5.3 快捷函数 render() 36
2.5.4 抛出 404 错误 37
2.5.5 使用模板系统 38
2.5.6 删除模板中硬编码的 URLs 38
2.5.7 URL names 的命名空间 38
2.6 编写一个简单的表单 39
2.7 用通用视图：减少重复代码 42
2.7.1 改良 URLconf 42
2.7.2 修改视图 43
2.8 静态文件 44
2.8.1 使用 CSS 自定义应用的风格 44
2.8.2 静态文件命名空间 45
2.8.3 添加一个背景图 45
2.9 重新设计后台 46
2.9.1 自定义后台表单 46
2.9.2 添加关联对象 47
2.9.3 定制实例的列表页面 51
2.9.4 定制 admin 整体界面 53
2.9.5 定制 admin 首页 54

第 3 章 视图层 55
3.1 视图层介绍 56
3.1.1 分析 View 视图的作用 56
3.1.2 实战演练：使用简易 View 视图文件实例 57
3.2 URL 调度器 59
3.2.1 URL 调度器介绍 59
3.2.2 Django URL 调度器的工作原理 61
3.2.3 路径转换器 61
3.2.4 URLconf 匹配 URL 65
3.2.5 设置视图参数的默认值 65

3.2.6 自定义错误页面 66
3.2.7 实战演练：使用 Django 框架
实现 URL 参数相加 67
3.3 编写 View 视图 69
3.3.1 一个简单的视图 70
3.3.2 返回错误信息 70
3.3.3 实战演练：在线文件上传
系统 72
3.4 异步视图 76
3.4.1 异步视图介绍 76
3.4.2 异步中间件 77
3.4.3 实战演练：使用异步视图展示
两种货币的交易数据 78

第 4 章 Django 数据库操作 83

4.1 Model 模型 84
4.1.1 Model 模型基础 84
4.1.2 META 内部类 86
4.1.3 实战演练：在 Django 框架中
创建 SQLite3 数据库 88
4.2 使用 QuerySet API 89
4.2.1 QuerySet API 基础 90
4.2.2 生成新的 QuerySet 对象的
方法 91
4.2.3 不返回 QuerySet 的方法 96
4.2.4 字段查找 99
4.2.5 实战演练：使用 QuerySet API
操作 SQLite 数据库 100
4.3 实战演练：使用 QuerySet API 操作
MySQL 数据库 105

第 5 章 使用模板 111

5.1 模板基础 112
5.1.1 配置引擎 112
5.1.2 Django 模板的基础用法 113
5.1.3 实战演练：使用简易模板 114
5.2 模板标签 Tags 115

5.2.1 常用的模板标签 115
5.2.2 实战演练：在模板中使用 for
循环显示列表内容 120
5.3 模板过滤器 Filter 121
5.3.1 常用的内置过滤器 121
5.3.2 国际化标签和过滤器 130
5.3.3 其他标签和过滤器库 131
5.3.4 实战演练：使用过滤器提取
列表和字典中的内容 131
5.4 模板继承 133
5.4.1 模板继承介绍 133
5.4.2 实战演练：使用模板继承 135
5.5 自定义模板标签和过滤器 137
5.5.1 基本方法 137
5.5.2 自定义模板过滤器 138
5.5.3 自定义模板标签 139
5.5.4 实战演练：创建自定义模板
过滤器 142

第 6 章 表单 145

6.1 表单介绍 146
6.1.1 HTML 表单介绍 146
6.1.2 Django 中的表单 147
6.2 使用表单 148
6.2.1 使用表单类 Form 的方法 148
6.2.2 实战演练：第一个表单
程序 151
6.3 表单的典型应用 152
6.3.1 表单 forms 的设计与使用 152
6.3.2 实战演练：简易用户登录验证
系统 160
6.3.3 实战演练：文件上传系统 163

第 7 章 站点配置和管理 169

7.1 系统配置文件 170
7.1.1 配置文件的特性 170
7.1.2 基本配置 170

7.2 静态文件 172
    7.2.1 静态文件介绍 173
    7.2.2 实战演练：在登录表单中使用静态文件 174
7.3 Django Admin 管理 176
    7.3.1 Django Admin 基础 176
    7.3.2 实战演练：使用 Django Admin 系统 179

第8章 站点的安全性 183

8.1 Django 安全概述 184
    8.1.1 跨站脚本(XSS)防护 184
    8.1.2 跨站请求伪造(CSRF)防护 184
    8.1.3 SQL 注入保护 185
    8.1.4 点击劫持保护 185
    8.1.5 SSL/HTTPS 185
    8.1.6 Host 协议头验证 186
8.2 使用 Cookie 和 Session 186
    8.2.1 Django 框架中的 Cookie 186
    8.2.2 Django 框架中的 Session 190
8.3 点击劫持保护 194
    8.3.1 点击劫持的例子 194
    8.3.2 使用 X-Frame-Options 195
8.4 跨站请求伪造保护 196
    8.4.1 在 Django 中使用 CSRF 防护的方法 196
    8.4.2 装饰器方法 197
    8.4.3 实战演练：求和计时器 198
    8.4.4 实战演练：每日任务管理器 200
8.5 加密签名 203
8.6 中间件 206
8.7 实战演练：安全版的仿 CSDN 登录验证系统 208
    8.7.1 系统设置 208
    8.7.2 会员注册和登录验证模块 209
    8.7.3 博客发布模块 214

第9章 站点管理 219

9.1 Django Web 国际化 220
    9.1.1 Django 中 Python 程序的国际化 220
    9.1.2 Django 中模板的国际化 221
    9.1.3 Django 中 URL 模式的国际化 224
9.2 Django Web 本地化 227
    9.2.1 Message File(消息文件) 227
    9.2.2 编译消息文件 228
    9.2.3 本地格式化 228
9.3 国际化和本地化的应用 230
    9.3.1 实战演练：展示法语环境 230
    9.3.2 实战演练：创建多语言环境 233
9.4 网站地图 sitemap 236
    9.4.1 安装 sitemap 236
    9.4.2 sitemap 的初始化 236
    9.4.3 类 Sitemap 的成员 237
    9.4.4 快捷类 GenericSitemap 238
    9.4.5 静态视图的 Sitemap 239
    9.4.6 创建网站地图索引 240
    9.4.7 模板定制 240
    9.4.8 实战演练：在 Django 博客系统中创建网站地图 241

第10章 系统优化、调试和部署 245

10.1 Django 性能与优化 246
    10.1.1 什么是优化？ 246
    10.1.2 Django 中的性能优化技术 246
    10.1.3 实战演练：在 Django 博客系统中添加 django-debug-toolbar 面板 249
10.2 Django 缓存处理 254
    10.2.1 缓存的思路 254
    10.2.2 设置缓存 255
    10.2.3 站点级缓存 260

10.2.4 缓存单个 view 视图..............261
   10.2.5 在 URLconf 中指定视图
          缓存.................................262
   10.2.6 模板片段缓存.....................262
   10.2.7 实战演练：在上传系统中
          使用 Redis 缓存..................263
 10.3 日志系统..268
   10.3.1 在 Django 视图中使用
          logging 268
   10.3.2 在 Django 中配置 logging.....269
   10.3.3 自定义 logging 配置和
          禁用 logging 配置................272
   10.3.4 Django 对 logging 模块的
          扩展....................................272
   10.3.5 实战演练：在日志中记录
          用户的访问操作...................273

第 11 章 邮件发送模块..............................277

 11.1 实战演练：使用 smtplib 发送
      邮件...278
 11.2 使用 django.core.mail 发送邮件........280
   11.2.1 django.core.mail 基础............280
   11.2.2 实战演练：使用 django.core.mail
          实现一个邮件发送程序........282
 11.3 实战演练：使用邮箱发送验证码的
      用户注册、登录验证系统.................284

第 12 章 用户登录验证模块......................291

 12.1 使用 auth 实现登录验证系统...........292
   12.1.1 auth 模块基础......................292
   12.1.2 实战演练：带登录验证功能的
          简易新闻系统......................297
 12.2 使用 django-allauth 实现登录验证
      系统...301
   12.2.1 django-allauth 框架基础........301
   12.2.2 实战演练：在 django-allauth
          中使用百度账户实现用户
          登录系统.............................304

第 13 章 计数器模块..................................311

 13.1 实战演练：一个简单的网页
      计数器...312
 13.2 实战演练：使用数据库保存统计
      数据...313
   13.2.1 创建 Django 工程.................313
   13.2.2 实现数据库.........................313
   13.2.3 配置 URL............................314
   13.2.4 实现视图.............................315
   13.2.5 实现模板.............................315
   13.2.6 调试运行.............................316
 13.3 实战演练：使用第三方库实现访问
      计数器...316
   13.3.1 准备工作.............................316
   13.3.2 配置 URL............................317
   13.3.3 实现数据库.........................317
   13.3.4 实现视图.............................318
   13.3.5 实现模板.............................319
   13.3.6 调试运行.............................321

第 14 章 Ajax 模块....................................323

 14.1 Ajax 技术的原理..............................324
 14.2 实战演练：无刷新计算器................325
 14.3 Ajax 上传和下载系统......................326
   14.3.1 实现文件上传功能..............326
   14.3.2 实现文件下载功能..............331

第 15 章 分页模块......................................335

 15.1 类 Paginator 和类 Page.....................336
   15.1.1 类 Paginator.........................336
   15.1.2 类 Page...............................337
   15.1.3 实战演练：实现简单的
          分页....................................337
 15.2 实战演练：自定义的美观的分页
      程序...339
 15.3 实战演练：使用分页显示网络
      信息...343

- 15.3.1 创建工程 344
- 15.3.2 设计视图 344
- 15.3.3 设计 URL 导航 344
- 15.3.4 实现模板文件 345

第 16 章 富文本编辑器模块 347

- 16.1 第三方库 django-mdeditor 348
  - 16.1.1 django-mdeditor 介绍 348
  - 16.1.2 实战演练：使用 django-mdeditor 实现富文本编辑器 348
- 16.2 第三方库 django-ckeditor 352
  - 16.2.1 django-ckeditor 介绍 352
  - 16.2.2 实战演练：在博客系统中使用 django-ckeditor 富文本编辑器 353

第 17 章 综合实战：民宿信息可视化 357

- 17.1 系统背景介绍 358
- 17.2 爬虫抓取信息 358
  - 17.2.1 系统配置 359
  - 17.2.2 Item 处理 359
  - 17.2.3 具体爬虫 360
  - 17.2.4 破解反扒字体加密 360
  - 17.2.5 下载器中间件 363
  - 17.2.6 保存爬虫信息 367
- 17.3 数据可视化 370
  - 17.3.1 数据库设计 370
  - 17.3.2 视图显示 373

第 1 章

初步认识 Django

　　Django 是一个开放源代码的 Web 应用框架，由 Python 语言写成。Django 遵守 BSD 版权，初次发布于 2005 年 7 月，并于 2008 年 9 月发布了第一个正式版本 1.0。Django 采用了 MVC 的软件设计模式，即模型 M、视图 V 和控制器 C。在本章的内容中，将详细讲解使用 Django 框架开发 Web 程序的基础知识。

1.1 Django Web 开发基础

Django 自称是"能够很好地应对应用上线期限的 Web 框架"。其最初在 21 世纪初发布，由 Lawrence Journal-World 报业的在线业务的 Web 开发者创建。2005 年正式发布，引入了以"新闻业的时间观开发应用"的方式。

扫码观看本节视频讲解

1.1.1 Web 开发和 Web 框架介绍

Django 是一款开发 Web 程序的 Python 框架，那么什么是 Web 开发呢？Web 开发指的是开发基于 B/S(浏览器/服务器)的架构，通过前后端的配合，将后台服务器的数据在浏览器上展现给前台用户的应用。比如将电子购物网站的商品数据在浏览器上展示给客户，在基于浏览器的学校系统管理平台上管理学生的数据，监控机房服务器的状态并将结果以图形化的形式展现出来等。

使用 Python 语言开发 Web 应用程序，最简单、原始和直接的办法是使用 CGI 标准，它是如何做的呢？ 以使用 Python CGI 脚本显示数据库中最新添加的 10 件商品为例，可以用下面的代码实现。

```
import pymysql

print("Content-Type: text/html\n")
print("<html><head><title>products</title></head>")
print("<body>")
print("<h1>products</h1>")
print("<ul>")

connection = pymysql.connect(user='连接数据库的用户名', passwd='连接数据库的密码', db='数据库的名字')
cursor = connection.cursor()
cursor.execute("SELECT name FROM products ORDER BY create_date DESC LIMIT 10")

for row in cursor.fetchall():
    print("<li>%s</li>" % row[0])

print("</ul>")
print("</body></html>")

connection.close()
```

上述 CGI 方案的运作过程是这样的：

(1) 客户端浏览器用户请求 CGI，脚本代码打印 Content-Type 行等一些 HTML 的起始标签。

(2) 连接数据库并执行一些查询操作，获取最新的 10 件商品的相关数据。在遍历这些商品的同时，生成一个商品的 HTML 列表项，然后输出 HTML 的结束标签并且关闭数据库连接。

(3) 将生成的 HTML 代码保存到一个.cgi 文件中，然后上传到网络服务器，用户通过浏览器即可访问。

上述流程在实际应用中会存在一些问题，例如：

- 如果在应用中有多处需要连接数据库会怎样呢？每个独立的 CGI 脚本，不应该重复编写数据库连接相关的代码。
- 前端、后端工程师以及数据库管理员集于一身，无法分工配合。
- 欠缺代码重用功能。
- 可扩展性一般，例如今天是取 10 个商品，明天我要删除 10 个商品怎么办？

以上问题是显而易见的，为了解决上述问题，简化 Web 开发的流程，很多聪明的开发者为自己发明了一个 Web 模块，这个模块可以实现常用的功能，并且能够解决上面的问题，这就是 Web 框架的最初模型。

Web 框架致力于解决一些共同的问题，为 Web 应用提供通用的架构模板，让开发者专注于网站应用业务逻辑的开发，而无须处理网络应用底层的协议、线程、进程等方面的问题。这样能大大提高开发者的效率和 Web 应用程序的质量。例如常用 Web 框架的架构如图 1-1 所示。

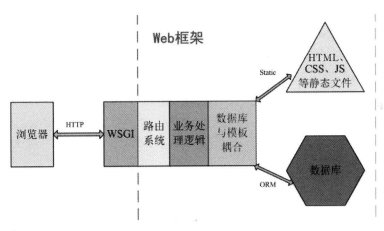

图 1-1　常用 Web 框架的架构

Django 是一个由 Python 语言编写的具有完整架站能力的开源 Web 框架。通过使用 Django，Python 开发者只需要编写很少的代码，就可以轻松地完成一个专业的企业级网站。

1.1.2　Django 框架介绍

Django 诞生于 2003 年，2006 年加入了 BSD 许可证，成为开源的 Web 框架。Django 这一词语是根据比利时的爵士音乐家 Django Reinhardt 命名的，有希望 Django 能够优雅地演奏(开发)各种乐曲(Web 应用)的美好含义。

Django 最初是由美国堪萨斯(Kansas)州 Lawrence 城中的一个新闻开发小组开发出来的。当时 Lawrence Journal-World 报纸的程序员 Adrian Holovaty 和 Simon Willison 用 Python 语

言编写 Web 新闻应用，他们的 World Online 小组制作并维护了当地的几个新闻站点。新闻界独有的特点是快速迭代，从开发到上线，通常只有几天或几个小时的时间。为了能在截止时间前完成工作，Adrian 和 Simon 打算开发一种通用的高效的网络应用开发框架，也就是 Django。

2005 年的夏天，当这个框架开发完成时，它已经用来制作了很多个 World Online 的站点。不久，小组中的 Jacob Kaplan-Moss 决定把这个框架发布为一个开源软件，于是短短数年，Django 项目就有了数以万计的用户和贡献者，在世界范围内广泛传播。 原来的 World Online 的两个开发者(Adrian 和 Jacob)仍然掌握着 Django，但是其发展方向受社区团队的影响更大。

1.1.3　Django 框架的特点

- 功能完善、要素齐全：Django 提供了大量的特性和工具，无须用户自己定义、组合、增删及修改。
- 完善的技术文档：经过十多年的发展和完善，Django 拥有广泛的实践经验和完善的在线文档，当开发者遇到问题时可以搜索在线文档寻求解决方案。
- 强大的数据库访问组件：Django 的 Model 层自带数据库 ORM 组件，使得开发者无须学习其他数据库访问技术(SQL、PyMySQL、SQLALchemy 等)。当然也可以不用 Django 自带的 ORM 组件，而是使用其他访问技术，比如 SQLALchemy。
- 灵活的 URL 映射：Django 使用正则表达式管理 URL 映射，灵活性高。
- 丰富的 Template 模板语言：类似 Jinjia 模板语言，不但原生功能丰富，还可以自定义模板标签。
- 自带免费的后台管理系统：只需要通过简单的几行配置和代码就可以实现一个完整的后台数据管理控制平台。
- 完整的错误信息提示：在开发调试过程中如果出现运行错误或者异常，Django 可以提供非常完整的错误信息帮助定位问题。

1.2　Django 的设计模式

在目前基于 Python 语言的几十个 Web 开发框架中，几乎所有的全栈框架都强制或引导开发者使用 MVC 设计模式。所谓全栈框架，是指除了封装网络和线程操作，还提供 HTTP、数据库读写管理、HTML 模板引擎等一系列功能的 Web 框架，比如 Django、Tornado 和 Flask。

扫码观看本节视频讲解

1.2.1　MVC 设计模式介绍

MVC 即 Model-View-Controller(模型-视图-控制器) 模式，是一种优秀的软件设计典范。

MVC 使用业务逻辑、数据、界面显示分离的方法组织代码，将业务逻辑聚集到一个部件里面，在改进和个性化定制界面及用户交互的同时，不需要重新编写业务逻辑。

- Model (模型)：简而言之即数据模型。模型不是数据本身(比如数据库里的数据)，而是抽象的描述数据的构成和逻辑关系。通常模型包括了数据表的各个字段(比如人的年龄和出生日期)和相互关系(单对单、单对多关系等)。数据库里的表会根据模型的定义来生成创建。
- View (视图)：主要用于显示数据，用来展示用户可以看到的内容或提供给用户可以输入或操作的界面。数据来源于哪里？当然是数据库了。那么用户输入的数据给谁？当然是给控制器了。
- Controller(控制器)：应用程序中处理用户交互的部分。通常控制器负责从视图读取数据，控制用户输入，并向模型发送数据(比如增加或更新数据表)。

MVC 最大的优点是实现了软件或网络应用开发过程中数据、业务逻辑和界面的分离，使软件开发更清晰，也使维护变得更容易。这与静态网页设计中使用 HTML 和 CSS 实现内容和样式的分离是同一个道理。

Django 框架借鉴了 MVC 设计模式，Django Web 项目的 4 大核心模块是：Model(模型)、URL(链接)、View(视图) 和 Template(模板)，这 4 大模块与 MVC 设计模式的对应关系如下。

- Django 中的 Model(模型)模块：此模块与 MVC 模式下的 Model 差不多。
- Django 中的 URL 和 View(视图)模块：这两个合起来与 MVC 模式下的 Controller 十分相似，可以联合使用 Django 中的 URL 和 View 向 Template(模板)传递正确的数据。另外，用户输入的数据也需要 Django 中的 View 来处理。
- Django 中的 Template(模板)：此模块与经典 MVC 模式下的 View 一致，Django 中的 Template(模板)用于显示 Django View 传来的数据，这决定了用户界面的显示外观。另外，在 Template 中也包含表单，可以用来搜集用户的输入。

1.2.2 MTV 设计模式介绍

Django 对传统的 MVC 设计模式进行了修改，将视图分成 View 模块和 Template 模块两部分，将动态的逻辑处理与静态的页面展现分离开。而 Model 采用了 ORM 技术，将关系型数据库表抽象成面向对象的 Python 类，将表操作转换成类操作，避免了复杂的 SQL 语句编写。MTV 和 MVC 本质上是一样的。

- 模型(Model)：和 MVC 中的定义一样。
- 模板(Template)：将数据与 HTML 语言结合起来的引擎。
- 视图(View)：负责实际的业务逻辑实现。

总而言之，Django 中的 MTV 设计模式是根据 MVC 模式演变过来的，使用 MVC 或 MTV 设计模式的最大好处是解耦合(降低程序间的依赖关系，改动项目中的一个程序文件不会影响项目中的其他文件)。

1.3 搭建 Django 开发环境

因为 Django 是由 Python 语言编写的 Web 框架，依赖于 Python 环境，所以在安装 Django 之前，必须先安装 Python。在本节的内容中，将详细讲解搭建 Django 开发环境的知识。

1.3.1 搭建 Python 环境

扫码观看本节视频讲解

Python 可以在 Windows、Linux 和 Mac 当今这三大主流的计算机系统中运行，接下来将讲解在 Windows 系统中下载并安装 Python 的过程。

（1）登录 Python 官方网站 https://www.python.org，单击顶部导航中的 Downloads | Windows 链接，如图 1-2 所示。

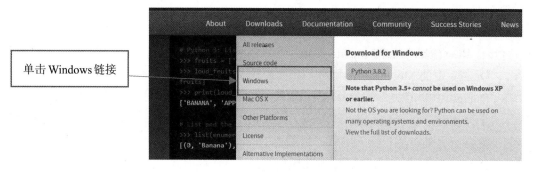

图 1-2 Python 下载页面

（2）此时会弹出如图 1-3 所示的 Windows 版下载界面，此时(作者写作本书时)最新的版本是 Python 3.8.2。

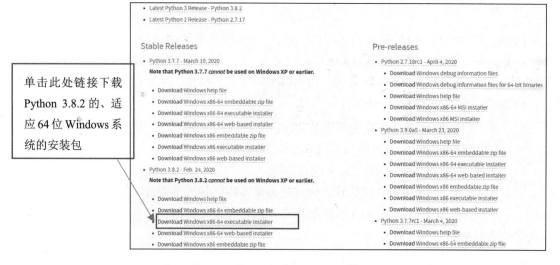

图 1-3 Windows 版 Python 下载界面

在图1-3所示的页面中列出的都是Windows系统平台的安装包，其中x86适合32位操作系统，x86-64适合64位操作系统。并且可以通过如下3种途径获取Python。

- web-based installer：需要通过联网完成安装。
- executable installer：通过可执行文件(*.exe)安装。
- embeddable zip file：这是嵌入式版本，可以集成到其他应用程序中。

（3）因为笔者的计算机是64位操作系统，所以需要选择一个64位的安装包，当前(笔者写稿时)最新版本为"Windows x86-64 executable installer"。弹出如图1-4所示的下载进度对话框。

（4）下载成功后得到一个".exe"格式的可执行文件，双击此文件开始安装。在第一个安装界面中勾选界面下方的两个复选框，然后单击Install Now链接，如图1-5所示。

图1-4　下载对话框　　　　　　　　　图1-5　第一个安装界面

注　意

勾选Add Python 3.8 to PATH复选框的目的是把Python的安装路径添加到系统路径下面。勾选这个选项后，以后在执行cmd命令时，输入python就会调用python.exe。如果不勾选这个选项，在cmd下输入python时会报错。

（5）弹出如图1-6所示的安装进度界面。

（6）安装完成后的界面效果如图1-7所示，单击Close按钮完成安装。

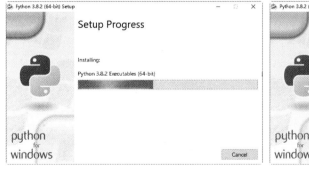

图1-6　安装进度界面　　　　　　　　图1-7　安装完成界面

（7）选择"开始"|"运行"命令，输入cmd后打开DOS命令界面，然后输入"python"

验证是否安装成功。弹出如图 1-8 所示的界面表示安装成功。

图 1-8 表示安装成功

1.3.2 搭建 Django 环境

在当今技术环境下,有多种安装 Django 框架的方法,下面对这些安装方法按难易程度进行排序,其中越靠前的越简单。

- Python 包管理器。
- 操作系统包管理器。
- 官方发布的压缩包。
- 源码库。

最简单的下载和安装方式是使用 Python 包管理工具,建议读者使用这种安装方式。例如可以使用 Setuptools 中的 easy_install(http://packages.python.org/distribute/easy_install.html),或 pip(http://pip.openplans.org)。目前在所有的操作系统平台上都可使用这两个工具,对于 Windows 用户来说,在使用 Setuptools 时需要将 easy_install.exe 文件放在 Python 安装目录下的 Scripts 文件夹中。此时只需在 DOS 命令行窗口中使用一条命令就可以安装 Django,例如,可以使用如下的 easy_install 命令安装当前的最新版本。

```
easy_install django
```

也可以使用如下的 pip 命令安装当前的最新版本。

```
pip install django
```

我们也可以指定要安装的版本,例如使用如下命令安装 Django 3.1:

```
pip install django==3.1
```

本书使用的版本是 3.1,在命令提示符界面中的安装过程如下所示:

```
Collecting Django==3.1
  Downloading Django-3.1-py3-none-any.whl (7.8 MB)
     |████████████████████████████████| 7.8 MB 24 kB/s
Successfully installed Django-3.1 asgiref-3.2.10
```

为了验证是否安装成功,进入 Python 交互式环境,输入如下命令查看安装的版本:

```
>>> import django
>>> print(django.get_version())
3.1
```

1.3.3 常用的 Django 命令

接下来讲解 Django 框架中常用的一些基本命令，读者需要打开 Linux 或 MacOS 的 Terminal(终端)，直接在终端中输入这些命令(不是 Python 的命令行界面中)。如果读者使用的是 Windows 系统，则在 CMD 上输入操作命令。

(1) 新建一个 Django 工程。

```
django-admin.py startproject project-name
```

"project-name"表示工程名字，一个工程是一个项目。在 Windows 系统中需要使用如下命令创建工程。

```
django-admin startproject project-name
```

> **注 意**
>
> 在给项目命名的时候必须避开 Django 和 Python 的保留关键字，比如 django 和 test 等，否则会引起冲突和莫名的错误。对于 mysite 的放置位置，不建议放在传统的 /var/wwww 目录下，它会具有一定的数据暴露危险，因此 Django 建议你将项目文件放在类似 /home/mycode 的位置。

(2) 新建 app(应用程序)。

```
python manage.py startapp app-name
```

或：

```
django-admin.py startapp app-name
```

通常一个项目有多个 app，当然通用的 app 也可以在多个项目中使用。

> **注 意**
>
> app 应用与 Project 项目的区别：
> - 一个 app 实现某个功能，比如博客、公共档案数据库或者简单的投票系统。
> - 一个 Project 是配置文件和多个 app 的集合，这些 app 组合成整个站点。
> - 一个 Project 可以包含多个 app。
> - 一个 app 可以属于多个 Project。
> - app 的存放位置可以是任何地点，但是通常都将它们放在与文件 manage.py 同级的目录下，这样方便导入文件。

(3) 同步数据库。

```
python manage.py syncdb
```

读者需要注意，在 Django 1.7.1 及以上的版本中需要使用以下命令：

```
python manage.py makemigrations
python manage.py migrate
```

这种方法可以创建表，当在 models.py 中新增类时，运行它就可以自动在数据库中创建表了，不用手动创建。

(4) 使用开发服务器。

开发服务器，即在开发时使用，在修改代码后会自动重启，这会方便程序的调试和开发。但是由于性能问题，建议只用来测试，不要用在生产环境中。

```
python manage.py runserver
# 当提示端口被占用的时候，可以用其他端口：
python manage.py runserver 8001
python manage.py runserver 999
#也可以关闭占用端口的进程

# 监听所有可用 IP (电脑可能有一个或多个内网 IP，一个或多个外网 IP，即有多个 IP 地址)
python manage.py runserver 0.0.0.0:8000
# 如果是在外网或者局域网电脑上运行，则需要使用这些设备的 IP 地址和端口来运行，比如
http://172.16.20.2: 8000
```

(5) 清空数据库。

```
python manage.py flush
```

此命令会询问是 yes 还是 no，选择 yes 会把数据全部清空，只留下空表。

(6) 创建超级管理员。

```
python manage.py createsuperuser
# 按照提示输入用户名和对应的密码就好了，邮箱可以留空，用户名和密码必填
# 修改用户密码可以用：
python manage.py changepassword username
```

(7) 通过如下命令导出数据和导入数据。

```
python manage.py dumpdata appname > appname.json
python manage.py loaddata appname.json
```

(8) 通过如下命令打开 Django 项目环境终端。

```
python manage.py shell
```

如果安装了 bpython 或 ipython，会自动调用它们的界面，推荐安装 bpython。这个命令和直接运行 Python 或 bpython 进入 shell 的区别是：可以在这个 shell 里面调用当前项目的 models.py 中的 API。

(9) 通过如下命令开启数据库操作命令行界面。

```
python manage.py dbshell
```

Django 会自动进入在 settings.py 中设置的数据库，如果是 MySQL 或 postgreSQL，会要求输入数据库用户密码。在这个终端可以执行数据库的 SQL 语句。如果对 SQL 比较熟悉，可以使用这种方式。

(10) 启动 Django Web 程序。

```
python manage.py runserver
```

1.4 实现第一个 Django Web 项目

为了使读者对 Django Web 有一个清晰的认识,接下来将通过一个具体实例的实现过程,详细讲解创建并运行第一个 Django 工程的方法。

扫码观看本节视频讲解

1.4.1 实战演练:使用 Django 命令创建 Django Web 项目

在接下来的内容中,将介绍使用命令创建一个 Django Web 项目的方法。

 源码路径:**daima\2\mysite**

(1) 在"命令提示符"中定位到目录"E:\123\Django-daima\2\",然后通过如下命令创建一个"mysite"目录作为"project(工程)"。

```
django-admin startproject mysite
```

注意,如果不能使用 django-admin,请用 django-admin.py,即使用下面的命令:

```
django-admin.py startproject mysite
```

创建成功后会看到如下所示的目录样式。

```
mysite
├── manage.py
└── mysite
    ├── __init__.py
    ├── settings.py
    ├── urls.py
    └── wsgi.py
```

也就是说,在"E:\123\Django-daima\2\"目录中新建了一个名为"mysite"的子目录,在子目录"mysite"的里面还有一个名为"mysite"的子目录,这个子目录 mysite 中是一些项目的设置文件 settings.py、总的 urls 配置文件 urls.py,以及部署服务器时用到的 wsgi.py 文件,而文件__init__.py 是 python 包的目录结构必需的,与调用有关。

- mysite:项目的容器,保存整个工程。
- manage.py:一个实用的命令行工具,可让你以各种方式与该 Django 项目进行交互。
- mysite/__init__.py:一个空文件,告诉 Python 该目录是一个 Python 包。
- mysite/settings.py:该 Django 项目的"设置/配置"文件。
- mysite/urls.py:该 Django 项目的 URL 声明,就像网站的"目录"。
- mysite/wsgi.py:一个 WSGI 兼容的 Web 服务器的入口,以便运行项目。

(2) 在"命令提示符"中定位到 mysite 目录下(注意,不是 mysite 中的 mysite 目录),然后通过如下命令新建一个应用(app),名称叫 learn。

```
E:\123\Django-daima\2\mysite>python manage.py startapp learn
```

此时可以看到在主 mysite 目录中多出了一个 learn 文件夹，在里面有如下所示的文件。

```
learn/
├── __init__.py
├── admin.py
├── apps.py
├── models.py
├── tests.py
└── views.py
```

（3）为了将新定义的 app 添加到 settings.py 文件的 INSTALLED_APPS 中，需要对文件 mysite/mysite/settings.py 进行如下修改。

```
INSTALLED_APPS = [
    'django.contrib.admin',
    'django.contrib.auth',
    'django.contrib.contenttypes',
    'django.contrib.sessions',
    'django.contrib.messages',
    'django.contrib.staticfiles',
    'learn',
]
```

这一步的目的是将新建的程序"learn"添加到 INSTALLED_APPS 中，如果不这样做，Django 就不能自动找到 app 中的模板文件(app-name/templates/下的文件)和静态文件 (app-name/static/中的文件)。

（4）定义视图函数，用于显示访问页面时的内容。在 learn 目录中打开文件 views.py，然后进行如下所示的修改。

```
#coding:utf-8
from django.http import HttpResponse
def index(request):
    return HttpResponse(u"欢迎光临，Python 工程师欢迎您！")
```

对上述代码的具体说明如下所示。

- ▶ 第 1 行：声明编码为 utf-8，因为我们在代码中用到了中文，如果不声明就会报错。
- ▶ 第 2 行：引入 HttpResponse，用来向网页返回内容。就像 Python 中的 print 函数一样，只不过 HttpResponse 是把内容显示到网页上。
- ▶ 第 3~4 行：定义一个 index()函数，第一个参数必须是 request，与网页发来的请求有关。在 request 变量里面包含 get/post 的内容、用户浏览器和系统等信息。函数 index()返回一个 HttpResponse 对象，可以经过一些处理，最终显示几个字到网页上。

现在问题来了，用户应该访问什么网址才能看到刚才写的这个函数呢？如何让网址和函数关联起来呢？接下来需要定义和视图函数相关的 URL 网址。

（5）开始定义视图函数相关的 URL 网址，对文件 mysite/mysite/urls.py 进行如下所示的修改。

```
from django.conf.urls import url
from django.contrib import admin
```

```
from learn import views as learn_views  # new

urlpatterns = [
    url(r'^$', learn_views.index),  # new
    url(r'^admin/', admin.site.urls),
]
```

(6) 最后在终端上输入如下命令运行上面创建的 Django Web 项目。

```
python manage.py runserver
```

运行成功后显示如下所示的提示信息。

```
C:\WINDOWS\system32>cd E:\123\Django-daima\2\mysite

E:\123\Django-daima\2\mysite>python manage.py runserver
Watching for file changes with StatReloader
Performing system checks...

System check identified no issues (0 silenced).

You have 18 unapplied migration(s). Your project may not work properly until
you apply the migrations for app(s): admin, auth, contenttypes, sessions.
Run 'python manage.py migrate' to apply them.
August 19, 2020 - 10:25:19
Django version 3.1, using settings 'mysite.settings'
Starting development server at http://127.0.0.1:8000/
Quit the server with CTRL-BREAK.
```

根据上面显示的提示信息，启动 Django Web 项目的 URL 是：

```
http://127.0.0.1:8000/
```

在浏览器中输入 http://127.0.0.1:8000/后的效果如图 1-9 所示。

图 1-9　执行效果

> **注　意**
>
> 　　在默认情况下，runserver 命令会将服务器设置为监听本机内部 IP 的 8000 端口。如果你想更换服务器的监听端口，请使用命令行参数。举个例子，下面的命令会使服务器监听 8080 端口。
>
> ```
> python manage.py runserver 8080
> ```
>
> 　　如果你想要修改服务器监听的 IP，只需在 python manage.py runserver 命令后输入新的 IP 即可。比如，为了监听所有服务器的公开 IP(这在你运行 Vagrant 或想要向网络上的其他电脑展示你的成果时很有用)，则使用：
>
> ```
> python manage.py runserver 0:8000
> ```
>
> 　　其中 0 是 0.0.0.0 的简写。

1.4.2 实战演练：使用 PyCharm 创建 Django Web 项目

对于开发者来说，创建 Django Web 项目的最简单方式是使用集成开发工具 PyCharm。利用 2018 年以后发布的 Pycharm，可以非常方便地创建 Django Web 项目，并且可以很方便地基于虚拟环境创建 Django 工程。

源码路径：**daima\2\untitled**

（1）打开 PyCharm，选择 File | New Project 命令，在弹出的对话框的左侧选择 Django 选项，如图 1-10 所示。

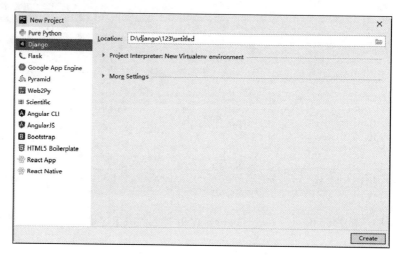

图 1-10　选择 Django 选项

- Location：设置保存当前 Django 工程的目录位置。
- Project Interpreter: New Virtualenv environment：设置当前工程所使用的虚拟 Python 环境，我们可以在 Base interpreter 下拉列表框中选择已经安装的 Python 环境作为虚拟环境。在一般情况下，在安装 Python 时会创建一个默认的 Python 环境。对于大多数开发者来说，可能已经安装了好几个版本的 Python 环境或虚拟环境，此时可以在 Base interpreter 下拉列表框中选择要使用的 Python 环境。例如有的开发者安装了 Anaconda3，那么就会在 PyCharm 的 Project Interpreter 选项组中显示这个 Python 环境。例如我的电脑中有两个 Python 环境，如图 1-11 所示。

> **注意**
> 建议大家使用 New Virtualenv environment 方式创建一个虚拟环境，这样不会影响已经安装的 Python 环境。在此处选择 Virtualenv(这可能需要你提前执行 pip install virtualenv 命令进行虚拟工具 virtualenv 的安装)。在通常情况下，虚拟环境会以 venv 的名字自动在工程目录下生成。

- Existing interpreter：如果不想使用虚拟环境，而是想使用现成的已经安装的 Python

环境或虚拟环境,请选择这个选项。

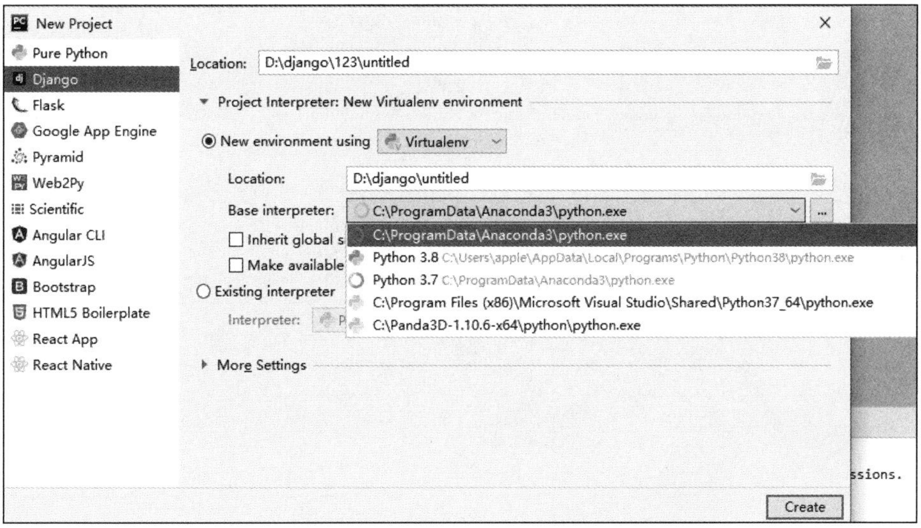

图 1-11 两个 Python 环境

例如我们按照如图 1-12 所示的参数,基于新建的虚拟环境创建一个名字为"untitled"的 Django Web 工程。

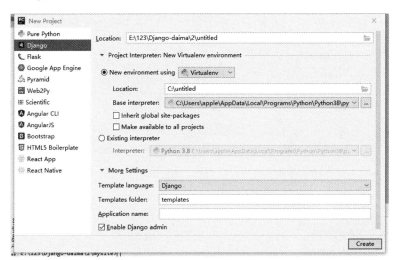

图 1-12 基于新建的虚拟环境创建一个 Django Web 工程

单击 More Settings 选项后可以进行更多设置。
- Template language:选择使用的模板语言,默认 Django 就行,可选 Jinjia。
- Templates folder:PyCharm 提供给我们的功能,额外创建一个工程级别的模板文件的保存目录,可以不设置,空着,这里使用默认设置。
- Enable Django admin:设置是否启用 Django Admin 功能,一般勾选此复选框。

(2) 单击右下角的 Create 按钮开始创建虚拟目录和 Django Web 工程,创建完成之后会在工程目录中显示 PyCharm 自动为我们创建的各个 Django 目录文件夹和默认的 Python 程序文件，如图 1-13 所示。

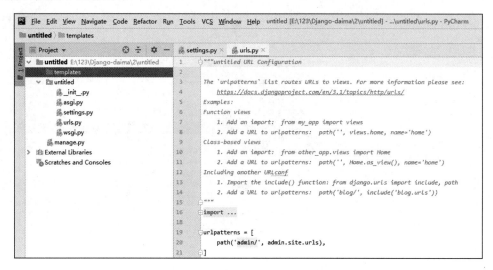

图 1-13　PyCharm 自动创建的 Django 目录和程序文件

选择 PyCharm 的 File | Settings | Project 命令进入 Settings 界面，此时可以看到在当前的 Python 环境中已经安装的库，其中包括最新版本的 Django，如图 1-14 所示。如果需要指定使用其他版本，比如 2.1、1.11 等，就不能这么操作了，需要在命令行下自己创建虚拟环境并安装 Django。或者在这里先删除 Django，然后再安装想要的指定版本。

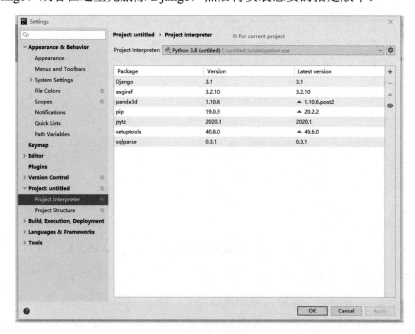

图 1-14　当前 Django Web 工程所使用的 Python 环境

（3）单击 PyCharm 顶部绿色三角形按钮 ▶ 启动当前 Django Web 工程，启动成功后会在 PyCharm 的 Run 面板中显示对应的提示信息，如图 1-15 所示。

在浏览器中输入 URL 为 http://127.0.0.1:8000/即可查看当前 Django Web 工程主页的执行效果，如图 1-16 所示。单击 PyCharm 顶部中的红色正方形按钮■可以停止当前 Django Web 工程的运行，在不用这个项目时请务必停止当前 Django Web 工程的运行。

图 1-15　运行成功后的提示信息

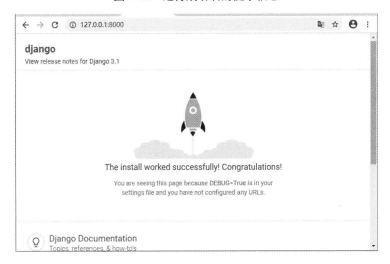

图 1-16　当前 Django Web 主页的执行效果

（4）也可以单击 PyCharm 左下角中的 Terminal 标签，在弹出的界面中会自动将命令行定位到当前 Django Web 工程的根目录，如图 1-17 所示。

图 1-17　定位到当前 Django Web 工程的根目录

然后在后面输入如下命令也可以启动运行当前的 Django Web 工程，运行成功后也会在 Run 面板中显示对应的提示信息，如图 1-18 所示。按下键盘上的 Ctrl+C 组合键可以停止当前 Django Web 工程的运行，在不用这个项目时请务必停止当前 Django Web 工程的运行。

```
python manage.py runserver
```

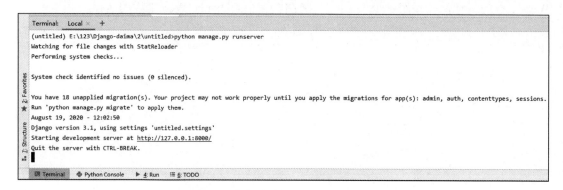

图 1-18　运行成功后的提示信息

第 2 章

分析 Django 项目的结构

在本章的内容中,将以创建一个在线 Web 投票系统为例,向大家展示使用 Django 开发典型的 Web 应用程序的方法,目的是让大家对 Django 的设计理念、功能模块、体系架构、基本用法有个初步的印象。

2.1 实战演练：在线投票系统

本在线投票系统具有如下所示的功能。
- 可以让用户针对某个主题进行投票。
- 可以查看投票结果。
- 可以在后台实现投票信息的增加、删除、修改和查找功能。

扫码观看本节视频讲解

源码路径：daima\2\first

开始创建我们的项目，首先使用如下命令新建一个 Django 工程。

```
django-admin startproject mysite
```

这将在目录下生成一个 mysite 目录，也就是这个 Django 工程的根目录。它包含一系列自动生成的目录和文件，具备各自专有的用途。

然后进入 mysite 项目根目录，确保与 manage.py 文件处于同一级，输入如下命令创建一个名为"polls"的 app。

```
python manage.py startapp polls
```

此时系统会自动生成 polls 应用的目录，其结构如下。

```
polls/
    __init__.py
    admin.py
    apps.py
    migrations/
    models.py
    tests.py
    views.py
```

2.2 编写第一个视图

在自动生成的视图文件 polls/views.py 中编写如下代码。

扫码观看本节视频讲解

```
from django.http import HttpResponse

def index(request):
    return HttpResponse("这是投票系统主页")
```

为了调用上述视图，我们还需要编写 URLconf，也就是路由路径。在 polls 目录中新建文件 urls.py，在其中输入如下代码。

```
from django.urls import path

from . import views
```

```
urlpatterns = [
    path('', views.index, name='index'),
]
```

然后在项目根目录的主 urls.py 文件中添加 urlpatterns 条目，指向我们刚才建立的 polls 这个 app 独有的 urls 文件，并且还需要导入 include 模块。打开文件 mysite/urls.py，代码如下。

```
from django.contrib import admin
from django.urls import include, path

urlpatterns = [
    path('polls/', include('polls.urls')),
    path('admin/', admin.site.urls),
]
```

在上述代码中，函数 include() 允许引用其他 URLconfs。每当 Django 遇到 :func:~django.urls.include 时，它会截断与此项匹配的 URL 的部分，并将剩余的字符串发送到 URLconfs 以供进一步处理。

我们设计 include() 的理念是使其可以即插即用。因为投票应用有它自己的 URLconf(polls/urls.py)，它们能够被放在 "/polls/"、"/fun_polls/"、"/content/polls/"或者其他任何路径下，这个应用都能够正常工作。

如果此时使用如下命令启动我们创建的 Django 项目，在浏览器中输入 http://127.0.0.1:8000/polls/后会显示视图文件 polls/views.py 中的内容，如图 2-1 所示。

```
python manage.py runserver
```

图 2-1 polls 应用的主页

在上面的文件 mysite/urls.py 中，函数 path()有四个参数，其中 route 和 view 是必须设置的参数，kwargs 和 name 为可选参数，各个参数的具体含义如下。

- route：是一个匹配 URL 的准则(类似正则表达式)。当 Django 响应一个请求时，它会从 urlpatterns 的第一项开始，按顺序依次匹配列表中的项，直到找到匹配的项。这些准则不会匹配 GET 和 POST 参数或域名。例如，URLconf 在处理请求 https://www.example.com/myapp/ 时，它会尝试匹配 myapp/。处理请求 https://www.example.com/myapp/?page=3 时，也只会尝试匹配 myapp/。

- view：当 Django 找到了一个匹配的准则时，就会调用这个特定的视图函数，并传入一个 HttpRequest 对象作为第一个参数，被"捕获"的参数以关键字参数的形式传入。稍后，我们会给出一个例子。

- kwargs：任意个关键字参数可以作为一个字典传递给目标视图函数。

- name：对 URL 进行命名，从而能够在 Django 的任意处，尤其是模板内显式地引用它。这是一个非常强大的功能，相当于给 URL 取了个全局变量名，这样不会将 url 匹配地址写死，使网址更具有灵活性。

2.3 实现数据库

数据库是动态 Web 程序的核心，接下来将详细讲解实现 Django 数据库功能的过程。

2.3.1 配置数据库

扫码观看本节视频讲解

打开配置文件 mysite/settings.py，这是整个 Django 项目的设置中心。Django 默认使用 SQLite 数据库，因为 Python 内置支持 SQLite 数据库，所以无须安装任何程序就可以直接使用 SQLite。当然，如果创建一个实际的项目，可以使用类似 PostgreSQL 的数据库，避免以后数据库迁移的相关问题。在配置文件 mysite/settings.py 中，下面是默认使用 SQLite 数据库的代码，这表示本项目使用的数据库文件是 db.sqlite3。

```
DATABASES = {
  'default': {
    'ENGINE': 'django.db.backends.sqlite3',
    'NAME': os.path.join(BASE_DIR, 'db.sqlite3'),
  }
}
```

如果想使用其他类型的数据库，请先安装相应的数据库操作模块，并将文件 settings 中 DATABASES 位置的'default'的键值进行相应的修改，用于连接数据库。其中：

- ENGINE(引擎)：可以是 django.db.backends.sqlite3、django.db.backends.postgresql、django.db.backends.mysql、django.db.backends.oracle，当然其他的也行。
- NAME(名称)：类似 MySQL 数据库管理系统中用于保存项目内容的数据库的名字。如果你使用的是默认的 SQLite，那么数据库将作为一个文件存放在你的本地机器内，此时的 NAME 应该是这个文件的完整绝对路径(包括文件名)，默认值为 os.path.join(BASE_DIR, 'db.sqlite3')，将把该文件储存在你的项目目录下。

> **注 意**
>
> 如果在 Django 项目中使用了 SQLite 以外的数据库，请确认在使用前已经创建了数据库。可以通过在数据库交互式命令行中使用 "CREATE DATABASE database_name;" 命令来完成这件事。另外，还要确保该数据库用户具有 create database 权限。这使得自动创建的 test database 能被以后的教程使用。如果使用的是 SQLite 数据库，那么不需要在使用前做任何事——数据库会在需要的时候自动创建。

在修改配置文件 settings 时，请顺便将 TIME_ZONE 设置为国内所在的时区：Asia/Shanghai。同时，请注意 settings 文件中顶部的 INSTALLED_APPS 设置项。它列出了所有的项目中被激活的 Django 应用(app)。必须将我们自定义的 app 在 INSTALLED_APPS 中注册。每个应用可以被多个项目使用,并且可以打包和分发给其他人在他们的项目中使用。

在默认情况下,INSTALLED_APPS 会自动包含下列条目,它们都是 Django 自动生成的。

- django.contrib.admin:admin 管理后台站点。
- django.contrib.auth:身份认证系统。
- django.contrib.contenttypes:内容类型框架。
- django.contrib.sessions:会话框架。
- django.contrib.messages:消息框架。
- django.contrib.staticfiles:静态文件管理框架。

上面的一些应用也需要建立一些数据库表,所以在使用它们之前我们要在数据库中创建这些表。可使用下面的命令创建数据表。

```
python manage.py migrate
```

migrate 命令将遍历 INSTALLED_APPS 设置中的所有项目,在数据库中创建对应的表,并打印出每一条动作信息。

2.3.2 创建数据库模型

接下来开始定义模型 model,模型本质上就是数据库表的布局,再附加一些元数据。Django 通过自定义 Python 类的形式来定义具体的模型,每个模型的物理存在方式就是一个 Python 的类 Class,每个模型代表数据库中的一张表,每个类的实例代表数据表中的一行数据,类中的每个变量代表数据表中的一列字段。Django 通过模型,将 Python 代码和数据库操作结合起来,实现对 SQL 查询语言的封装。也就是说,你即使不会管理数据库,不会 SQL 语言,也同样能通过 Python 的代码进行数据库的操作。Django 通过 ORM 对数据库进行操作,奉行代码优先的理念,将 Python 程序员和数据库管理员进行分工解耦。

在我们这个在线投票应用中,将创建两个模型:Question 和 Choice。Question 包含一个问题和一个发布日期。Choice 包含两个字段:该选项的文本描述和该选项的投票数。每一条 Choice 都关联到一个 Question。这些都由 Python 的类来体现,编写的全是 Python 的代码,不接触任何 SQL 语句。现在,编辑模型文件 polls/models.py,具体代码如下。

```python
from django.db import models
class Question(models.Model):
    question_text = models.CharField(max_length=200)
    pub_date = models.DateTimeField('date published')

class Choice(models.Model):
    question = models.ForeignKey(Question, on_delete=models.CASCADE)
    choice_text = models.CharField(max_length=200)
    votes = models.IntegerField(default=0)
```

在上述代码中,每一个类都是 django.db.models.Model 的子类。每一个字段都是类 Field 的一个实例,例如用于保存字符数据的 CharField 和用于保存时间类型的 DateTimeField,它们告诉 Django 每一个字段保存的数据类型。

每一个 Field 实例的名字就是字段的名字(例如 question_text 或者 pub_date)。在 Python 代码中会使用这个值，数据库也会将这个值作为表的列名。

也可以在每个 Field 中使用一个可选的第一位置参数来提供一个可读的字段名，让模型更友好、更易读，并且将被作为文档的一部分来增强代码的可读性。

一些 Field 类必须提供某些特定的参数。例如 CharField 需要指定 max_length。这不仅是数据库结构的需要，同样也用于数据验证功能。

有必填参数，当然就会有可选参数，比如在 votes 里我们将其默认值设置为 0。

最后还需要注意，我们使用 ForeignKey 定义了一个外键关系。它告诉 Django，每一个 Choice 关联到一个对应的 Question(注意要将外键写在"多"的一方)。Django 支持通用的数据关系：一对一、多对一和多对多。

2.3.3 启用模型

虽然上面的模型文件 polls/models.py 的代码看着有点少，但是包含了大量的信息。通过这个模型文件，Django 会做下面两件事。

- 创建该 app 对应的数据库表结构。
- 为 Question 和 Choice 对象创建基于 Python 的数据库访问 API。

但是，首先我们得先告诉 Django 项目，我们要使用投票 app。要将应用添加到项目中，需要在 INSTALLED_APPS 设置中增加指向该应用的配置文件的链接。对于本例的投票应用，它的配置类文件 PollsConfig 是 polls/apps.py，所以它的用点"."引用的路径为 polls.apps.PollsConfig。我们需要在 INSTALLED_APPS 中，将该路径添加进去：

```
INSTALLED_APPS = [
'polls.apps.PollsConfig',
'django.contrib.admin',
'django.contrib.auth',
'django.contrib.contenttypes',
'django.contrib.sessions',
'django.contrib.messages',
'django.contrib.staticfiles',
]
```

实际上，在多数情况下，我们简写成'polls'就可以了：

```
INSTALLED_APPS = [
'django.contrib.admin',
'django.contrib.auth',
'django.contrib.contenttypes',
'django.contrib.sessions',
'django.contrib.messages',
'django.contrib.staticfiles',
'polls'
]
```

接下来运行下一个命令：

```
python manage.py makemigrations polls
```

你会看到类似下面的提示：

```
Migrations for 'polls':
  polls/migrations/0001_initial.py:
    - Create model Choice
    - Create model Question
    - Add field question to choice
```

通过运行 makemigrations 命令，Django 会检测我们对模型文件的修改，也就是告诉 Django 你对模型有改动，并且你想把这些改动保存为一个"迁移(migration)"。

migrations 是 Django 保存模型修改记录的文件，这些文件保存在磁盘上。在本例中，它就是 polls/migrations/0001_initial.py，你可以打开它看看，里面保存的都是可读并且可编辑的内容，方便我们随时手动修改内容。

接下来使用 migrate 命令对数据库执行真正的迁移动作。但是在此之前，让我们先看看在使用 migration 进行数据迁移的时候实际执行的 SQL 语句是什么。有一个叫作 sqlmigrate 的命令可以展示 SQL 语句，例如：

```
python manage.py sqlmigrate polls 0001
```

此时你将会看到如下类似的文本(经过适当的格式调整，以方便阅读)：

```sql
BEGIN;
--
-- Create model Choice
--
CREATE TABLE "polls_choice" (
    "id" serial NOT NULL PRIMARY KEY,
    "choice_text" varchar(200) NOT NULL,
    "votes" integer NOT NULL
);
--
-- Create model Question
--
CREATE TABLE "polls_question" (
    "id" serial NOT NULL PRIMARY KEY,
    "question_text" varchar(200) NOT NULL,
    "pub_date" timestamp with time zone NOT NULL
);
--
-- Add field question to choice
--
ALTER TABLE "polls_choice" ADD COLUMN "question_id" integer NOT NULL;
ALTER TABLE "polls_choice" ALTER COLUMN "question_id" DROP DEFAULT;
CREATE INDEX "polls_choice_7aa0f6ee" ON "polls_choice" ("question_id");
ALTER TABLE "polls_choice"
  ADD CONSTRAINT
"polls_choice_question_id_246c99a640fbbd72_fk_polls_question_id"
    FOREIGN KEY ("question_id")
    REFERENCES "polls_question" ("id")
    DEFERRABLE INITIALLY DEFERRED;

COMMIT;
```

> **注意**
> - 实际的输出内容将根据你使用的数据库会有所不同。上面的是 PostgreSQL 的输出。表名是自动生成的，通过组合应用名 (polls) 和小写的模型名组成的，例如 polls_question 和 polls_choice。(你可以重写此行为。)
> - 主键 (IDs) 是自动添加的。(你也可以重写此行为。)
> - 按照惯例，Django 会在外键字段名上附加"_id"。(你仍然可以重写此行为。)
> - 生成 SQL 语句时针对所使用的数据库，会为你自动处理特定于数据库的字段，例如 auto_increment (MySQL)、serial (PostgreSQL) 或 integer primary key (SQLite)。在引用字段名时也是如此，比如使用双引号或单引号。
> - 这些 SQL 命令并没有在你的数据库中实际运行，它只是在屏幕上显示出来，以便让你了解 Django 真正执行的是什么。

如果你感兴趣，也可以运行如下命令检查项目中的错误，这并不会实际进行迁移或者链接数据库的操作。

```
python manage.py check
```

接下来可以运行 migrate 命令，在数据库中进行真正的表操作了。

```
python manage.py migrate
```

migrate 命令会对所有还未实施的迁移记录进行操作，本质上就是将对模型的修改体现到数据库中具体的表上面。Django 通过一张叫作 django_migrations 的表，记录并跟踪已经实施的 migrate 动作，通过对比获得哪些 migrations 尚未提交。

migration 命令的功能非常强大，允许随时修改模型，而不需要删除或者新建数据库或数据表，在不丢失数据的同时，实时动态更新数据库。我们将在后面的章节对此进行深入的阐述，但是现在，只需要记住修改模型时的操作分如下 3 步。

(1) 在 models.py 中修改模型。
(2) 运行 python manage.py makemigrations，为改动创建迁移记录。
(3) 运行 python manage.py migrate，将操作同步到数据库。

> **注意**
> 数据库迁移被分解成生成和应用两个命令，是为了让用户能够在代码控制系统上提交迁移数据并使其能在多个应用里使用。这不仅仅会让开发更加简单，也给别的开发者和生产环境中的使用带来方便。

2.3.4 使用模型的 API

接下来进入 Python 交互环境，尝试使用 Django 提供的数据库访问 API。要进入 Python 的 shell 交互环境，请输入下面的命令：

```
python manage.py shell
```

相比较直接输入"python"命令的方式进入 Python 环境，调用 manage.py 参数能将

DJANGO_SETTINGS_MODULE 环境变量导入，它将自动按照 mysite/settings.py 中的设置，配置好 python shell 环境，这样，就可以导入和调用任何项目内的模块了。

或者也可以先进入一个纯净的 python shell 环境，然后启动 Django，具体如下。

```
>>> import django
>>> django.setup()
```

进入 shell 后，尝试使用下面的 API：

```
>>> from polls.models import Question, Choice  # 导入我们写的模型类
# 现在系统内还没有 questions 对象
>>> Question.objects.all()
<QuerySet []>

# 创建一个新的 question 对象
# Django 推荐使用 timezone.now() 代替 python 内置的 datetime.datetime.now()
# 这个 timezone 就来自于 Django 的依赖库 pytz
from django.utils import timezone
>>> q = Question(question_text="What's new?", pub_date=timezone.now())

# 你必须显式地调用 save() 方法，才能将对象保存到数据库内
>>> q.save()

# 默认情况，你会自动获得一个自增的名为 id 的主键
>>> q.id
1

# 通过 python 的属性调用方式，访问模型字段的值
>>> q.question_text
"What's new?"
>>> q.pub_date
datetime.datetime(2012, 2, 26, 13, 0, 0, 775217, tzinfo=<UTC>)

# 通过修改属性来修改字段的值，然后显式地调用 save 方法进行保存
>>> q.question_text = "What's up?"
>>> q.save()

# objects.all() 用于查询数据库内的所有 questions
>>> Question.objects.all()
<QuerySet [<Question: Question object>]>
```

其中，上面的<Question: Question object>是一个不可读的内容展示，我们无法从中获得任何直观的信息，为此我们需要一点小技巧，让 Django 在打印对象时显示一些我们指定的信息。

返回到模型文件 polls/models.py，在类 Question 和类 Choice 中分别添加一个方法，代码如下。这个技巧不仅仅能给你在使用命令行时带来方便，而且在 Django 自动生成的 admin 里也可以使用这个方法来表示对象。

```
from django.db import models

class Question(models.Model):
    # ...
```

```
    def __str__(self):
        return self.question_text

class Choice(models.Model):
    # ...
```

另外，也可以自定义一个模型方法 was_published_recently()，用于判断这个投票主题是否为最近时间段发布的。

```
import datetime

from django.db import models
from django.utils import timezone


class Question(models.Model):
    # ...
    def was_published_recently(self):
        return self.pub_date >= timezone.now() - datetime.timedelta(days=1)
```

请注意上面分别导入了两个关于时间的模块，一个是 Python 内置的 datetime，另一个是 Django 工具包提供的 timezone。

保存上面的修改后，重新启动一个新的 python shell，再来看看其他 API：

```
>>> from polls.models import Question, Choice

# 先看看__str__()的效果，直观多了吧？
>>> Question.objects.all()
<QuerySet [<Question: What's up?>]>

# Django 提供了大量的关键字参数查询 API
>>> Question.objects.filter(id=1)
<QuerySet [<Question: What's up?>]>
>>> Question.objects.filter(question_text__startswith='What')
<QuerySet [<Question: What's up?>]>

# 获取今年发布的问卷
>>> from django.utils import timezone
>>> current_year = timezone.now().year
>>> Question.objects.get(pub_date__year=current_year)
<Question: What's up?>

# 查询一个不存在的 ID，会弹出异常
>>> Question.objects.get(id=2)
Traceback (most recent call last):
...
DoesNotExist: Question matching query does not exist.

# Django 为主键查询提供了一个缩写：pk。下面的语句和 Question.objects.get(id=1)效果一样
>>> Question.objects.get(pk=1)
<Question: What's up?>

# 看看我们自定义的方法用起来怎么样
>>> q = Question.objects.get(pk=1)
>>> q.was_published_recently()
```

```
True
# 让我们试试主键查询
>>> q = Question.objects.get(pk=1)

# 显示所有与 q 对象有关系的 choice 集合，目前是空的，还没有任何关联对象
>>> q.choice_set.all()
<QuerySet []>

# 创建 3 个 choice
>>> q.choice_set.create(choice_text='Not much', votes=0)
<Choice: Not much>
>>> q.choice_set.create(choice_text='The sky', votes=0)
<Choice: The sky>
>>> c = q.choice_set.create(choice_text='Just hacking again', votes=0)

# Choice 对象可通过 API 访问和它们关联的 Question 对象
>>> c.question
<Question: What's up?>

# 同样的，Question 对象也可通过 API 访问关联的 Choice 对象
>>> q.choice_set.all()
<QuerySet [<Choice: Not much>, <Choice: The sky>, <Choice: Just hacking again>]>
>>> q.choice_set.count()
3

# API 会自动进行连表操作，通过双下划线分割关系对象。连表操作可以无限多级，一层一层地连接
# 下面是查询所有的 Choice，它所对应的 Question 的发布日期是今年。(重用了上面的
current_year 结果)
>>> Choice.objects.filter(question__pub_date__year=current_year)
<QuerySet [<Choice: Not much>, <Choice: The sky>, <Choice: Just hacking again>]>

# 使用 delete 方法删除对象
>>> c = q.choice_set.filter(choice_text__startswith='Just hacking')
>>> c.delete()
```

关于模型的使用就暂时先介绍这么多，这部分内容是 Django 项目的核心，也是动态网站与数据库交互的核心。

2.4 使用 admin 后台管理 Web

在很多时候，我们不光要开发针对客户使用的前端页面，还要给后台管理人员提供相应的管理界面。但是大多数时候，编写用于增加、修改和删除内容的后台管理站点都是一件非常乏味的工作并且没有多少创造性，而且也需要花费较多的时间和精力。Django 最大的优点之一，就是体贴地为你提供了一个基于项目 Model 创建的后台管理站点 admin。这个界面只给站点管理员使用，并不对大众开放。虽然 admin 的界面可能不是那么美观，功能不是那么强大，内容不一定符合你的要求，但是它是免费的、现成的，并且还是可定制的，有完善的帮助文档。

扫码观看本节视频讲解

2.4.1 创建管理员用户

首先，我们需要通过下面的命令，创建一个可以登录 admin 站点的用户。

```
python manage.py createsuperuser
```

此时要求输入用户名，例如输入 admin：

```
Username: admin
```

此时要求输入邮箱地址，例如输入 xxx@xxx.xxx：

```
Email address: xxx@xxx.xxx
```

此时要求输入密码和确认密码：

```
Password: **********
Password (again): *********
Superuser created successfully.
```

注 意

Django 1.10 版本后，超级用户的密码要求具备一定的复杂性，如果密码强度不够，Django 会提醒你，但是可以强制通过，如图 2-2 所示。

```
(mysite) E:\123\Django-daima\2\first\mysite>python manage.py createsuperuser
Username (leave blank to use 'apple'): admin
Email address: xxx@123.com
Password:
Password (again):
The password is too similar to the username.
This password is too common.
Bypass password validation and create user anyway? [y/N]: y
Superuser created successfully.
```

图 2-2 提醒密码强度不够

2.4.2 启动 Web 项目

使用如下命令启动我们的 Django 项目：

```
python manage.py runserver
```

服务器启动后，在浏览器访问 http://127.0.0.1:8000/admin，就能看到 admin 的登录界面，如图 2-3 所示。

在现实应用中，为了提高站点的安全性，我们一般不能将管理后台的 url 随便暴露给他人，不能用/admin/这么简单的路径。可以打开根 url 路由文件 mysite/urls.py，修改其中 admin.site.urls 对应的表达式，比如：

图 2-3 后台登录界面

```
from django.contrib import admin
from django.urls import path

urlpatterns = [
```

```
    path('polls/', include('polls.urls')),
    path('control/', admin.site.urls),
]
```

此时我们必须访问 http://127.0.0.1:8000/control/ 才能进入 admin 界面。

2.4.3 进入 admin 站点

利用刚才建立的 admin 账户登录后台，将会看到如图 2-4 所示的界面效果。

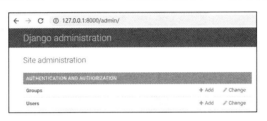

图 2-4　后台主页

由此可见，当前只有两个可编辑的内容：Groups 和 Users，它们是 django.contrib.auth 模块提供的身份认证框架。

2.4.4 在 admin 中注册投票应用

现在还无法看到投票应用，必须先在 admin 中进行注册，告诉 admin 站点。请将 polls 的模型加入站点内，接受站点的管理。打开文件 polls/admin.py，在里面添加下面的内容。

```
from django.contrib import admin
from .models import Question

admin.site.register(Question)
```

2.4.5 体验便捷的管理功能

在文件 polls/admin.py 中注册 question 模型后，刷新 admin 页面就能看到 Questions 栏目。如图 2-5 所示。

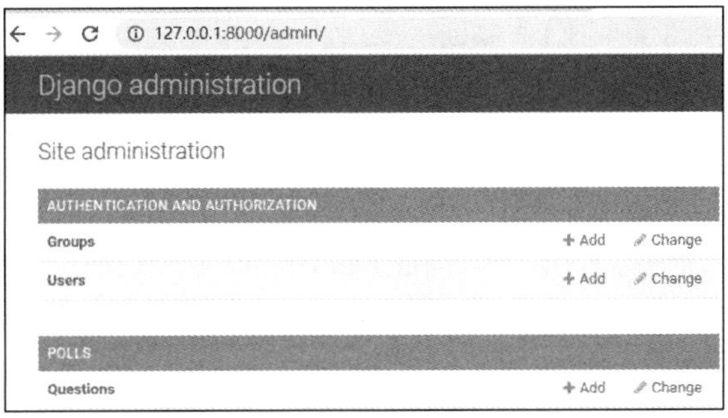

图 2-5　出现 Question 栏目

单击 Questions 选项，进入修改列表页面。这个页面会显示所有的数据库内的 questions 对象，你可以在这里对它们进行修改，如图 2-6 所示。

图 2-6　Questions 列表界面

看到下面的"What's new?"了吗？它就是我们先前创建的一个 question 对象，并且通过__str__方法的帮助，显示了较为直观的信息，而不是一个冷冰冰的对象类型名称。单击 What's new?可以进入编辑界面，单击 Add 按钮后来到 Add question 页面，在此页面中可以添加新的投票主题，如图 2-7 所示。我们可以添加一个主题"你喜欢的球队是谁？"。

图 2-7　添加新的投票主题

页面中的表单是由 Question 模型自动生成的，不同的模型字段类型(DateTimeField、CharField)会表现为不同的 HTML input 框类型。每一个 DateTimeField 都会自动生成一个可单击链接。日期是 Today，且有一个日历弹出框。时间是 Now，且有一个通用的时间输入列表框。

在页面的底部是一些可选项按钮，具体说明如下。

- delete：弹出一个删除确认页面。
- Save and add another：保存当前修改，并加载一个新的空白的当前类型对象的表单。
- Save and continue editing：保存当前修改，并重新加载该对象的编辑页面。
- SAVE：保存修改，返回当前对象类型的列表页面。

如果 Date published 字段的值和在前面 modds.py 文件中的不一致，可能是因为你没有正确地配置 TIME_ZONE，在国内，通常是 8 个小时的时间差别。修改 TIME_ZONE 配置并重新加载页面，就能显示正确的时间了。在页面的右上角，单击 HISTORY 按钮，你会看到对当前对象的所有修改操作都在这里有记录，包括修改时间和操作人员，如图 2-8 所示。

Change history: 你喜欢的球队是谁?

DATE/TIME	USER	ACTION
Aug. 19, 2020, 10:46 a.m.	admin	Added.

图 2-8　修改记录页面

2.5　视图和模板

一个视图就是一个页面，通常提供特定的功能，使用特定的模板实现。例如在一个博客应用中，你可能会看到如下所示的视图。

扫码观看本节视频讲解

- 博客主页：显示最新发布的一些内容。
- 每篇博客的详细页面：博客的永久链接。
- 基于年的博客页面：显示指定年内的所有博客文章。
- 基于月的博客页面：显示指定月内的所有博客文章。
- 基于天的博客页面：显示指定日内的所有博客文章。
- 发布评论：处理针对某篇博客发布的评论。

在我们的在线投票应用中，将建立下面的视图。

- 问卷“index”页：显示最新的一些问卷。
- 问卷“detail”页面：显示一个问卷的详细文本内容，没有调查结果，但是有一个投票或调查表单。
- 问卷“results”页面：显示某个问卷的投票或调查结果。
- 投票动作页面：处理针对某个问卷的某个选项的投票动作。

在 Django 中，网页和其他一些内容都是通过视图来处理的。视图其实就是一个简单的 Python 函数(在基于类的视图中称为方法)。Django 通过对比请求的 URL 地址来选择对应的视图。在我们上网的过程中，很可能看见过类似下面这样的 URL：

```
ME2/Sites/dirmod.asp?sid=&type=gen&mod=Core+Pages&gid=A6CD4967199A42D9B65B1B
```

请大家不要担心，Django 里的 URL 规则要比这简洁明了得多！

一个 URL 模式定义了一种 URL 的基本格式，比如：/newsarchive/<year>/<month>/。为了将 URL 和视图关联起来，Django 使用了 'URLconfs' 来配置。URLconf 将 URL 模式映射到视图。

2.5.1　编写视图

打开视图文件 polls/views.py，输入下列代码。

```
def detail(request, question_id):
   return HttpResponse("You're looking at question %s." % question_id)

def results(request, question_id):
```

```python
    response = "You're looking at the results of question %s."
    return HttpResponse(response % question_id)

def vote(request, question_id):
    return HttpResponse("You're voting on question %s." % question_id)
```

然后在路径导航文件 polls/urls.py 中加入下面的 URL 模式，将其映射到我们上面新增的视图。

```python
from django.urls import path

from . import views

urlpatterns = [
    # ex: /polls/
    path('', views.index, name='index'),
    # ex: /polls/5/
    path('<int:question_id>/', views.detail, name='detail'),
    # ex: /polls/5/results/
    path('<int:question_id>/results/', views.results, name='results'),
    # ex: /polls/5/vote/
    path('<int:question_id>/vote/', views.vote, name='vote'),
]
```

现在去浏览器中访问/polls/34/(注意：这里省略了域名。另外，使用了二级路由后，URL 中都要添加字符串 polls 前缀，参考前面的章节)，它将调用 detail()函数，然后在页面中显示你在 URL 里提供的 ID。访问/polls/34/results/和/polls/34/vote/，将分别显示预定义的伪结果和投票页面，如图 2-9 所示。

上面访问的路由的过程是什么呢？当有人访问/polls/34/地址时，Django 将首先加载 mysite.urls 模块，因为它是 settings 文件里设置的根 URL 配置文件。在该

图 2-9　访问/polls/34/

文件里，Django 发现了 urlpatterns 变量，于是在其内按顺序进行匹配。当它匹配上 polls/时，就裁去 URL 中匹配的文本 polls/，然后将剩下的文本"34/"传递给 polls.urls 进行下一步的处理。在 polls.urls 中，又匹配到<int:question_id>/，最终结果就是调用该模式对应的 detail()视图，也就是下面的函数：

```
detail(request=<HttpRequest object>, question_id=34)
```

函数中的 question_id=34 参数，是由<int:question_id>/而来。使用尖括号"捕获"这部分 URL，且以关键字参数的形式发送给视图函数。上述字符串中的 question_id 部分定义了将被用于区分匹配模式的变量名，而 int 则作为一个转换器决定了应该以哪种变量类型匹配这部分的 URL 路径。

2.5.2　编写一个真正有用的视图

每个视图至少要做两件事之一：返回一个包含请求页面的 HttpResponse 对象或者弹出一个类似 Http404 的异常，而其余的功能可以根据项目需求继续添加。下面是一个新的 index()

主页视图，用于替代先前无用的 index，它会根据发布日期显示最近的 5 个投票主题。本实例目前暂时只有一个主题，访问 http://127.0.0.1:8000/polls/ 的效果如图 2-10 所示。

```python
from django.http import HttpResponse
from .models import Question

def index(request):
    latest_question_list = Question.objects.order_by('-pub_date')[:5]
    output = ', '.join([q.question_text for q in latest_question_list])
    return HttpResponse(output)

# Leave the rest of the views (detail, results, vote) unchanged
# 省略了那些没改动过的视图(detail, results, vote)
```

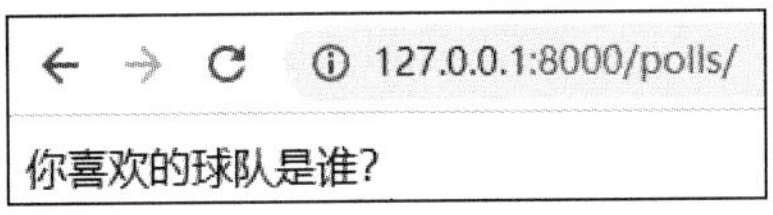

图 2-10　根据发布日期显示最近的 5 个投票主题

这里有个非常重要的问题：在当前视图中的 HTML 页面是硬编码的。如果想改变页面的显示内容，就必须修改这里的 Python 代码。为了解决这个问题，需要使用 Django 提供的模板系统，解耦视图和模板之间的硬连接。

首先，在 polls 目录下创建一个新的 templates 目录，Django 会在它里面查找模板文件。在 Django 项目的配置文件 settings.py 中，在 TEMPLATES 配置选项中描述了 Django 如何载入和渲染模板。默认的设置文件设置了 DjangoTemplates 后端，并将 APP_DIRS 设置成了 True。这一选项将会让 DjangoTemplates 在每个 INSTALLED_APPS 文件夹中寻找 templates 子目录。这就是为什么尽管我们没有像在 2.4 中那样修改 DIRS 设置，Django 也能正确找到 polls 模板位置的原因。

在 templates 目录中，再创建一个新的子目录名叫 polls，进入该子目录，创建一个新的 HTML 文件 index.html。换句话说，你的模板文件应该是 polls/templates/polls/index.html。因为 Django 会寻找对应的 app_directories，所以你只需要使用 polls/index.html 就可以引用这一模板了。

> **注　意**
>
> **模板命名空间**
>
> 也许有人会想：为什么不把模板文件直接放在 polls/templates 目录下，而是费劲地再建个子目录 polls 呢？设想这么个情况，有另外一个 app，它也有一个名叫 index.html 的文件，当 Django 在搜索模板时，有可能就找到它，然后退出搜索，这就命中了错误的目标，不是我们想要的结果。解决这个问题的最好办法就是在 templates 目录下再建立一个与 app 同名的子目录，将自己所属的模板都放到里面，从而达到独立命名空间的作用，不会再出现引用错误。

现在在文件 polls/templates/polls/index.html 中写入下列代码。

```
{% if latest_question_list %}
    <ul>
    {% for question in latest_question_list %}
        <li><a href="/polls/{{ question.id }}/">{{ question.question_text }}</a></li>
    {% endfor %}
    </ul>
{% else %}
    <p>No polls are available.</p>
{% endif %}
```

然后，修改视图文件 polls/views.py，让新的 HTML 文件 index.html 生效：

```
from django.http import HttpResponse
from django.template import loader

from .models import Question

def index(request):
    latest_question_list = Question.objects.order_by('-pub_date')[:5]
    template = loader.get_template('polls/index.html')
    context = {
        'latest_question_list': latest_question_list,
    }
    return HttpResponse(template.render(context, request))
```

上面的代码会加载 polls/index.html 文件，并传递给它一个参数。这个参数是一个字典，包含模板变量名和 Python 对象之间的映射关系。

在浏览器中通过访问/polls/，你可以看到一个列表，此时会包含"你喜欢的球队是谁？"主题，以及连接到其对应详细内容页面的链接点，如图 2-11 所示。

新的主页　　　　　　　　　　　单击链接后

图 2-11　新的主页和链接页面

> **注意**
> 如果显示的是 No polls are available.，说明你在前面没有添加 Questions 对象。前面的大量手动 API 操作你没有做。没关系，我们在 admin 中追加对象就可以。

2.5.3　快捷函数 render()

在实际运用中，加载模板、传递参数、返回 HttpResponse 对象是一整套再常用不过的操作了，为了减少操作，Django 提供了一个快捷方式：render()函数。我们可以将视图文件 polls/views.py 修改成下面的代码。

```
from django.shortcuts import render
from .models import Question
```

```python
def index(request):
    latest_question_list = Question.objects.order_by('-pub_date')[:5]
    context = {'latest_question_list': latest_question_list}
    return render(request, 'polls/index.html', context)
```

此时我们不再需要导入 loader 和 HttpResponse，而是从 django.shortcuts 导入 render。函数 render() 的第一个位置参数是请求对象(就是 view 函数的第一个参数)，第二个位置参数是模板。还可以有一个可选的第三参数——一个字典，包含需要传递给模板的数据。最后 render 函数返回一个经过字典数据渲染过的模板封装而成的 HttpResponse 对象。

2.5.4 抛出 404 错误

现在开始在视图文件 polls/views.py 中编写返回具体问卷文本内容：

```python
from django.http import Http404
from django.shortcuts import render

from .models import Question
def detail(request, question_id):
    try:
        question = Question.objects.get(pk=question_id)
    except Question.DoesNotExist:
        raise Http404("Question does not exist")
    return render(request, 'polls/detail.html', {'question': question})
```

通过上述代码，如果请求的问卷 ID 不存在，那么会弹出一个 404 错误。如果你想试试上面这段代码是否能正常工作，可以新建模板 polls/detail.html 文件，暂时写入下面的代码：

```
{{ question }}
```

另外，还可以使用函数 get_object_or_404() 获取错误处理，此函数的用法跟函数 render() 一样。例如在文件 polls/views.py 中使用如下代码。

```python
from django.shortcuts import get_object_or_404, render

from .models import Question
# ...
def detail(request, question_id):
    question = get_object_or_404(Question, pk=question_id)
    return render(request, 'polls/detail.html', {'question': question})
```

和使用函数 render() 一样，也需要从 Django 内置的快捷方式模块中导出 get_object_or_404()。方法 get_object_or_404() 将一个 Django 模型作为第一个位置参数，后面可以跟上任意个数的关键字参数，如果对象不存在则弹出 Http404 错误。同样，还有一个 get_list_or_404() 方法，和上面的 get_object_or_404() 类似，只不过是用来替代 filter() 函数，当查询列表为空时弹出 404 错误。(filter 是模型 API 中用来过滤查询结果的函数，它的结果是一个列表集。而 get 则是查询一个结果的方法，和 filter 是一个和多个的区别！)

为什么我们使用辅助函数 get_object_or_404()，而不是自己捕获 ObjectDoesNotExist 异常呢？还有，为什么模型 API 不直接抛出 ObjectDoesNotExist，而是抛出 Http404 呢？因为这样做会增加模型层和视图层的耦合性。指导 Django 设计的最重要的思想之一就是要保证松散耦合。一些受控的耦合将会被包含在 django.shortcuts 模块中。

2.5.5 使用模板系统

在上面的 detail()视图中，向模板传递了上下文变量 question。下面是 polls/detail.html 模板里正式的代码：

```
<h1>{{ question.question_text }}</h1>
<ul>
{% for choice in question.choice_set.all %}
    <li>{{ choice.choice_text }}</li>
{% endfor %}
</ul>
```

在模板系统中圆点"."是万能的魔法师，我们可以用它访问对象的属性。例如代码：{{ question.question_text }}，Django 首先会在 question 对象中尝试查找一个字典，如果失败则尝试查找属性，如果再失败，则尝试作为列表的索引进行查询。

在 {% for %}循环中的方法调用——question.choice_set.all，其实就是 Python 的代码 question.choice_set.all()，它将返回一组可迭代的 Choice 对象，并用在{% for %}标签中。

2.5.6 删除模板中硬编码的 URLs

在模板文件 polls/index.html 中还有一部分硬编码代码，也就是 href 里的 "/polls/" 部分：

```
<li><a href="/polls/{{ question.id }}/">{{ question.question_text }}</a></li>
```

它对于代码修改非常不利，假设想在文件 urls.py 中修改路由表达式，那么在我们所有的模板中对这个 urls 的引用都需要修改，这是无法接受的！

在前面我们给 urls 定义了一个 name 别名，可以用它来解决这个问题。具体代码如下：

```
<li><a href="{% url 'detail' question.id %}">{{ question.question_text }}</a></li>
```

Django 会在 polls.urls 文件中查找 name='detail'的 url，具体的就是下面这行代码：

```
path('<int:question_id>/', views.detail, name='detail'),
```

举个例子，如果你想将 polls 的 detail 视图的 URL 更换为 polls/specifics/12/，那么我们不需要在模板中重新修改 url 地址，仅仅只需要在文件 polls/urls.py 中将对应的正则表达式改成下面这样就行了，所有模板中对它的引用都会自动修改成新的链接。

```
# 添加新的单词'specifics'
path('specifics/<int:question_id>/', views.detail, name='detail'),
```

2.5.7 URL names 的命名空间

在本实例中只有一个 app，也就是 polls，但是在实际工作项目中可能会有 5 个、10 个甚至更多的 app 同时存在。Django 如何区分这些 app 之间的 URL name 呢？

答案是使用 URLconf 的命名空间。在 polls/urls.py 文件的开头部分，添加一个 app_name 变量来指定该应用的命名空间：

```
from django.urls import path

from . import views

app_name = 'polls'
urlpatterns = [
    path('', views.index, name='index'),
    path('<int:question_id>/', views.detail, name='detail'),
    path('<int:question_id>/results/', views.results, name='results'),
    path('<int:question_id>/vote/', views.vote, name='vote'),
]
```

现在，让我们将代码修改得更严谨一点，将 polls/templates/polls/index.html 中的：

```
<li><a href="{% url 'detail' question.id %}">{{ question.question_text }}</a></li>
```

修改为：

```
<li><a href="{% url 'polls:detail' question.id %}">
{{ question.question_text }}</a></li>
```

注意引用方法是冒号，不是圆点，也不是斜杠。

2.6 编写一个简单的表单

接下来我们将继续编写投票应用，并专注于简单的表单处理，以及精简我们的代码。让我们更新一下在 2.5 节中编写的投票详细页面的模板("polls/detail.html")，让它包含一个 HTML 表单元素<form>：

扫码观看本节视频讲解

```
<h1>{{ question.question_text }}</h1>

{% if error_message %}<p><strong>{{ error_message }}</strong></p>{% endif %}

<form action="{% url 'polls:vote' question.id %}" method="post">
{% csrf_token %}
{% for choice in question.choice_set.all %}
    <input type="radio" name="choice" id="choice{{ forloop.counter }}" value="{{ choice.id }}">
    <label for="choice{{ forloop.counter }}">{{ choice.choice_text }}</label><br>
{% endfor %}
<input type="submit" value="Vote">
</form>
```

对上述代码的具体说明如下。

- 上面的模板在 Question 的每个 Choice 前添加一个单选按钮。每个单选按钮的 value 属性是对应的各个 Choice 的 ID。每个单选按钮的 name 是 "choice"。这意味着，当有人选择一个单选按钮并提交表单时，它将发送一个 POST 数据

- choice=#，其中#为选择的 Choice 的 ID。这是 HTML 表单的基本概念。
- 我们设置表单的 action 为{% url 'polls:vote' question.id %}，并设置 method="post"。使用 method="post"(与其相对的是 method="get")是非常重要的，因为这个提交表单的行为会改变服务器端的数据。无论何时，当你需要创建一个改变服务器端数据的表单时，请使用 method="post"。这不是 Django 的特定技巧；这是优秀的网站开发技巧。
- forloop.counter 指示 for 标签已经循环多少次。
- 创建一个 POST 表单(它具有修改数据的作用)，因为表单容易引起跨站点请求伪造的问题，所以我们必须保证这个表单的安全性。我们比较幸运的是，Django 内置了一个模块用来防御跨站点请求伪造问题。简而言之，所有针对内部 URL 的 POST 表单都应该使用 {% csrf_token %} 模板标签。

接下来创建一个 Django 视图来处理提交的数据，在前面为投票应用创建了一个 URLconf，其中包含下面这一行代码：

```
path('<int:question_id>/vote/', views.vote, name='vote'),
```

我们还创建了一个 vote() 函数的虚拟实现，接下来创建一个真实的版本。将下面的代码添加到视图文件 polls/views.py 中。

```
from django.http import HttpResponse, HttpResponseRedirect
from django.shortcuts import get_object_or_404, render
from django.urls import reverse

from .models import Choice, Question
# ...
def vote(request, question_id):
    question = get_object_or_404(Question, pk=question_id)
    try:
        selected_choice = question.choice_set.get(pk=request.POST['choice'])
    except (KeyError, Choice.DoesNotExist):
        # Redisplay the question voting form.
        return render(request, 'polls/detail.html', {
            'question': question,
            'error_message': "You didn't select a choice.",
        })
    else:
        selected_choice.votes += 1
        selected_choice.save()
        # 在成功处理 POST 数据后，始终返回 HttpResponseRedirect。
        # 这样可以防止在用户单击后退按钮时两次发布数据。
        # user hits the Back button.
        return HttpResponseRedirect(reverse('polls:results', args=(question.id,)))
```

对上述代码的具体说明如下。
- request.POST 是一个类似字典的对象，允许通过键名访问提交的数据。本例中，request.POST['choice']返回被选项的 ID，并且值的类型永远是 string 字符串，哪怕它看起来像数字！同样地，你也可以用类似的方法获取 GET 请求发送过来的数据。

- request.POST['choice']有可能触发一个 KeyError 异常,如果你的 POST 数据里没有提供 choice 键值,上面的代码会返回表单页面并给出错误提示。
- 在选择计数器加 1 后,返回的是一个 HttpResponseRedirect,而不是先前我们常用的 HttpResponse。HttpResponseRedirect 需要一个参数:重定向的 URL。当你成功处理 POST 数据后,应当保持一个良好的习惯,始终返回一个 HttpResponseRedirect。这不仅仅是对 Django 而言,它是一个良好的 Web 开发习惯。
- 我们在上面 HttpResponseRedirect 的构造器中使用了一个 reverse()函数。它能帮助我们避免在视图函数中硬编码 URL。它首先需要一个我们在 URLconf 中指定的 name,然后是传递的数据。例如'/polls/3/results/',其中的 3 是某个 question.id 的值。重定向后将进入 polls:results 对应的视图,并将 question.id 传递给它。直白地说,就是把任务扔给另外一个路由对应的视图去做。

当有人对某个主题投票后,vote()视图将重定向到问卷的结果显示页面。下面我们来写这个处理结果页面的视图(polls/views.py):

```
from django.shortcuts import get_object_or_404, render

def results(request, question_id):
    question = get_object_or_404(Question, pk=question_id)
    return render(request, 'polls/results.html', {'question': question})
```

同样还需要写一个新的模板文件 polls/templates/polls/results.html,代码如下。

```
<h1>{{ question.question_text }}</h1>

<ul>
{% for choice in question.choice_set.all %}
    <li>{{ choice.choice_text }} -- {{ choice.votes }}
vote{{ choice.votes|pluralize }}</li>
{% endfor %}
</ul>

<a href="{% url 'polls:detail' question.id %}">Vote again?</a>
```

现在可以到浏览器中访问/polls/1/进行投票,此时会看到一个结果页面,每投一次,它的内容就更新一次。如果你提交的时候没有选择项目,则会得到一个错误提示。

如果在你的投票主题中没有创建 choice 选项对象,那么可以按下面的操作进行补充。

```
E:\123\Django-daima\2\first\mysite>python manage.py shell
Python 3.6.1 (v3.6.1:69c0db5, Mar 21 2017, 18:41:36) [MSC v.1900 64 bit (AMD64)]
on win32
Type "help", "copyright", "credits" or "license" for more information.
(InteractiveConsole)
>>> from polls.models import Question
>>> q = Question.objects.get(pk=1)
>>> q.choice_set.create(choice_text='Not much', votes=0)
<Choice: Choice object>
>>> q.choice_set.create(choice_text='The sky', votes=0)
<Choice: Choice object>
```

```
>>> q.choice_set.create(choice_text='Just hacking again', votes=0)
<Choice: Choice object>
```

此时访问/polls/1/会显示这个投票主题的选项，如图 2-12 所示。

图 2-12　某个投票主题

2.7　用通用视图：减少重复代码

上面实现的 detail、index 和 results 视图的代码非常相似，这有点冗余，这是一个专业开发者所不能忍受的。它们都具有类似的业务逻辑，实现类似的功能：都是从 URL 传递过来的参数去数据库查询数据，然后加载一个模板，利用刚才的数据渲染模板，并返回这个模板。由于这个过程是如此的常见，Django 很善解人意地帮你想办法偷懒，于是它提供了一种快捷方式，名为"通用视图"。

扫码观看本节视频讲解

接下来我们将原来的代码改为使用通用视图的方式，整个过程分为如下三步。

- 修改 URLconf 设置。
- 删除一些旧的无用的视图。
- 采用基于类视图的新视图。

> **注　意**
>
> 为什么要重构代码？
>
> 一般来说，当编写一个 Django 应用时，你应该先评估一下通用视图是否可以解决你的问题。你应该在一开始就使用它，而不是进行到一半时再重构代码。本书目前为止是有意将重点放在以"艰难的方式"编写视图，就是将重点放在核心概念上。这就像在使用计算器之前，你需要掌握基础数学一样。

2.7.1　改良 URLconf

打开文件 polls/urls.py，将原来代码中的<question_id>修改为<pk>：

```
from django.urls import path

from . import views

app_name = 'polls'
urlpatterns = [
```

```python
    path('', views.IndexView.as_view(), name='index'),
    path('<int:pk>/', views.DetailView.as_view(), name='detail'),
    path('<int:pk>/results/', views.ResultsView.as_view(), name='results'),
    path('<int:question_id>/vote/', views.vote, name='vote'),
]
```

2.7.2 修改视图

打开视图文件 polls/views.py，删除 index、detail 和 results 视图，替换成 Django 的通用视图，代码如下。

```python
from django.http import HttpResponseRedirect
from django.shortcuts import get_object_or_404, render
from django.urls import reverse
from django.views import generic

from .models import Choice, Question


class IndexView(generic.ListView):
    template_name = 'polls/index.html'
    context_object_name = 'latest_question_list'

    def get_queryset(self):
        """Return the last five published questions."""
        return Question.objects.order_by('-pub_date')[:5]


class DetailView(generic.DetailView):
    model = Question
    template_name = 'polls/detail.html'


class ResultsView(generic.DetailView):
    model = Question
    template_name = 'polls/results.html'


def vote(request, question_id):
    ... # 同前面的一样，不需要修改
```

在上述代码中，使用了两种通用视图 ListView 和 DetailView(它们是作为父类被继承的)。这两者分别代表"显示一个对象的列表"和"显示特定类型对象的详细页面"的抽象概念。每一种通用视图都需要知道它要作用在哪个模型上，这通过 model 属性提供。DetailView 需要从 url 捕获到称为"pk"的主键值，因此我们在 URL 文件中将 2 条目和 3 条目的<question_id>修改成了<pk>。

在默认情况下，通用视图 DetailView 使用一个称作 <app_name>/<model_name>_detail.html 的模板。在本例中，实际使用的是 polls/detail.html。属性 template_name 就是用来指定这个模板名的，用于代替自动生成的默认模板名(一定要仔细观察上面的代码，对号入座，注意细节)。同样地，在 results 列表视图中，指定 template_name 为

'polls/results.html'，这样就确保了虽然 results 视图和 detail 视图同样继承了 DetailView 类，使用了同样的 model:Question，但它们依然会显示不同的页面。

类似地，通用视图 ListView 使用一个默认模板：<app name>/<model name>_list.html。我们也使用 template_name 这个变量来告诉 ListView 使用我们已经存在的 "polls/index.html" 模板，而不是使用它自己默认的那个。

在前面的步骤中，我们给模板提供了一个包含 question 和 latest_question_list 的上下文变量。而对于 DetailView，question 变量会被自动提供，因为我们使用了 Django 的模型(Question)，Django 会智能地选择合适的上下文变量。然而，对于 ListView，自动生成的上下文变量是 question_list。为了覆盖它，我们提供了 context_object_name 属性，表示我们希望使用 latest_question_list 而不是 question_list。

现在可以运行 Django 服务器，然后试试基于类视图的应用程序了。类视图是 Django 比较高级的一种用法，初学者可能不太好理解，我们先有个印象，详情在后面讲解。

2.8 静态文件

除了前面介绍的 HTML 模板文件外，在 Django Web 中还可以使用图片文件、JavaScript 脚本文件和 CSS 样式表文件。因为这些文件的内容基本是固定不变的，不需要动态生成，所以通常将这些文件统称为"静态文件"。

扫码观看本节视频讲解

对于小型项目来说，可以将静态文件放在任何 Web 服务器能够找到的地方。但是对于大型项目来说，尤其是那些包含多个 app 在内的项目，处理那些由 app 带来的多套不同的静态文件是非常麻烦的。这时候可以考虑使用 django.contrib.staticfiles，它可以收集每个应用(和任何你指定的地方)的静态文件到一个统一指定的地方，并且易于访问。

2.8.1 使用 CSS 自定义应用的风格

首先在 polls 目录下创建一个名为 static 的目录。Django 将在该目录下查找静态文件，这种方式和 Diango 在 polls/templates/目录下查找 template 模板的方式类似。

在 Django 的 STATICFILES_FINDERS 设置中包含了一系列的查找器，它们知道去哪里找到 static 文件。AppDirectoriesFinder 是默认查找器中的一个，它会在每个 INSTALLED_APPS 中指定的应用的子文件中寻找名称为 static 的特定文件夹，就像我们在 polls 中刚创建的那个一样。管理后台采用相同的目录结构管理它的静态文件。

在刚创建的 static 文件夹中创建一个名为 polls 的文件夹，然后再在 polls 文件夹中创建一个名为 style.css 的文件。也就是说，这个样式文件的路径是 polls/static/polls/style.css。因为 AppDirectoriesFinder 的存在，我们可以在 Django 中简单地以 polls/style.css 的形式引用此文件，类似引用模板路径的方式。

2.8.2 静态文件命名空间

虽然我们可以像管理模板文件一样，把 static 文件直接放入 polls/static ——而不是创建另一个名为 polls 的子文件夹，但是这实际上是一个很麻烦的做法。Django 只会使用第一个找到的静态文件。如果你在其他应用中有一个相同名字的静态文件，Django 将无法区分它们。我们需要指引 Django 选择正确的静态文件，而最简单的方式就是把它们放入各自的命名空间。也就是说，需要把这些静态文件放入另一个与应用名相同的目录中。

在样式表文件 polls/static/polls/style.css 中编写如下代码。

```
li a {
    color: green;
}
```

接下来在模板文件 polls/templates/polls/index.html 的开头添加以下代码。

```
{% load static %}
<link rel="stylesheet" type="text/css" href="{% static 'polls/style.css' %}">
```

在上述代码中，模板标签{% static %}会生成静态文件的绝对路径。

最后在系统配置文件 settings.py 中设置引用 static，代码如下。

```
STATIC_URL = '/static/'
STATICFILES_DIRS = (
os.path.join(BASE_DIR, "static"),
)
```

此时通过如下命令启动本项目：

```
python manage.py runserver
```

重新浏览 http://localhost:8000/polls/ 会发现有问题的链接是绿色的（这是 Django 自己的问题标注方式），这意味着你追加的样式表起作用了，如图 2-13 所示。

图 2-13　绿色的超级链接颜色

2.8.3 添加一个背景图

创建一个用于存放图像的目录，在 polls/static/polls 目录下创建一个名为 images 的子目录。在这个目录中，存放一张名为 background.jpg 的图片。换言之，在目录 polls/static/polls/images/中存放一张图片。然后在样式文件 polls/static/polls/style.css 中添加如下代码。

```
body {
    background: white url("images/background.jpg ") no-repeat;
}
```

此时在浏览器重载 http://localhost:8000/polls/，将在屏幕的左上角见到这张背景图，如图 2-14 所示。

图 2-14　显示背景图

2.9　重新设计后台

在前面我们使用的是 Django 自动创建的后台系统，整个界面比较简单，并且没有为某个投票主题设置投票选项的功能，这是不符合现实投票系统需求的。在接下来的内容中，将详细讲解重新设计后台页面的方法。

扫码观看本节视频讲解

2.9.1　自定义后台表单

在前面的内容中，通过使用 admin.site.register(Question)语句，我们在 admin 后台中注册了 Question 模型，Django 会自动生成一个该模型的默认表单页面。如果想自定义该页面的外观和工作方式，可以在注册对象的时候告诉 Django 我们创建的自定义选项。我们可以通过修改文件 polls/admin.py 中代码的方式，达到修改 admin 表单默认排序方式的目的。例如：

```
from django.contrib import admin
from .models import Question

class QuestionAdmin(admin.ModelAdmin):
    fields = ['pub_date', 'question_text']

admin.site.register(Question, QuestionAdmin)
```

我们只需要创建一个继承 admin.ModelAdmin 的模型管理类，然后将它作为第二个参数传递给 admin.site.register()，第一个参数是 Question 模型的本身。通过上面的修改，让字段 Publication date 显示在字段 Question 的前面(默认是在后面)。此时执行后的效果如图 2-15 所示。

图 2-15　某个投票主题(一)

当表单中含有大量字段的时候，可以将表单划分为一些字段的集合。再次修改文件 polls/admin.py：

```python
from django.contrib import admin

from .models import Question


class QuestionAdmin(admin.ModelAdmin):
    fieldsets = [
        (None,               {'fields': ['question_text']}),
        ('Date information', {'fields': ['pub_date']}),
    ]
admin.site.register(Question, QuestionAdmin)
```

此时在字段集合 fieldsets 中每一个元组的第一个元素是该字段集合的标题，此时的执行效果如图 2-16 所示。

图 2-16　某个投票主题(二)

2.9.2　添加关联对象

虽然我们已经有了 Question(投票主题)的管理页面，但是在现实投票系统中，一个 Question 有多个 Choices(投票选项)，如果想在后台显示 Choices 的内容该如何实现呢？可以使用如下两个方法解决这个问题。

（1）像 Question 一样将 Choices 注册到 admin 站点，这很容易，只需修改文件 polls/admin.py，增加下面的内容：

```
from django.contrib import admin
from .models import Choice, Question

# ...
admin.site.register(Choice)
```

此时重新执行本实例，就可以在后台主页中看到 Choices 条目了，效果如图 2-17 所示。

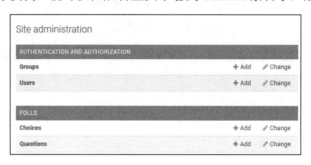

图 2-17　后台主页

单击右边的 Add 按钮，进入 Add choice 表单页面，在这个页面中可以为某投票主题添加投票选项，如图 2-18 所示。

在 Add Choice 表单页面中，Question 字段是一个 select 选择框，包含当前数据库中所有的 Question 实例。在 Django 的 admin 站点中，自动地将所有的外键关系展示为一个 select 选择框。在我们的例子中，目前只有一个 question 对象存在。请注意图中的绿色加号➕，它连接到 Question 模型，每一个包含外键关系的对象都会有这个绿色加号。单击➕会弹出一个新增 Question 的表单，类似 Question 自己的添加表单。填入相关信息后单击 SAVE 按钮，Django 自动将该 Question 保存在数据库，并作为当前 Choice 的关联外键对象。也就是说，新建一个 Question 并作为当前 Choice 的外键。

图 2-18　添加投票选项

（2）上一种在后台显示 Choices 内容的方式的效率一般，更好的做法是在创建 Question 对象的时候就直接添加一些 Choice，这就是我们要说的第二种方法。首先，在文件 polls/admin.py 中删除 Choice 模型对 register()方法的调用。然后，编辑 Question 的内容，最后整个文件的代码应该如下。

```python
from django.contrib import admin

from .models import Choice, Question

class ChoiceInline(admin.StackedInline):
    model = Choice
    extra = 3

class QuestionAdmin(admin.ModelAdmin):
    fieldsets = [
        (None,               {'fields': ['question_text']}),
        ('Date information', {'fields': ['pub_date'], 'classes': ['collapse']}),
    ]
    inlines = [ChoiceInline]

admin.site.register(Question, QuestionAdmin)
```

通过上述代码，告诉 Django 系统 Choice 对象将在 Question 管理页面进行编辑。在默认情况下，会提供 3 个 Choice 对象(投票选项)的编辑区域。此时重新运行本实例，Add question 页面的执行效果如图 2-19 所示。

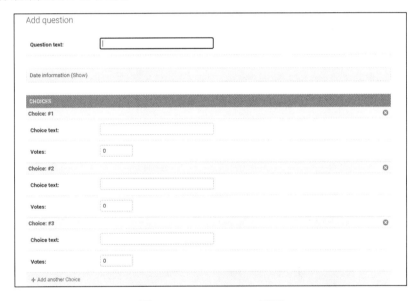

图 2-19　Add question 页面

并且编辑修改某个投票主题的页面(例如 http://127.0.0.1:8000/admin/polls/question/1/change/)变成如图 2-20 所示的效果。

在 Add question 页面的最后有一个 Add another Choice 链接，单击此链接后可弹出一个新的表单选项，用于为当前投票主题添加一个新的投票选项。如果你想删除新增的表单选项，单击它最右边的灰色图标 ⊗ 即可。但是，不可以删除默认生成的 3 个 Choice 对象(投票选项)。

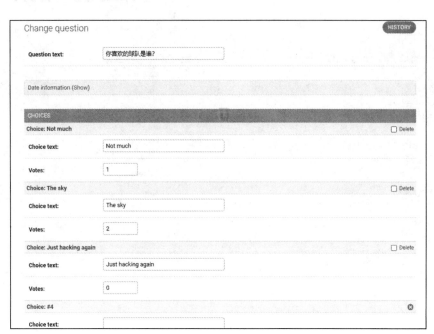

图 2-20　Change question 页面

此时我们的程序还有一个小小的问题，上面页面中每个投票选项下面的表单需要占据大块的页面空间，查看起来很不方便。为此，Django 提供了一种扁平化的显示方式，仅仅只需在文件 polls/admin.py 中修改 ChoiceInline 继承的类为 admin.TabularInline，替代先前的 StackedInline 类(其实，从类名上你就能看出两种父类的区别)即可，代码如下。

```
class ChoiceInline(admin.TabularInline):
    #...
```

此时刷新一下页面，会看到用规整的、类似表格的方式显示 Add question 页面，如图 2-21 所示。

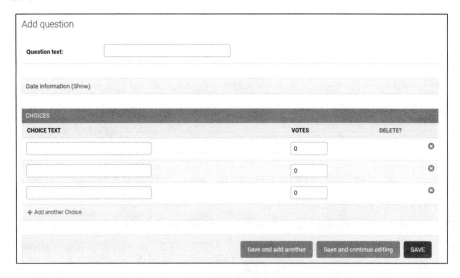

图 2-21　规整的显示方式

2.9.3 定制实例的列表页面

接下来开始装饰"实例列表"(change list)页面,该页面显示了当前系统中所有的 questions(投票主题)实例。在默认情况下,显示效果如图 2-22 所示。

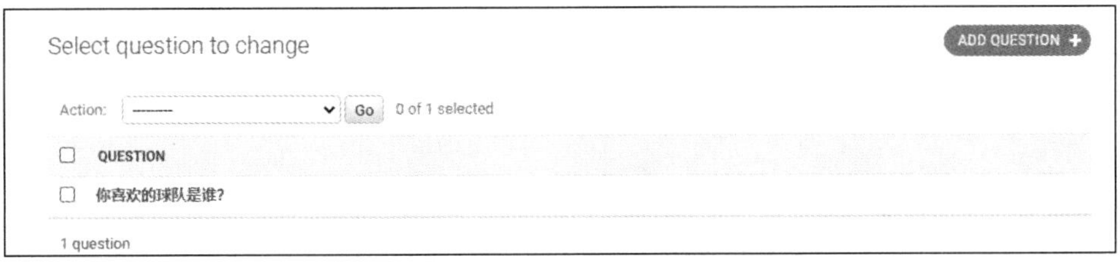

图 2-22 投票主题列表页面(一)

通常,Django 只会显示方法 __str()__ 指定的内容。但是在很多时候,我们可能要同时显示一些别的内容,此时需要借助 list_display 属性来实现。list_display 是一个由字段组成的元组,其中的每一个字段都会按顺序显示在 change list 页面上。修改文件 polls/admin.py 的代码:

```
class QuestionAdmin(admin.ModelAdmin):
    # ...
    list_display = ('question_text', 'pub_date', 'was_published_recently')
```

通过上述代码,把方法 was_published_recently()的结果也显示出来了。此时投票主题列表页面的执行效果如图 2-23 所示。

图 2-23 投票主题列表页面(二)

你可以单击每一列的标题,来根据这列的内容进行排序。但是 was_published_recently 列除外,它不支持这种根据函数输出结果进行排序的方式。同时需要注意,was_published_recently 这一列的列标题默认是方法的名字,内容是输出的字符串表示形式。

我们可以通过给方法提供一些属性的方法来改进输出的样式,例如修改文件 polls/models.py,增加最后面三行代码:

```
class Question(models.Model):
    # ...
    def was_published_recently(self):
        now = timezone.now()
        return now - datetime.timedelta(days=1) <= self.pub_date <= now
    was_published_recently.admin_order_field = 'pub_date'
```

```
was_published_recently.boolean = True
was_published_recently.short_description = 'Published recently?'
```

此时在投票主题列表页面中新增了"PUBLISHED RECENTLY?",执行效果如图2-24所示。

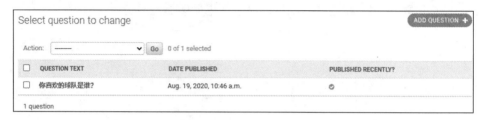

图 2-24　投票主题列表页面(三)

另外,我们还可以对显示结果进行过滤。通过使用 list_filter 属性,在文件 polls/admin.py 的 QuestionAdmin 中添加如下代码。

```
list_filter = ['pub_date']
```

此时在投票主题列表页面的右侧多出了一个基于 pub_date 的过滤面板,执行效果如图 2-25 所示。

图 2-25　右侧新增的基于 pub_date 的过滤面板

根据我们选择的不同的过滤条件,Django 会在面板中添加不同的过滤选项。因为 pub_date 是一个 DateTimeField,所以 Django 自动添加了一些选项,如 Any date、Today、Past 7 days、This month、This year。

在文件 polls/admin.py 的 QuestionAdmin 中,添加如下代码可以实现搜索功能。

```
search_fields = ['question_text']
```

这会在页面的顶部增加一个搜索框,如图 2-26 所示。当输入搜索关键字后,Django 会在 question_text 字段内进行搜索。只要你愿意,可以使用任意多个搜索字段,Django 在后台使用的都是 SQL 查询语句的 LIKE 语法,但是有限制的搜索字段有助于提高后台数据库的查询效率。

其实本页面还提供分页功能,默认每页显示 100 条,只是现在我们实例中只有一个投票主题,所以看到分页链接。

图 2-26　在顶部新增的搜索表单

2.9.4　定制 admin 整体界面

在每一个项目的 admin 页面顶端都显示 Django administration 是不合理的，可利用 Django 的模板系统快速修改它。

(1) 在文件 manage.py 的同级目录下创建一个 templates 目录，然后打开设置文件 mysite/settings.py，在 TEMPLATES 条目中添加一个 DIRS 选项：

```
TEMPLATES = [
    {
        'BACKEND': 'django.template.backends.django.DjangoTemplates',
        'DIRS': [os.path.join(BASE_DIR, 'templates')],  # 要有这一行，如果已经存在
请保持原样
        'APP_DIRS': True,
        'OPTIONS': {
            'context_processors': [
                'django.template.context_processors.debug',
                'django.template.context_processors.request',
                'django.contrib.auth.context_processors.auth',
                'django.contrib.messages.context_processors.messages',
            ],
        },
    },
]
```

DIRS 是一个文件系统目录的列表，是模板的搜索路径。当加载 Django 模板时，会在 DIRS 中进行查找。

虽然和静态文件一样，我们可以把所有的模板都放在一起，形成一个大大的模板文件夹，但是不推荐这么做。正确的做法是，让每一个模板都存放在它所属应用的模板目录内(例如 polls/templates)，而不是整个项目的模板目录(templates)，因为这样每个应用才可以被方便和正确地重用。只有对整个项目有作用的模板文件才放在根目录的 templates 中，比如 admin 界面。

(2) 在刚才创建的 templates 目录中再创建一个 admin 目录，然后将模板文件 admin/base_site.html 复制到该目录内。这个 HTML 文件来自 Django 源码，它位于 django/contrib/admin/templates 目录内。例如在笔者的 Windows 系统中，位于 C:\Python39\Lib\site-packages\django\contrib\admin\templates\admin 目录中。

当然我们也可以自定义设计一个后台页面，例如：

```
{% extends "admin/base.html" %}

{% block title %}{{ title }} | {{ site_title|default:_('Django site admin') }}{% endblock %}

{% block branding %}
<h1 id="site-name"><a href="{% url 'admin:index' %}">投票站点管理界面</a></h1>
{% endblock %}

{% block nav-global %}{% endblock %}
```

在上述代码中，使用硬编码方式强行改名为"投票站点管理界面"。但是在实际的项目中，可以使用django.contrib.admin.AdminSite.site_header属性，方便地对这个页面title进行自定义。此时后台主页的执行效果如图2-27所示。

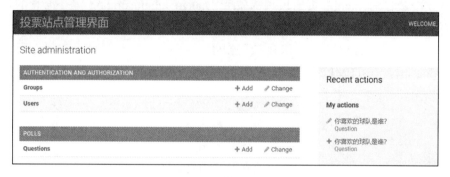

图 2-27　后台主页

> **注　意**
>
> 所有 Django 默认的 admin 模板都可以被重写，类似刚才重写模板文件 base_site.html 的方法，从源代码目录将 HTML 文件复制至你自定义的目录内，然后修改文件。

2.9.5　定制 admin 首页

在默认情况下，admin 首页会显示所有 INSTALLED_APPS 中在 admin 应用中注册过的 app，以字母顺序进行排序。要想自定义设计 admin 首页，需要重写模板文件 admin/index.html，就像前面修改模板文件 base_site.html 的方法一样，从源码目录复制到你指定的目录内。编辑该文件，你会看到文件内使用了一个 app_list 模板变量。该变量包含了所有已经安装的 Django 应用。你可以硬编码链接到指定对象的 admin 页面，使用任何你认为好的方法，来替代这个 app_list。

第 3 章

视 图 层

　　Django Web 开发遵循经典软件设计开发的 MVC 模式。View(视图)模块的主要功能是根据用户的请求返回数据,用于展示用户可以看到的内容(比如网页、图片),也可以用来处理用户提交的数据,比如保存到数据库中。在本章的内容中,将详细讲解 Django Web 视图层的知识。

3.1 视图层介绍

Django 视图层用于处理用户的请求并返回响应，可以对外接收用户请求，对内调度模型层和模板层，调度数据库和前端，最后根据业务逻辑，将处理好的数据与前端结合并返回给用户。

3.1.1 分析 View 视图的作用

扫码观看本节视频讲解

在 Django 框架中，当服务器收到用户通过浏览器发来的请求后，会根据在文件 urls.py 中设置的 URL 导航去视图 View 中查找与请求对应的处理方法，将找到的内容返回给客户端 HTTP 页面数据。

1. 一个最简单的视图 View

例如在下面的视图文件 views.py 中，设置当用户发来一个请求 request 时，通过 HttpResponse 打印输出文本"人生苦短，我用 Python!"。

```
from django.http import HttpResponse
def index(request):
    return HttpResponse("人生苦短，我用Python!")
```

在 Django Web 程序中，View 模块的功能是不仅从数据库提取数据，而且还设置要显示数据库内容的模板，并提供模板渲染页面所需的内容对象(Context Object)。

2. 一个博客的例子

下面举一个新闻博客的例子，假设设置/blog/这个 URL 展示博客中所有的文章列表，而在/blog/articles/<int:id>/这个 URL 展示具体文章的详细内容。那么我们可以首先通过如下文件 blog/urls.py 设置 URL。

```
from django.urls import path

from . import views

urlpatterns = [
    path('blog/', views.index, name='index'),
    path('blog/articles/<int:id>/', views.articles_detail,
name='articles_detail'),
]
```

在上述演示代码中，如果用户在浏览器输入/blog/，当 URL 收到客户端请求后会调用视图文件 views.py 中的 index 方法，这样可以列表显示博客中所有的文章。如果用户在浏览器输入/blog/articles/<int:id>/，URL 不仅会调用文件 views.py 中的 articles 方法，而且还会把表示博客文章编号的参数 id 通过<int:id>的形式传递给视图中的 articles_detail 方法。

接下来可以通过如下文件 blog/views.py 实现视图界面。

```python
from django.shortcuts import render, get_object_or_404
from .models import Articles
# 展示所有文章
def index(request):
    latest_articless = Article.objects.all().order_by('-pub_date')
    return render(request, 'blog/articles_list.html', {"latest_articles": latest_articles})
# 展示所有文章
def articles_detail(request, id):
    article = get_object_or_404(Articles, pk=id)
    return render(request, 'blog/articles_detail.html', {"articles": articles})
```

在上述代码中用到了如下所示的 4 个方法。

(1) 方法 index：提取要展示的数据对象列表 latest_articles，然后通过方法 render 传递给模板 blog/articles_list.html，这样可以列表展示在数据库中保存的文章。

(2) 方法 article_detail：通过 id 获取某篇博客文章的对象 articles，然后通过 render 方法传递给模板 blog/articles_detail.html 显示，这样可以显示某篇博客文章的详细信息。

(3) 方法 render：功能是返回页面内容，此方法有如下所示的 3 个参数。

- 参数 request：用于生成此响应的请求对象。
- 参数'blog/articles_detail.html'：表示模板的名称和位置。
- 参数"articles": articles：表示需要传递给模板的内容，也被称为 context object。

(4) 方法 get_object_or_404：功能是调用 Django 的 get 方法查询要访问的对象，如果不存在会抛出一个 HTTP 404 异常。

接下来看模板文件，在模板中可以直接调用通过上面视图文件传递过来的内容。例如，下面是模板文件 blog/article_list.html 的实现代码。

```
{% block content %}
{% for articles in latest_articles %}
    {{ article.title }}
    {{ article.pub_date }}
{% endfor %}
{% endblock %}
```

例如，下面是模板文件 blog/articles_detail.html 的实现代码。

```
{% block content %}
{{ article.title }}
{{ article.pub_date }}
{{ article.body }}
{% endblock %}
```

3.1.2 实战演练：使用简易 View 视图文件实例

在下面的实例代码中，演示了在 Django 框架中使用表单计算数字求和的过程。

> 源码路径：**daima\3\myshitu**

(1) 通过如下命令新建一个项目(project)，设置名称为 myshitu。

```
django-admin.py startproject myshitu
```

(2) 通过如下命令新建一个应用(app)，设置名称为 shitu。

```
python manage.py startapp shitu  # shitu 是一个 app 的名称
```

(3) 把新定义的 app 加到设置文件 settings.py 中，修改文件 myshitu\myshitu\settings.py 中的 INSTALLED_APPS 部分为：

```
INSTALLED_APPS = [
    'django.contrib.admin',
    'django.contrib.auth',
    'django.contrib.contenttypes',
    'django.contrib.sessions',
    'django.contrib.messages',
    'django.contrib.staticfiles',

    'shitu',
]
```

如果不将新建的 app 添加到 INSTALLED_APPS 中，Django 就不能自动找到 app 中的模板文件(app-name/templates/中的文件)和静态文件(app-name/static/中的文件)。

(4) 定义视图函数(访问页面时的内容)。

在 shitu 目录中打开文件 views.py，修改为如下所示的源代码。

```python
# coding:utf-8
from django.shortcuts import render

from django.http import HttpResponse


def index(request):
    return HttpResponse(u"欢迎光临 Python 架构师大舞台！")
```

在上述代码中通过使用 HttpResponse 向网页返回内容，就像 Python 中的 print 一样，只不过 HttpResponse 是把内容显示到网页上。还定义了 index()函数，其参数必须是 request，与网页发来的请求有关，在 request 变量中包含 get 或 post 的内容、用户浏览器、系统等信息。函数 index()返回一个 HttpResponse 对象，最终在网页中输出显示文本"欢迎光临 Python 架构师大舞台！"。

(5) 定义和视图函数相关的 URL(网址)文件 myshitu\myshitu\url.py，设置访问指定 URL 地址时对应的视图内容。文件 url.py 的代码如下所示。

```python
from django.contrib import admin
from django.urls import path
from shitu import views as shitu_views  # new


urlpatterns = [
```

```
    path('', shitu_views.index),  # new
    path('admin/', admin.site.urls),
]
```

(6) 通过如下命令运行程序，执行效果如图 3-1 所示。

```
python manage.py runserver
```

图 3-1 执行效果

3.2 URL 调度器

URL 是 Web 服务的入口，用户通过浏览器发送过来的任何请求，都是发送到一个指定的 URL 地址，然后被响应。在 Django Web 项目中编写路由，就是向外暴露我们接收哪些 URL 的请求，除此之外的任何 URL 都不被处理，也没有返回。通俗地理解，URL 路由是我们的 Web 服务对外暴露的 API。对于高质量的 Web 应用来说，使用简洁、优雅的 URL 模式是一个非常值得重视的细节。Django 允许自由地设计 URL，而不受框架束缚。

扫码观看本节视频讲解

3.2.1 URL 调度器介绍

为了给一个应用程序设计 URL，需要创建一个 Python 模块，通常被称为 URLconf(URL configuration)。这个模块是纯粹的 Python 代码，包含 URL 模式(简单的正则表达式)到 Python 函数(你的视图)的简单映射。具体映射可短可长，也可以引用其他映射。而且，因为它是纯粹的 Python 代码，所以可以动态构造。

URL 路由在 Django 项目中的体现就是文件 urls.py，在一个项目中可以有很多个 urls.py 文件，但是绝对不会在同一目录下。实际上 Django 提倡项目有个根 urls.py，每个 app 下分别有自己的一个 urls.py，既集中又分治，是一种解耦的模式。

我们随便新建一个 Django 项目，会自动默认为创建一个/project_name/urls.py 文件，并且自动生成下面的代码，这就是项目的根 URL。

```
from django.contrib import admin
from django.urls import path

urlpatterns = [
    path('admin/', admin.site.urls),
]
```

在上述代码中，Django 会默认导入 path 方法和 admin 模块，然后有一条指向 admin 后台的 url 路径。当一个用户请求 Django Web 中的一个网页时，下面是 Django 系统决定执行哪个 Python 代码的算法。

（1）Django 确定使用根 URLconf 模块。通常，这是 ROOT_URLCONF 设置的值，但是如果传入的 HttpRequest 对象拥有 urlconf 属性(通过中间件设置)，它的值将被用来代替 ROOT_URLCONF 设置。也就是说，此时可以自定义项目入口 URL 是哪个文件。

（2）加载这个模块的程序并寻找可用的 urlpatterns，这通常是 django.urls.path()或者 django.urls.re_path()实例的一个列表。

（3）依次匹配每个 URL 模式，在与请求的 URL 相匹配的第一个模式停下。也就是说，URL 匹配是从上往下的短路操作，正因如此，URL 在列表中的位置非常关键。

（4）导入并调用匹配行中给定的视图，该视图是一个简单的 Python 函数(被称为视图函数)，或基于类的视图。视图将获得如下参数。

- 一个 HttpRequest 实例。
- 如果匹配的表达式返回了未命名的组，那么匹配的内容将作为位置参数提供给视图。
- 关键字参数由表达式匹配的命名组成，但是可以被 django.urls.path()的可选参数 kwargs 覆盖。

如果没有匹配到任何表达式，或者在匹配过程中抛出异常，则 Django 会调用一个适当的错误处理视图。

例如，下面是一个使用 URLconf 的文件 urls.py。

```
from . import views
urlpatterns = [
    path('articles/2003/', views.special_case_2003),
    path('articles/<int:year>/', views.year_archive),
    path('articles/<int:year>/<int:month>/', views.month_archive),
    path('articles/<int:year>/<int:month>/<slug:slug>/',
views.article_detail),
]
```

通过上述代码可知：

- 要捕获 url 中的值，需要使用尖括号，而不是圆括号。
- 可以转换捕获到的值为指定类型，比如上述代码中的 int。在默认情况下，捕获到的结果保存为字符串类型，不包含"/"这个特殊字符。
- 匹配模式的开头不需要添加斜杠"/"，因为在默认情况下，在每个 url 的前面都带一个斜杠"/"，既然是大家都有的部分，就不用再特别写一个了。

对应上述代码，下面是不同的 URL 访问请求的匹配例子。

- /articles/2005/03/：将匹配第三条，并调用 views.month_archive(request, year=2005, month=3)。
- /articles/2003/：将匹配第一条，并调用 views.special_case_2003(request)。
- /articles/2003：一条都匹配不上，因为它最后少了一个斜杠，而列表中的所有模式中都以斜杠结尾。
- /articles/2003/03/building-a-django-site/：将匹配最后一个，并调用函数 views.article_detail(request, year=2003, month=3, slug="building-a-django-site")。

3.2.2 Django URL 调度器的工作原理

在 Django 框架中,因为 URL 和 View(视图)模块与 MVC 模式下的 Controller 相对应,所以 URL 通常被用在视图(View)模块程序中。那么视图是什么呢?视图是当 Web 服务器收到用户的请求后,将请求结果返回给用户看的内容。

当服务器端收到用户请求后,会根据文件 urls.py 中的关系条目,在视图 View 中查找与请求对应的处理方法,找到后返回给客户端 http 页面数据。有其他 Web 开发框架经验的读者会发现,上述过程和其他 Web 开发模块中的路由机制(Router)的原理类似。

假设有一个博客系统,其 URL 文件 blog/urls.py 的代码如下所示。

```
from django.urls import path

from . import views

urlpatterns = [
   path('blog/', views.index),
    path('blog/article/<int:id>/', views.article),
]

# blog/views.py
def index(request):
    # 展示所有文章

def article(request, id):
    # 展示某篇具体文章
```

上述代码的工作流程如下所示。

(1) 如果用户在浏览器中输入/blog/,URL 在收到这个请求后会调用视图文件 views.py 中的 index 方法,可以通过方法 index 展示所有的文章。

(2) 如果用户在浏览器中输入/blog/article/<int:id>/,URL 不仅会调用文件 views.py 中的 article 方法,而且还会把表示文章 id(编号)的参数通过尖括号<>的形式传递给视图。此处的参数 int 表示只传递整数,传递的参数名字是 id。

在配置 URL 时需要把 Web 程序的 urls(例如 blog 的 urls:blog.urls)加入项目中的 URL 配置中(mysite/urls.py),例如如下所示的配置代码。

```
from django.conf.urls import url, include

urlpatterns = [
   url(r'^/', include('blog.urls')),
]
```

3.2.3 路径转换器

在 Django 框架中主要有三种传递 URL 参数的方法,分别是 path 路径转换器、自定义 path 路径转换器和_re_path,功能是通过 URL 把参数传递给视图 View。从 Django 2.0 开始,

Django 官方迫于压力和同行的影响，不得不将原来的正则匹配表达式改为更加简单的 path 表达式，但依然通过 re_path()方法保持对 1.x 版本的兼容。

1. 路径转换器 path

实现正常参数传递，有如下两种传递格式：

```
<变量类型:变量名>
<变量名>
```

例如<int:id>，<slug:slug>或<username>。在默认情况下，Django 内置了下面的路径转换器。

- str：匹配任何非空字符串，但不含斜杠 "/"，如果你没有专门指定转换器，那么这个是默认使用的。
- int：匹配 0 和正整数，返回一个 int 类型。
- slug：可理解为注释、后缀、附属等概念，通常位于 url 的最后，是一些解释性字符。该转换器匹配任何 ASCII 字符以及连接符和下划线，比如 building-your-1st-django-site。
- uuid：匹配一个 uuid 格式的对象。为了防止冲突，规定必须使用破折号，所有字母必须小写，例如 075194d3-6885-417e-a8a8-6c931e272f00。返回一个 UUID 对象。
- path：匹配任何非空字符串，重点是可以匹配包含路径分隔符 "/" 的字符串。这个转换器可以帮助你匹配整个 url 而不是一段一段的 url 字符串。要区分 path 转换器和 path()方法。

2. 注册自定义的路径转换器

对于更复杂的匹配需求，需要我们定义自己的路径转换器。将转换器定义为一个类，在其中包含下面的内容。

- 类属性 regex：一个字符串形式的正则表达式属性。
- 方法 to_python(self, value)：用来将匹配到的字符串转换为我们想要的数据类型，并传递给视图函数。如果转换失败，它必须弹出 ValueError 异常。
- 方法 to_url(self, value)：将 Python 数据类型转换为一段 url 的方法，是上面方法的反向操作。

例如，在 urlconf 的同目录下新建文件 converters.py，然后编写下面的类。

```
class FourDigitYearConverter:
    regex = '[0-9]{4}'

    def to_python(self, value):
        return int(value)

    def to_url(self, value):
        return '%04d' % value
```

然后在 URLconf 中注册上述类 FourDigitYearConverter，并使用它注册 yyyy。

```
from django.urls import register_converter, path

from . import converters, views

register_converter(converters.FourDigitYearConverter, 'yyyy')

urlpatterns = [
    path('articles/2003/', views.special_case_2003),
    path('articles/<yyyy:year>/', views.year_archive),
    ...
]
```

3. re_path 方法

虽然从 Django 2.0 开始修改了 URL 配置方法，但依然兼容老版本。而这个兼容的办法，就是用 re_path()方法代替 path()方法。使用正则表达式 regex 匹配，采用命名组的方式传递参数，具体格式如下所示：

(?P<变量名>表达式)

其实 re_path()方法就是以前的 url()方法，只不过导入的位置变了。例如在下面的演示文件 blog/urls.py 中，通过使用上述 path 方法和 re_path 方法传递文章 id 给视图，这两种方式的功能是完全一样的。

```
from django.urls import path, re_path

from . import views

urlpatterns = [
    path('blog/article/<int:id>/', views.article, name = 'article'),
    re_path(r'^blog/article/(?P<id>\d+)/$', views.article, name='article'),
]

# View (in blog/views.py)
def article(request, id):
    # 展示某篇文章
```

在上述代码中，在 re_path 中的单引号前面有一个小写字母 r，表示引号里为正则表达式，请忽略'\'不要转义，^代表开头，$代表结尾，\d+代表正整数。

在上述演示代码中还给 URL 取了一个名字 'article'，这相当于给 URL 取了一个全局变量的名字。它能够让你在 Django 的任意处，尤其是模板内显式地引用它。假设你需要在模板中通过链接指向一篇具体文章，我们可以使用如下两种方法实现。

- 第一种方法：使用命名 URL 方法实现。

```
<a href="{% url 'article' id %}">Article</a>
```

- 第二种方法：使用常规 URL 方法实现。

```
<a href="blog/article/id">Article</a>
```

很明显使用第一种方法是最好的选择，假设需要把全部模板链接由 blog/article/id 改为

blog/articles/id，第二种方法需要修改所有模板，而第一种方法只要修改 URL 配置中的一个字母，所以使用第一种方法会更高效。

命名 URL 一般只能在模板里使用，而不能直接在视图中使用。如果我们有了命名的 URL，如何把它转化为常规的 URL 在视图里使用呢？在 Django 框架中可以使用内置方法 reverse()实现。假设在不同的 Web 项目(比如 news 或 blog)中都有 article 这个命名 URL，我们怎么区分呢？ 我们只需要在 article 前面加上 blog 或 news 这个命名空间即可，例如下面的演示代码：

```
from django.urls import reverse
reverse('blog:article', args=[id])
```

4. 嵌套参数

在使用正则表达式时允许嵌套参数，Django 将处理这些嵌套参数并传递给视图。当转换时，Django 将试着填充给所有外部捕捉参数，忽略任何嵌套捕捉参数。请看下面可选的带有页面参数的 URL 模式：

```
from django.urls import re_path

urlpatterns = [
    re_path(r'^blog/(page-(\d+)/)?$', blog_articles),                    # bad
    re_path(r'^comments/(?:page-(?P<page_number>\d+)/)?$', comments),   # good
]
```

在上述代码中用到了两个使用嵌套参数的模式。例如，blog/page-2/ 将匹配给 blog_articles 并带有两个参数: page-2/和 2。第二个模式为 comments 匹配 comments/page-2/ 并带有设置为 2 的关键参数 page_number。在上述代码中，外部参数是一个非捕捉参数"?"。

视图 blog_articles 需要反转最外层捕捉的参数，page-2/在这里不需要参数，而 comments 可以在没有参数或 page_number 值的情况下反转。

嵌套捕捉参数在视图参数和 URL 之间创建一个强耦合，例如 blog_articles 视图接收部分 URL (page-2/) 而不只是视图要的值。当反转时这种耦合更明显，因为反转视图我们需要传递一段 URL 而不是 page_number。只有当正则表达式需要一个参数但视图忽略它时，才捕捉该视图需要的值并使用非捕捉参数。

5. URL 指向基于类的视图(View)

在目前的 Django 框架中，path 和 re_path 都只能指向视图 view 中的一个函数或方法，而不能指向一个基于类的视图。Django 额外提供了一个 as_view()方法，可以将一个类伪装成方法，这一点在使用 Django 自带的 view 类或自定义的类时会非常重要。例如，在下面的演示文件 blog/urls.py 中使用 as_view()方法。

```
from django.urls import path, re_path

from . import views
```

```
urlpatterns = [
   path('', views.ArticleList.as_view(), name='article_list'),
   path('blog/article/<int:id>/', views.article, name='article'),
   re_path(r'^blog/article/(?P<id>\d+)/$', views.article, name='article'),
]

# View (in blog/views.py)
from django.views.generic import ListView
from .views import Article

class ArticleList(ListView):

   queryset = Article.objects.filter(date__lte=timezone.now()).order_by('date')[:5]
   context_object_name = 'latest_article_list'
   template_name = 'blog/article_list.html'

def article(request, id):
   # 展示某篇文章
```

6. 通过 URL 方法传递额外的参数

在配置 URL 时可以通过字典的形式传递额外的参数给视图，而无需在链接中写入这个参数。例如，在下面的演示文件 blog/urls.py 中使用了这一用法。

```
from django.urls import path, re_path

from . import views

urlpatterns = [

   path('', views.ArticleList.as_view(), name='article_list', {'blog_id': 3}),
   re_path(r'^blog/article/(?P<id>\d+)/$', views.article, name='article'),
]
```

3.2.4 URLconf 匹配 URL

我们可以将请求 URL 看作一个普通的 Python 字符串，URLconf 在其上面进行查找并匹配，在匹配时不包括 GET 或 POST 请求方式的参数以及域名。例如，在请求 https://www.example.com/myapp/ 中，URLconf 将查找 myapp/。在请求 https://www.example.com/myapp/?page=3 中，URLconf 也会查找 myapp/。

URLconf 不检查使用何种 HTTP 请求方法，所有请求方法 POST、GET、HEAD 等都将路由到同一个 URL 的同一个视图。在视图中，才根据具体请求方法的不同，进行不同的处理。

3.2.5 设置视图参数的默认值

我们可以设置视图参数的默认值，例如下面是一个 URLconf 和视图的例子。

```python
from django.urls import path

from . import views

urlpatterns = [
    path('blog/', views.page),
    path('blog/page<int:num>/', views.page),
]

# View (in blog/views.py)
def page(request, num=1):
    # 根据 num 在页面中输出显示相应条目的 blog.
    ...
```

在上述代码中，两个 URL 模式指向同一个视图 views.page。但是第一个模式不会从 URL 中捕获任何值。如果第一个模式匹配，page()函数将使用 num 参数的默认值"1"。如果第二个模式匹配，page()将使用捕获的 num 值。

3.2.6　自定义错误页面

当 Django 找不到与请求匹配的 URL 时或当抛出一个异常时，将调用一个错误处理视图。Django 默认自带的错误视图包括 400、403、404 和 500，分别表示请求错误、拒绝服务、页面不存在和服务器错误。它们分别位于：

- handler400：django.conf.urls.handler400。
- handler403：django.conf.urls.handler403。
- handler404：django.conf.urls.handler404。
- handler500：django.conf.urls.handler500。

这些值可以在根 URLconf 中进行设置，在其他 app 中的二级 URLconf 中设置这些变量无效。Django 有内置的 HTML 模板，用于返回错误页面给用户，但是这些 403、404 页面效果不够美观，通常需要开发者自定义设计错误页面。例如，首先在根 URLconf 中额外增加下面的条目，并导入 views 模块。

```python
from django.contrib import admin
from django.urls import path
from app import views

urlpatterns = [
    path('admin/', admin.site.urls),
]

# 增加的条目
handler400 = views.bad_request
handler403 = views.permission_denied
handler404 = views.page_not_found
handler500 = views.error
```

然后在文件 app/views.py 中增加 4 个处理错误的视图：

```python
def bad_request(request):
    return render(request, '400.html')

def permission_denied(request):
    return render(request, '403.html')

def page_not_found(request):
    return render(request, '404.html')

def error(request):
    return render(request, '500.html')
```

然后根据项目需求，创建对应的 400、403、404、500 四个页面文件就可以了。另外还需要注意模板文件的引用方式、视图的放置位置等。

3.2.7 实战演练：使用 Django 框架实现 URL 参数相加

在下面的实例代码中，演示了使用 Django 框架实现参数相加功能的过程。

源码路径：daima\3\zqxt_views

(1) 在命令提示符界面中使用命令定位到 H 盘，然后通过如下命令创建一个"zqxt_views"目录作为"project(工程)"目录。

```
django-admin startproject zqxt_views
```

也就是说在 H 盘中新建了一个 zqxt_views 目录，其中还有一个相同名字的子目录 zqxt_views，在这个子目录 zqxt_views 中保存了项目的设置文件 settings.py、总的 urls 配置文件 urls.py，以及部署服务器时用到的 wsgi.py 文件，文件 __init__.py 是 python 包的目录结构必需的，与调用有关。

(2) 在 CMD 中定位到 zqxt_views 目录下(注意,不是 zqxt_views 中的 zqxt_views 目录)，然后通过如下命令新建一个应用(app)，名称叫 calc。

```
cd zqxt_views
python manage.py startapp calc
```

此时自动生成的目录结构大致如下所示。

```
zqxt_views/
├── calc
│   ├── __init__.py
│   ├── admin.py
│   ├── apps.py
│   ├── models.py
│   ├── tests.py
│   └── views.py
├── manage.py
└── zqxt_views
    ├── __init__.py
    ├── settings.py
    ├── urls.py
    └── wsgi.py
```

(3) 为了将新定义的 app 添加到 settings.py 文件的 INSTALL_APPS 中，需要对文件 zqxt_views/zqxt_views/settings.py 进行如下修改。

```
INSTALLED_APPS = [
    'django.contrib.admin',
    'django.contrib.auth',
    'django.contrib.contenttypes',
    'django.contrib.sessions',
    'django.contrib.messages',
    'django.contrib.staticfiles',
    'calc',
]
```

这一步的目的是将新建的程序"calc" 添加到 INSTALL_APPS 中，如果不这样做，Django 就不能自动找到 app 中的模板文件(app-name/templates/ 中的文件)和静态文件(app-name/static/中的文件)。

(4) 定义视图函数，用于显示访问页面时的内容。对文件 calc/views.py 中的代码进行如下所示的修改。

```
from django.shortcuts import render
from django.http import HttpResponse

def add(request):
    a = request.GET['a']
    b = request.GET['b']
    c = int(a)+int(b)
    return HttpResponse(str(c))
```

在上述代码中，request.GET 类似于一个字典，当没有传递 a 的值时，a 的默认值为 0。

(5) 开始定义视图函数相关的 URL 网址，添加一个网址来对应我们刚才新建的视图函数。对文件 zqxt_views/zqxt_views/urls.py 进行如下所示的修改。

```
from django.conf.urls import url
from django.contrib import admin
from learn import views as learn_views  # new

urlpatterns = [
    url(r'^$', learn_views.index),  # new
    url(r'^admin/', admin.site.urls),
]
```

(6) 最后在终端上运行如下命令进行测试。

```
python manage.py runserver
```

在浏览器中输入"http://localhost:8000/add/"后的执行效果如图 3-2 所示。

如果在 URL 中输入数字参数，例如在浏览器中输入"http://localhost:8000/add/?a=4&b=5"，执行后会显示这两个数字(4 和 5)的和，执行效果如图 3-3 所示。

```
MultiValueDictKeyError at /add/
"'a'"
   Request Method: GET
      Request URL: http://localhost:8000/add/
   Django Version: 1.10.4
   Exception Type: MultiValueDictKeyError
  Exception Value: "'a'"
Exception Location: C:\Program Files\Python36\lib\site-packages\django-1.10.4-py3.6.egg\django\utils\datastructures.py in __getitem__, line 85
 Python Executable: C:\Program Files\Python36\python.exe
    Python Version: 3.6.0
       Python Path: ['H:\\zqxt_views',
                    'C:\\Program Files\\Python36\\python36.zip',
                    'C:\\Program Files\\Python36\\DLLs',
                    'C:\\Program Files\\Python36\\lib',
                    'C:\\Program Files\\Python36',
                    'C:\\Program Files\\Python36\\lib\\site-packages',
                    'C:\\Program Files\\Python36\\lib\\site-packages\\flask-0.12-py3.6.egg',
                    'C:\\Program Files\\Python36\\lib\\site-packages\\click-6.6-py3.6.egg',
                    'C:\\Program Files\\Python36\\lib\\site-packages\\itsdangerous-0.24-py3.6.egg',
                    'C:\\Program Files\\Python36\\lib\\site-packages\\jinja2-2.8.1-py3.6.egg',
                    'C:\\Program Files\\Python36\\lib\\site-packages\\werkzeug-0.11.13-py3.6.egg',
                    'C:\\Program Files\\Python36\\lib\\site-packages\\markupsafe-0.23-py3.6-win-amd64.egg',
                    'C:\\Program Files\\Python36\\lib\\site-packages\\tornado-4.4.2-py3.6-win-amd64.egg',
                    'C:\\Program Files\\Python36\\lib\\site-packages\\django-1.10.4-py3.6.egg']
       Server time: Sat, 31 Dec 2016 12:05:23 +0800
```

图 3-2　执行效果(一)

9

图 3-3　执行效果(二)

在 Python 程序中，也可以采用"/add/3/4/"这样的方式对 URL 中的参数进行求和处理。这时需要修改文件 calc/views.py 的代码，在里面新定义一个求和函数 add2()，具体代码如下所示。

```
def add2(request, a, b):
    c = int(a) + int(b)
    return HttpResponse(str(c))
```

接着修改文件 zqxt_views/urls.py 的代码，再添加一个新的 URL，具体代码如下所示。

```
url(r'^add/(\d+)/(\d+)/$', calc_views.add2, name='add2'),
```

此时可以看到网址中多了"\d+"，正则表达式中的"\d"代表一个数字，"+"代表一个或多个前面的字符，写在一起"\d+"就表示一个或多个数字，用括号括起来的意思是保存为一个子组(更多知识请参见 Python 正则表达式)，每一个子组将作为一个参数，被文件 views.py 中的对应视图函数接收。此时输入如下网址执行后就可以看到和图 3-3 同样的执行效果。

```
http://localhost:8000/add/?add/4/5/
```

3.3　编写 View 视图

一个视图函数(或简称为视图)是一个 Python 函数，能够接受 Web 请求并返回一个 Web 响应。这个响应可以是 Web 页面的 HTML 内容，或者是重定向，或者是 404 错误，或者是 XML 文档，或是一个图片……或是任何内容。视图本身包含返回响应所需的任何逻辑。这个代码可以存在任何地方，只要它在你的 Python 路径上就

扫码观看本节视频讲解

行。为了将代码放置在某处，约定将视图放在名为 views.py 的文件里，这个文件放置在项目或应用目录里。

3.3.1 一个简单的视图

下面是一个以 HTML 文档形式返回当前日期和时间的视图。

```
from django.http import HttpResponse
import datetime

def current_datetime(request):
    now = datetime.datetime.now()
    html = "<html><body>It is now %s.</body></html>" % now
    return HttpResponse(html)
```

对上述代码的具体说明如下。
- 首先，我们从 django.http 模块导入类 HttpResponse，以及 Python 的 datetime 库。
- 然后，我们定义一个名为 current_datetime 的函数。这是一个视图函数。每个视图函数都将 HttpRequest 对象作为第一个参数，通常名为 request。
- 上述视图返回一个包含生成的响应的 HttpResponse 对象，每个视图函数都要返回 HttpResponse 对象。

注意

视图函数名称无关紧要；它不需要以特定的名称来让 Django 识别它。我们在这里命名 current_datetime，因为这个名字可以清楚地表示它的用途。

3.3.2 返回错误信息

Django 提供了返回 HTTP 错误的内置接口。HttpResponse 的子类除了处理 200 错误外，还提供了很多常见的 HTTP 状态代码。我们可以在 request/response 文档中找到所有可用子类的列表。返回这些子类中某个子类的实例而不是 HttpResponse 来表示错误。比如：

```
from django.http import HttpResponse, HttpResponseNotFound

def my_view(request):
    # ...
    if foo:
        return HttpResponseNotFound('<h1>Page not found</h1>')
    else:
        return HttpResponse('<h1>Page was found</h1>')
```

并不是每个可用的 HTTP 响应代码都有对应的子类，因为很多响应并不是很常见的。我们也可以将 HTTP 状态代码传递给 HttpResponse 的构造函数，这样就可以为任何状态代码创建返回类。比如：

```
from django.http import HttpResponse

def my_view(request):
```

```
# ...
# Return a "created" (201) response code.
return HttpResponse(status=201)
```

因为 404 错误是最常见的 HTTP 错误，所以可以考虑使用更简单的方法来处理 404 错误。

(1) Http404 异常。

```
class django.http.Http404
```

这是一个 Django 内置的异常类，可以在需要的地方用到它，Django 会捕获它，并且带上 HTTP404 错误码返回当前 app 的标准错误页面或者自定义错误页面。例如下面的用法：

```
from django.http import Http404
from django.shortcuts import render
from polls.models import Poll

def detail(request, poll_id):
    try:
        p = Poll.objects.get(pk=poll_id)
    except Poll.DoesNotExist:
        raise Http404("Poll does not exist")
    return render(request, 'polls/detail.html', {'poll': p})
```

为了在 Django 返回 404 时显示自定义的 HTML，我们可以创建名为 404.html 的 HTML 模板，并将其放置在模板树顶层。这个模板将在 DEBUG 设为 False 时提供，当 DEBUG 为 True 时，你可以提供 Http404 信息，并且在标准的 404 调试模板里显示。我们可以使用这些信息来调试，它们通常不适合在生产环境下的 404 模板。

(2) 自定义报错视图。

虽然在 Django 中的默认报错视图能够满足大部分的 Web 应用，但是我们也可以很方便地自定义创建报错视图。例如，在下面用 handler404 覆盖 page_not_found() 视图。

```
handler404 = 'mysite.views.my_custom_page_not_found_view'
```

也可以用 handler500 覆盖 custom_error()视图：

```
handler500 = 'mysite.views.my_custom_error_view'
```

也可以用 handler403 覆盖 permission_denied()视图：

```
handler403 = 'mysite.views.my_custom_permission_denied_view'
```

也可以用 handler400 覆盖 bad_request()视图：

```
handler400 = 'mysite.views.my_custom_bad_request_view'
```

为了测试自定义报错处理的响应，可以适当地在测试视图里引发异常。例如：

```
from django.core.exceptions import PermissionDenied
from django.http import HttpResponse
from django.test import SimpleTestCase, override_settings
from django.urls import path

def response_error_handler(request, exception=None):
```

```
       return HttpResponse('Error handler content', status=403)

def permission_denied_view(request):
    raise PermissionDenied

urlpatterns = [
    path('403/', permission_denied_view),
]
handler403 = response_error_handler

# ROOT_URLCONF must specify the module that contains handler403 = ...
@override_settings(ROOT_URLCONF=__name__)
class CustomErrorHandlerTests(SimpleTestCase):
    def test_handler_renders_template_response(self):
        response = self.client.get('/403/')
        # Make assertions on the response here. For example:
        self.assertContains(response, 'Error handler content', status_code=403)
```

3.3.3 实战演练：在线文件上传系统

在本实例中实现两种文件上传模式，第一种是简易上传系统，没有使用表单；第二种使用了表单。当文件被提交到服务器时，文件数据最终被放入 request.FILES 中。当使用表单时，必须在 HTML 表单中正确设置属性 enctype="multipart/form-data"，否则 request.FILES 将是空的。另外，必须使用 POST 方法提交表单。

Django 内置了两个模型字段来处理上传的文件：FileField 和 ImageField，上传到 FileField 或 ImageField 的文件不存储在数据库中，而是存储在文件系统中。FileField 和 ImageField 在数据库(通常是 VARCHAR)中作为字符串字段来创建，包含对实际文件的引用。如果删除包含 FileField 或 ImageField 的模型实例，Django 将不会删除物理文件，而只删除对该文件的引用。

源码路径：**daima\3\simple-file-upload**

1．基本配置

（1）在文件 settings.py 中设置使用 SQLite 数据库，设置保存上传文件的目录。代码如下：

```
DATABASES = {
    'default': {
        'ENGINE': 'django.db.backends.sqlite3',
        'NAME': os.path.join(BASE_DIR, 'db.sqlite3'),
    }
}
STATIC_URL = '/static/'
STATIC_ROOT = os.path.join(BASE_DIR, 'staticfiles')

MEDIA_URL = '/media/'
MEDIA_ROOT = os.path.join(BASE_DIR, 'media')
```

(2) 在文件 urls.py 中设置 URL 路径导航，分别设置主页、简易上传、表单上传和后台管理对应的 URL。代码如下：

```
urlpatterns = [
   path('', views.home, name='home'),
   path('uploads/simple/', views.simple_upload, name='simple_upload'),
   path('uploads/form/', views.model_form_upload, name='model_form_upload'),
   path('admin/', admin.site.urls),
]

if settings.DEBUG:
   urlpatterns += static(settings.MEDIA_URL, document_root=settings.MEDIA_ROOT)
```

2. 简易上传

在接下来介绍的简易文件上传系统中，使用 FileSystemStorage 实现文件上载功能。

(1) 在视图文件 views.py 中分别设置系统主页和 uploads/simple/对应的视图，使用 FileSystemStorage 将本地文件上传到服务器中。代码如下：

```
from django.shortcuts import render, redirect
from django.conf import settings
from django.core.files.storage import FileSystemStorage

from uploads.core.models import Document
from uploads.core.forms import DocumentForm


def home(request):
   documents = Document.objects.all()
   return render(request, 'core/home.html', { 'documents': documents })


def simple_upload(request):
   if request.method == 'POST' and request.FILES['myfile']:
       myfile = request.FILES['myfile']
       fs = FileSystemStorage()
       filename = fs.save(myfile.name, myfile)
       uploaded_file_url = fs.url(filename)
       return render(request, 'core/simple_upload.html', {
           'uploaded_file_url': uploaded_file_url
       })
   return render(request, 'core/simple_upload.html')
```

(2) 在上面的简易文件上传视图函数 simple_upload()中，对应的模板文件是 core/simple_upload.html，在 HTML 文件中设置使用 POST 方式提交信息，设置文件类型是 multipart/form-data。代码如下：

```
{% extends 'base.html' %}

{% load static %}

{% block content %}
 <form method="post" enctype="multipart/form-data">
```

```
    {% csrf_token %}
    <input type="file" name="myfile">
    <button type="submit">上传</button>
</form>

{% if uploaded_file_url %}
  <p>上传文件保存在: <a href="{{ uploaded_file_url }}">{{ uploaded_file_url }}</a></p>
{% endif %}

<p><a href="{% url 'home' %}">返回主页</a></p>
{% endblock %}
```

执行后的系统主页效果如图 3-4 所示，简易文件上传页面效果如图 3-5 所示，上传某个文件成功后的效果如图 3-6 所示。

图 3-4　系统主页　　　　　　　　　　图 3-5　简易上传

图 3-6　上传成功

3. 表单上传

接下来介绍的表单上传方式是一种更方便的方式，通过模型表单执行验证，自动生成上传文件的绝对路径，可以解决文件名冲突和其他常见功能。

（1）首先编写为数据库服务的模型文件 models.py。代码如下：

```
from django.db import models

class Document(models.Model):
    description = models.CharField(max_length=255, blank=True)
    document = models.FileField(upload_to='documents/')
    uploaded_at = models.DateTimeField(auto_now_add=True)
```

在上述代码中，注意 upload_to 的参数，文件将自动被上传到"MEDIA_ROOT/documents/"目录中。另外，也可以写为下面的形式，此时今天(2020 年 11 月 25 日)上传的文件将上传到"MEDIA_ROOT/documents/2020/11/25/"目录中。

```
document = models.FileField(upload_to='documents/%Y/%m/%d/')
```

另外，upload_to 也可以是返回字符串的可调用函数。这个可调用函数接受两个参数：

instance 和 filename。例如：

```
def user_directory_path(instance, filename):
    #文件将被上传到MEDIA_ROOT/user_<id>/<filename>
    return 'user_{0}/{1}'.format(instance.user.id, filename)

class MyModel(models.Model):
    upload = models.FileField(upload_to=user_directory_path)
```

(2) 然后编写 form 模型文件 forms.py，通过 fields 获取表单中的数据。代码如下：

```
from django import forms
from uploads.core.models import Document

class DocumentForm(forms.ModelForm):
    class Meta:
        model = Document
        fields = ('description', 'document', )
```

处理这个表单的视图将在 request.FILES 中收到文件数据，可以用 request.FILES['file'] 来获取上传文件的具体数据，其中的键值'file'是根据 file = forms.FileField()的变量名来的。

(3) 编写视图文件 views.py，通过 request.POST 和 request.FILES 分别获取数据提交方式和表单中的信息。代码如下：

```
def model_form_upload(request):
    if request.method == 'POST':
        form = DocumentForm(request.POST, request.FILES)
        if form.is_valid():
            form.save()
            return redirect('home')
    else:
        form = DocumentForm()
    return render(request, 'core/model_form_upload.html', {
        'form': form
    })
```

(4) 编写模板文件 model_form_upload.html，设计一个 HTML 表单。代码如下：

```
{% extends 'base.html' %}

{% block content %}
  <form method="post" enctype="multipart/form-data">
    {% csrf_token %}
    {{ form.as_p }}
    <button type="submit">上传</button>
  </form>

  <p><a href="{% url 'home' %}">返回主页</a></p>
{% endblock %}
```

通过如下命令更新数据库：

```
python manage.py makemigrations
python manage.py migrate
```

表单上传页面效果如图 3-7 所示，上传某个文件成功后的效果如图 3-8 所示，并且会自动为重名的上传文件重新命名。

图 3-7　表单上传页面

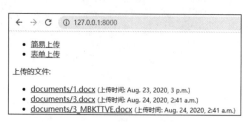

图 3-8　重名的文件

3.4　异步视图

异步视图是从 Django 3.1 开始提供的新特性。除了可以使用同步函数实现视图之外，视图也可以使用异步函数，这通常使用 Python 的 async def 语法定义。Django 将自动检测这些异步函数，并在异步上下文中运行它们。但是在使用异步视图时，需要使用基于 ASGI 的异步服务器来获得性能优势。

扫码观看本节视频讲解

3.4.1　异步视图介绍

在 Django 3.1 及其以后的版本中，要定义一个异步视图很简单，只需使用 Python 的 async def 语法进行定义，Django 会自动探测到它们，并在异步上下文中运行它们。异步视图有很多优点，比如能够在不使用 Python 线程的情况下为数百个连接提供服务，允许使用慢速流、长轮询和其他响应类型等。例如下面设计了一个简单的异步视图：

```
async def my_view(request):
    await asyncio.sleep(0.5)
    return HttpResponse('Hello, async world!')
```

在使用异步视图后，无论是运行在 WSGI 模式还是 ASGI 模式，都支持所有的异步特性。但是，在 WSGI 服务器下，异步视图将在其自有的一次性事件循环中运行。这意味着我们可以放心使用异步特性(例如并发异步 HTTP 请求)，但是你不会获得异步堆栈的好处。

我们可以随心所欲地混合异步和同步的视图、中间件和测试，Django 会确保我们始终获得正确的执行上下文。但是，Django 官方建议大多数时候依然使用同步视图，只有在真正有需求时使用异步视图，不过这完全取决于你的选择，你可以任性地都使用异步视图。

Django 对异步的支持完全向后兼容，对现有的同步代码没有速度限制，它不会对任何现有的 Django 项目产生明显的影响。例如下面是使用异步视图的另外一个例子：

```
import datetime
from django.http import HttpResponse

async def current_datetime(request):
```

```
    now = datetime.datetime.now()
    html = '<html><body><h1>欢迎访问 Django 教程: https://www.xample.com</h1>It is
now %s.</body></html>' % now
    return HttpResponse(html)
```

在 Django 3.1 版本中，Django 还没有完成异步 ORM 的功能开发，为了在异步视图中使用 ORM，需要将同步的代码转换为异步的代码，这就需要使用 asgiref 库，这个库已经作为安装依赖随 Django 一起安装。使用 asgiref 库的核心是使用 asgiref.sync 中的 sync_to_async 方法，具体使用方法有以下两种。

(1) 第一种以函数调用的方式，注意括号的位置：

```
from asgiref.sync import sync_to_async
results = sync_to_async(Blog.objects.get)(pk=123)
#注意圆括号，千万不要写成 results = sync_to_async(Blog.objects.get(pk=123))
```

(2) 第二种以装饰器的方式：

```
from asgiref.sync import sync_to_async

@sync_to_async
def get_blog(pk):
    return Blog.objects.select_related('author').get(pk=pk)
```

相应地，其实也有一个异步变同步的函数，用于在同步视图中包装异步调用：

```
from asgiref.sync import async_to_sync
async def get_data(...):
    ...
sync_get_data = async_to_sync(get_data)
@async_to_sync
async def get_other_data(...):
    ...
```

3.4.2 异步中间件

从 Django 3.1 开始，中间件可以支持同步和异步请求的任何组合。如果 Django 不能同时支持这两者，它将调整请求以满足中间件的需求，但是这样会降低性能。在默认情况下，Django 假设中间件只能处理同步请求。要更改这个假设，需要在中间件函数或类中设置以下属性。

- sync_capable：一个布尔值，指示中间件是否可以处理同步请求，默认为 True。
- async_capable：一个布尔值，指示中间件是否可以处理异步请求，默认为 False。

如果中间件同时具有 sync_capable=True 和 async_capable=True，那么 Django 将在不转换请求的情况下传递它。在这种情况下，可以使用 asyncio.iscoroutine function()检查传递给我们的 get_response 对象是否是一个协程函数，从而确定我们的中间件是否会接收异步请求。

既然同步和异步是可以共存的，那么如何将中间件设置为同步的，或者异步的，或者同步加异步的呢？在 django.utils.decorators 模块中包含三个装饰器：sync_only_middleware、async_only_middleware 和 sync_and_async_middleware，用于实现上面的功能。

中间件返回的可调用函数必须与 get_response 方法的 sync 或 async 性质匹配。如果有异步 get_response 响应，则必须返回一个协程函数(async def)。如果中间件提供了 process_view、process_template_response 和 process_exception 方法，则还应进行相应的调整以匹配同步/异步模式。如果你不这样做，Django 会根据需要对它们进行单独的调整，并产生性能损失。

下面是一个创建同时支持同步和异步功能的中间件的例子。

```python
import asyncio
from django.utils.decorators import sync_and_async_middleware

@sync_and_async_middleware
def simple_middleware(get_response):
    # One-time configuration and initialization goes here.
    if asyncio.iscoroutinefunction(get_response):
        async def middleware(request):
            # Do something here!
            response = await get_response(request)
            return response

    else:
        def middleware(request):
            # Do something here!
            response = get_response(request)
            return response

    return middleware
```

总地来说，Django 对同步/异步视图和同步/异步中间件之间的搭配组合有很好的适配能力，不会让我们的项目运行不起来。只不过如果搭配不当，会导致性能损失。

3.4.3　实战演练：使用异步视图展示两种货币的交易数据

本实例的功能是爬虫抓取网络中的数据，获取 Dinar(巴林第纳尔)和 Dirham(阿联酋迪拉姆)两种货币的实时交易数据。

源码路径：daima\3\django-async-views-example

（1）在文件 settings.py 中设置使用 SQLite 数据库，设置保存上传文件的目录。代码如下：

```python
DATABASES = {
    'default': {
        'ENGINE': 'django.db.backends.sqlite3',
        'NAME': os.path.join(BASE_DIR, 'db.sqlite3'),
    }
}
STATIC_URL = '/static/'
STATIC_ROOT = os.path.join(BASE_DIR, 'staticfiles')

MEDIA_URL = '/media/'
MEDIA_ROOT = os.path.join(BASE_DIR, 'media')
```

(2) 编写模型文件 models.py，分别创建 3 个数据库模型类 BaseKurs、Dinar 和 Dirham。因为 Django 3.1 不支持模型部分的异步视图，所以使用 asgiref 中的 sync_to_async 实现了"同步/异步"的匹配。代码如下：

```python
from django.db import models
from asgiref.sync import sync_to_async
class BaseKurs(models.Model):
    harga_jual = models.DecimalField(max_digits=12, decimal_places=2)
    harga_beli = models.DecimalField(max_digits=12, decimal_places=2)
    tanggal_dan_waktu = models.DateTimeField(auto_now_add=True)

    class Meta:
        abstract = True


class Dinar(BaseKurs):

    def __str__(self):
        return f'Dinar: {self.harga_jual}'


class Dirham(BaseKurs):

    def __str__(self):
        return f'Dirham: {self.harga_jual}'


dinar_get_all = sync_to_async(Dinar.objects.all)
dinar_create = sync_to_async(Dinar.objects.create)

dirham_get_all = sync_to_async(Dirham.objects.all)
dirham_create = sync_to_async(Dirham.objects.create)
```

(3) 编写文件 scrapper.py 实现网络爬虫功能。代码如下：

```python
import httpx
from bs4 import BeautifulSoup as bs4
from decimal import Decimal

URL = 'http://indonesiasoftware.com/dc_gold2012.php'

async def fetch_kurs() -> str:
    async with httpx.AsyncClient() as client:
        headers = {
            'User-Agent': ('Mozilla/5.0 (Windows NT 6.0; rv:2.0) '
                          'Gecko/20100101 Firefox/4.0 Opera 12.14')
        }
        response = await client.get(URL, headers=headers)
        return response.text


async def get_latest_kurs() -> dict:
    content = await fetch_kurs()
    soup = bs4(content, 'html.parser')
```

```python
    prices = soup.find_all('td', {'class': 'harga'})
    prices = [price.get_text().strip().replace(',', '') for price in prices]

    return {
        'dinar': {
            'jual': Decimal(prices[2]),
            'beli': Decimal(prices[3])
        },
        'dirham': {
            'jual': Decimal(prices[4]),
            'beli': Decimal(prices[5])
        }
    }
```

(4) 编写视图文件 view.py，使用关键字 async 定义了 3 个异步视图：主页、获取新数据和删除数据。在视图中，使用 asgiref 中的 sync_to_async 实现了"同步/异步"的匹配。代码如下：

```python
import asyncio
from asgiref.sync import sync_to_async
from django.shortcuts import render, redirect
from .scrapper import get_latest_kurs
from .models import (dirham_create, dinar_create, dirham_get_all, dinar_get_all,
Dinar, Dirham)


async def index_view(request):
    list_dinar, list_dirham = await asyncio.gather(
        dinar_get_all(),
        dirham_get_all()
    )
    context = {
        'list_dinar': list_dinar,
        'list_dirham': list_dirham,
    }
    return await sync_to_async(render)(request, 'index.html', context)


async def fetch_view(request):
    kurs = await get_latest_kurs()
    await asyncio.gather(
        dinar_create(**{
            'harga_jual': kurs['dinar']['jual'],
            'harga_beli': kurs['dinar']['beli']
        }),
        dirham_create(**{
            'harga_jual': kurs['dirham']['jual'],
            'harga_beli': kurs['dirham']['beli']
        })
    )
    return redirect('index')


async def remove_view(request):
```

```
    await asyncio.gather(
        sync_to_async(Dinar.objects.all().delete)(),
        sync_to_async(Dirham.objects.all().delete)(),
    )
    return redirect('index')
```

注 意

读者需要注意的是，在目前的 3.1 版本中，Django 还没有完全支持异步特性。有些部分仍然在同步运行，所以在本实例中使用函数 converte() 来转换数据。以便等未来 Django 完全支持异步特性后，可以将转换数据作为 wait() 调用的值，这样可以实现数据同步。

（5）最后编写模板文件 index.html。代码如下：

```html
<!DOCTYPE html>
<html>
  <head>
    <meta charset="utf-8">
    <meta name="viewport" content="width=device-width, initial-scale=1">
    <title>巴林第纳尔兑换阿联酋迪拉姆</title>
    <link rel="stylesheet" href="https://cdn.jsdelivr.net/npm/bulma@0.8.0/css/bulma.min.css">
    <script defer src="https://use.fontawesome.com/releases/v5.3.1/js/all.js"></script>
  </head>
  <body>
    <section class="section">
      <div class="container">
        <h1 class="title">汇率系统：Dinar(巴林第纳尔)兑换 Dirham(阿联酋迪拉姆)</h1>
      </div>
    </section>

    <section class="section">
      <div class="columns">
        <div class="column is-full">
          <a href="{% url 'remove' %}" class="button is-danger">全部删除</a>
          <a href="{% url 'fetch' %}" class="button is-primary">获取新数据</a>
        </div>
      </div>
      <div class="columns">
        <div class="column is-half">
          <h4 class="title is-4">Dinar</h4>
          <div class="columns">
            <div class="column is-full">
              <table class="table">
                <thead>
                  <th>日期</th>
                  <th>销售金额</th>
                  <th>购买金额</th>
                </thead>
                <tbody>
                  {% for dinar in list_dinar %}
                    <tr>
                      <td>{{ dinar.tanggal_dan_waktu }}</td>
                      <td>{{ dinar.harga_jual }}</td>
```

```html
          <td>{{ dinar.harga_beli }}</td>
        </tr>
        {% endfor %}
      </tbody>
    </table>
   </div>
  </div>
</div>
<div class="column is-half">
  <h4 class="title is-4">Dirham</h4>
  <div class="columns">
    <div class="column is-full">
      <table class="table">
        <thead>
          <th>日期</th>
          <th>销售金额</th>
          <th>购买金额</th>
        </thead>
        <tbody>
          {% for dirham in list_dirham %}
          <tr>
            <td>{{ dirham.tanggal_dan_waktu }}</td>
            <td>{{ dirham.harga_jual }}</td>
            <td>{{ dirham.harga_beli }}</td>
          </tr>
          {% endfor %}
        </tbody>
      </table>
    </div>
   </div>
  </div>
 </div>
</section>
</body>
</html>
```

执行后的效果如图 3-9 所示。

图 3-9　执行效果

第 4 章

Django 数据库操作

在动态 Web 应用程序中，数据库技术永远是核心技术。Django 框架是与数据库密切相关的，支持 SQLite3、MySQL 和 PostgreSQL 等数据库。在本章的内容中，将详细讲解使用 Django 框架开发动态数据库程序的知识。

4.1　Model 模型

在 Django 框架中,通常将与数据库相关的代码写在文件 models.py 中,此文件对应于 MVC 模式中的 Model(模型)。Django 框架支持 PostgreSQL、MySQL、SQLite、Oracle 等常用的数据库,在开发 Django 程序时只需在文件 settings.py 中进行配置即可,无须更改文件 models.py 的代码,这样极大地提高了开发效率。在 Django Web 程序中,通过如下所示的命令运行程序,会调用模型文件 models.py 创建对应的数据库表,数据库表的各个字段是在文件 models.py 中设置的。

扫码观看本节视频讲解

```
python manage.py syncdb
```

4.1.1　Model 模型基础

在模型文件 models.py 中,我们需要编写创建数据库表的程序代码,在代码中设置数据库表的各个字段。然后通过如下所示的命令运行程序,就可以根据文件 models.py 的代码自动创建数据库表。

```
python manage.py syncdb
```

Model (模型) 是指数据模型,是抽象的描述数据的构成和逻辑关系,不是数据库中的具体数据。每个 Django 模型实际上是一个类,这个类继承了 models.Model。每个 Model 包括属性、关系(比如一对一、一对多和多对多)和方法。当定义好 Model 模型后,Django 中的接口会自动在数据库生成相应的数据表(table)。这样就不用自己用 SQL 创建表格或在数据库中操作创建表格,大大提高了开发效率。

我们举一个和图书出版有关的案例,出版社都有自己的名字和注册地址。出版的每一本书有对应的书名、图书介绍和出版日期。因为一个出版社可以出版很多本不同名字的书,所以需要使用外键 ForeignKey 定义出版社与出版图书一对多的关系。此时我们可以定义如下所示的模型文件 models.py 来满足需求。

```python
from django.db import models

class Publisher(models.Model):
    name = models.CharField(max_length=30)
    address = models.CharField(max_length=60)

    def __str__(self):
        return self.name

class Book(models.Model):
    name = models.CharField(max_length=30)
    description = models.TextField(blank=True, default='')
    publisher = ForeignKey(Publisher, on_delete = models.CASCADE)
```

```
    add_date = models.DateField()

    def __str__(self):
        return self.name
```

对上述代码的具体说明如下所示。

(1) 类 Publisher：表示出版社，name 表示出版社的名字，address 表示出版社的地址。类名和数据库中的表名是对应的，例如类名 Publisher 说明会在数据库中创建一个名为"Publisher"的表。类 Publisher 中的属性 name 和 address 与数据库中的字段相对应，例如上述代码说明在数据库表 Publisher 中创建两个字段 name 和 address。

(2) 类 Book：表示出版社出版的某一本书，name 表示书名，description 表示图书简介，publisher 表示作者，add_date 表示出版日期。类名 Book 说明会在数据库中创建一个名为"Book"的表。类中的属性 name 和 description 表示会在数据库 Book 中创建字段 name 和 description。

(3) CharField：表示当前字段的数据类型是一个字符字段类型，包含如下所示的参数。

- max_length 的取值可以是一个数值或 None。
- 如果不是必填选项，可以设置 blank = True 和 default = ''。
- 如果用于 username，想使其唯一，可以设置 unique = True。
- 如果有 choice 选项，可以设置 choices = XXX_CHOICES。

(4) TextField：表示当前字段的数据类型是一个文本字段类型，包含如下所示的参数。

- max_length 的取值可以是一个数值。
- 如果不是必填项，可以设置为 blank = True 和 default = ''。

除了 CharField 和 TextField 外，Django 支持的常用数据类型如下所示。

(1) DateField()和 DateTimeField()：表示日期和时间字段类型。

- 一般将其取值设置为默认日期 default date。
- 对于上一次修改日期(last_modified date)，可以设置 auto_now=True。

(2) EmailField()：表示邮件字段类型。

- 如果不是必填项，可以设置为 blank = True 和 default = ''。
- 如果 Email 表示用户名，那么这个值应该是唯一的，建议在开发时设置为 unique = True。

(3) IntegerField()、SlugField()、URLField()、BooleanField()。

- 可以设置取值为 blank = True 或 null = True。
- 通常将 BooleanField 的值设置为 defautl = True/False。

(4) FileField(upload_to=None, max_length=100)：表示文件字段类型，通常用于保存上传的文件。

- upload_to = "/some folder/"。
- max_length = xxxx。

(5) ImageField(upload_to=None, height_field=None, width_field=None, max_length=100)：表示图像字段类型，通常用于保存上传的图像。

- upload_to = "/some folder/"。
- 其他选项是可选的。

(6) ForeignKey(to, on_delete, **options)：表示外键，用于表示一对多关系类型。
- 参数 to 必须指向其他模型，例如 Book 或 'self'。
- 必须设置 on_delete(删除选项)的值，例如 on_delete = models.CASCADE 或 on_delete = models.SET_NULL。
- 可以设置 default = xxx 或 null = True。
- 为了便于反向查询，可以设置 related_name = xxx。

(7) ManyToManyField(to, **options)：表示多对多关系类型。
- to 必须指向其他模型，例如 User 或'self'。
- 如果设置 symmetrical = False，则不符合多对多关系的规则。
- 设置 through = 'intermediary model'，表示建立中间模型来搜集更多信息。
- 设置 related_name = xxx 便于反向查询。

> **注意**
> 在上述出版社和图书的演示代码中，CharField 中的 max_length 和 ForeignKey 的 on_delete 选项是必须要设置的。例如设置这个表单在填写时是否可以为空(blank = True or False)，这个选项会影响到数据的完整性。

4.1.2 META 内部类

在 Django 框架中，模型类 META 是一个内部类，用于定义一些 Django 模型类的行为特性。在类 META 中主要包括如下所示的属性。

(1) 属性 abstract：用于设置当前的模型类是不是一个抽象类，抽象类不会拥有对应的数据库表。在具体项目中，通常使用关键字 abstract 来定义公共属性字段，然后创建子类来继承这些字段。

例如在下面的模型代码中，Person 是一个抽象类，Employee 是一个继承了 Person 的子类，在执行 syncdb 命令时，虽然不会生成 Person 表，但是会生成表 Employee，表 Employee 包括 Person 中继承来的字段。以后假设再加入一个 Customer 模型类，它能够同样继承 Person 的公共属性。

```
class Person(models.Model):
    name=models.CharField(max_length=100)
    GENDER_CHOICE=((u'M',u'Male'),(u'F',u'Female'),)
    gender=models.CharField(max_length=2,choices=GENDER_CHOICE,null=True)
    class Meta:
        abstract=True
class Employee(Person):
    joint_date=models.DateField()
class Customer(Person):
    first_name=models.CharField(max_length=100)
    birth_day=models.DateField()
```

通过使用上面的模型代码，运行 python manage.py syncdb 命令后的输出结果如下。从中可以看出不会创建表 Person。

```
$ python manage.py syncdb
Creating tables ...
Creating table myapp_employee
Creating table myapp_customer
Installing custom SQL ...
Installing indexes ...
No fixtures found.
app_label
```

（2）属性 app_label：这个选项仅仅在模型类不在默认的应用程序包下的 models.py 文件中时才会使用，这时必须指定这个模型类是哪个应用程序的。假设在其他地方编写了一个模型类，而这个模型类是属于名为"aaa"的 app 的，那么必须要指定为：

```
app_label='aaa'
db_table
```

（3）属性 db_table：用于设置数据库的表名。在 Django 框架中，有一套默认的依照一定规则生成数据模型的数据库表名。假设想使用自己定义的表名，就必须通过属性 db_table 设置这个表名，例如：

```
table_name='表名'
db_table
```

另外，有些数据库工具需要设置数据库表空间，比方 Oracle，此时通过属性 db_tablespace 来设置将这个模型对应的数据库表放在哪个数据库表空间。

（4）属性 get_latest_by：用于设置 latest() 是按照哪个字段进行选取的。

（5）属性 managed：如果不希望 Django 自己主动依据模型类生成映射的数据库表，可以把属性 managed 的值设置为 False。

（6）属性 order_with_respect_to：用于在多对多的关系中指向一个关联对象。在设置属性 order_with_respect_to 后，会得到一个类似于 get_XXX_order() 和 set_XXX_order() 样式的方法，通过这些方法能够设置或者获取排序的对象。

（7）属性 ordering：功能是告诉 Django 模型对象返回的记录结果集是按照哪个字段进行排序的。例如下面的代码演示了三种排序方法。

```
ordering=['order_date']          # 按升序排列
ordering=['-order_date']         # 按降序排列
ordering=['?order_date']         # 随机排序
```

（8）属性 permissions：用于在 Django Admin 管理模块下设置指定的权限。

（9）属性 proxy：用于实现代理模型。

（10）属性 unique_together：用于设置两个字段保持唯一性，假设想设置表 Person 中的字段 First_Name 和 Last_Name 的组合必须是唯一的，那么可以通过如下代码实现。

```
unique_together = (("first_name", "last_name"),)
verbose_name
```

(11) 属性 verbose_name：用于给模型类起一个可读的名字。例如：

```
verbose_name = "apple"
```

例如在下面的模式文件 models.py 中，演示了常见的 Django Model META 类选项的含义。

```python
from django.db import models

class Meta:
    # 按 Priority 降序, order_date 升序排列
    get_latest_by = ['-priority', 'order_date']
    # 自定义数据库里表格的名字
    db_table = 'music_album'
    # 按什么字段排序
    ordering = ['pub_date']
    # 定义 APP 的标签
    app_label = 'myapp'
    # 声明此类是否为抽象
    abstract = True
    # 添加授权
    permissions = (("can_deliver_pizzas", "Can deliver pizzas"),)
```

4.1.3 实战演练：在 Django 框架中创建 SQLite3 数据库

在下面的实例代码中，演示了在 Django 框架中创建 SQLite3 数据库信息的过程。

源码路径：daima\4\learn_models

(1) 首先新建一个名为 learn_models 的项目，然后进入 learn_models 文件夹新建一个名为 people 的 app。

```
django-admin.py startproject learn_models   # 新建一个项目
cd learn_models   # 进入该项目的文件夹
django-admin.py startapp people   # 新建一个 people 应用(app)
```

(2) 将新建的应用(people)添加到文件 settings.py 中的 INSTALLED_APPS 中，也就是告诉 Django 有这么一个应用。

```python
INSTALLED_APPS = (
    'django.contrib.admin',
    'django.contrib.auth',
    'django.contrib.contenttypes',
    'django.contrib.sessions',
    'django.contrib.messages',
    'django.contrib.staticfiles',
    'people',
)
```

(3) 打开文件 people/models.py，新建一个继承自类 models.Model 的子类 Person，此类中有姓名和年龄这两个字段。具体实现代码如下所示。

```python
from django.db import models
class Person(models.Model):
```

```
name = models.CharField(max_length=30)
age = models.IntegerField()
def __str__(self):
    return self.name
```

在上述代码中,因为双下划线"__"在 Django QuerySet API 中有特殊含义(用于表示数据之间的关系和包含,不区分大小写,以什么开头或结尾,日期的大于小于,正则等),所以在 name 和 age 这两个字段中不能有双下划线"__"。

> **注 意**
> 是在 name 和 age 这两个字段中不能有双下划线"__",函数__str__(self)则不受此限制影响。

(4) 开始同步数据库操作,在此使用默认数据库 SQLite3,无须进行额外配置。具体命令如下所示。

```
# 进入 manage.py 所在的那个文件夹下输入这个命令
python manage.py makemigrations
python manage.py migrate
```

通过上述命令可以创建一个数据库表,当在前面的文件 models.py 中新增类 people 时,运行上述命令就可以自动在数据库中创建对应的数据库表,不用开发者手动创建。在命令行运行后会发现 Django 生成了一系列的表,也生成了上面刚刚新建的表 people_person。命令运行界面效果如图 4-1 所示。

```
mac:learn_models tu$ python manage.py syncdb
Creating tables ...
Creating table django_admin_log
Creating table auth_permission
Creating table auth_group_permissions
Creating table auth_group
Creating table auth_user_groups
Creating table auth_user_user_permissions
Creating table auth_user
Creating table django_content_type
Creating table django_session
Creating table people_person
```

图 4-1 命令行运行界面效果

(5) 输入 CMD 命令进行测试,整个测试过程如下所示。

```
$ python manage.py shell
>>> from people.models import Person
>>> Person.objects.create(name="haoren", age=24)
<Person: haoren>
>>> Person.objects.get(name="haoren")
<Person: haoren>
```

4.2 使用 QuerySet API

在 Django 框架中提供了丰富的模型 API,通过这些 API 可以快速实现数据库建模和操作。一旦在 Django 程序中建立好数据模型,

扫码观看本节视频讲解

Django 会自动为开发者生成一套数据库抽象的 API(QuerySet 查询集方法)，使开发者快速创建、检索、更新和删除对象。在本节的内容中，将详细讲解使用 QuerySet API 的知识。

4.2.1　QuerySet API 基础

由前面的学习可知，Django 框架使用了一种直观的方式把数据库表中的数据表示成 Python 对象，用一个模型类表示数据库中的一个表，一个模型类的实例代表这个数据库表中的一条特定的记录。通过使用关键字参数，可以实例化某个模型实例来创建一个对象，然后调用 QuerySet API 中的方法 save()把对象保存到数据库中。

例如，下面是一个简单的多人博客模型示例。

```python
from django.db import models

# 标记用户博客简介
class Blog(models.Model):
    name = models.CharField(max_length=100)
    tagline = models.TextField()

    def __str__(self):
        return self.name

# 标记作者
class Author(models.Model):
    name = models.CharField(max_length=50)
    email = models.EmailField()

    def __str__(self):
        return self.name

# 标记用户博文
class Entry(models.Model):
    blog = models.ForeignKey(Blog, on_delete=models.CASCADE)
    headline = models.CharField(max_length=255)
    body_text = models.TextField()
    pub_date = models.DateField()
    authors = models.ManyToManyField(Author)
    n_comments = models.IntegerField(default=0)
    n_pingbacks = models.IntegerField(default=0)
    rating = models.IntegerField(default=0)

    def __str__(self):
        return self.headline
```

接下来可以在 Python 交互式窗口中，调用 QuerySet API 向数据库中插入一些数据进行测试：

```
>>> from polls.models import *
>>> b = Blog(name="Beatles Blog", tagline="All the latest Beatles news.")
>>> b.save()
>>> a1 = Author.objects.create(name="dkey", email="dkey@163.com")
```

```
>>> a2 = Author.objects.create(name="jerry", email="jerry@163.com")
>>> a3 = Author.objects.create(name="jerry", email="jerry@qq.com")
```

在上述测试过程中，在数据库表 Blog 中添加了一条数据，在数据库表 Author 中添加了 3 条数据。

4.2.2　生成新的 QuerySet 对象的方法

在 QuerySet API 中，如下所示的方法可以生成新的 QuerySet 对象。

(1) filter()：实现查询过滤功能，这是精确的查询，而不是通配匹配查询。例如：

```
>>> Author.objects.filter(name="dkey")
<QuerySet [<Author: dkey>]>
```

(2) exclude()：实现反向查询，返回查询条件相反的对象。例如：

```
>>> Author.objects.exclude(name="dkey")
<QuerySet [<Author: jerry>, <Author: jerry>]>
```

(3) order_by()：对结果集进行升序或降序处理，可以指定需要排序的字段。例如：

```
# 升序
>>> Author.objects.filter().order_by('name')
<QuerySet [<Author: dkey>, <Author: jerry>, <Author: jerry>]>

# 降序
>>> Author.objects.filter().order_by('-name')
<QuerySet [<Author: jerry>, <Author: jerry>, <Author: dkey>]>
```

如果想要随机排序查询结果，则可以使用"?"实现。例如：

```
>>> Author.objects.order_by('?')
<QuerySet [<Author: jerry>, <Author: dkey>, <Author: jerry>]>
```

> **注意**
> order_by('?')会根据你使用的数据库后端进程处理，查询过程可能会很慢。

(4) reverse()：用颠倒的顺序显示查询的数据元素，再次调用 reverse()会恢复到正常的显示顺序。

(5) values()：当 QuerySet 作为迭代器使用时返回一个字典，而不是模型实例。这些字典中的每一个都表示一个对象，其中的键与模型对象的属性名称相对应。例如：

```
>>> Author.objects.filter(name='jerry').values()
<QuerySet [{'id': 2, 'name': 'jerry', 'email': 'jerry@163.com'}, {'id': 3, 'name': 'jerry', 'email': 'jerry@qq.com'}]>
```

方法 values()采用可选的位置参数，参数 fields 指定 SELECT 应限制的字段名称。如果指定了字段，每个字典将仅包含指定字段的字段键/值。如果不指定字段，则每个字典将包含数据库表中每个字段的键和值。例如：

```
>>> Author.objects.values('name')
<QuerySet [{'name': 'dkey'}, {'name': 'jerry'}, {'name': 'jerry'}]>
```

方法 values() 还包含一个可选的关键字参数 expressions，被传递给 annotate()。例如：

```
>>> from django.db.models.functions import Lower
>>> Author.objects.values(lower=Lower('name'))
<QuerySet [{'lower': 'dkey'}, {'lower': 'jerry'}, {'lower': 'jerry'}]>
```

(6) values_list()：在迭代时返回元组，每个元组都包含传入到 values_list() 中的相应字段或表达式的值，因此第一个项目是第一个字段。例如：

```
>>> Author.objects.values_list('id', 'name')
<QuerySet [(1, 'dkey'), (2, 'jerry'), (3, 'jerry')]>
```

如果只传入单个字段，则还可以传入 flat 参数，如果设置为 True，意味着返回的结果是单值，而不是一元组。例如：

```
>>> Author.objects.values_list('name', flat=True).order_by('-id')
<QuerySet ['jerry', 'jerry', 'dkey']>
```

另外，也可以使用 named=True 以获得结果为 namedtuple() 的信息。例如：

```
>>> Author.objects.values_list('id','name', named=True)
<QuerySet [Row(id=1, name='dkey'), Row(id=2, name='jerry'), Row(id=3, name='jerry')]>
```

使用命名元组可能会使结果更具可读性，将结果转换为命名元组的代价很小。如果不传递任何 values_list() 中的数据，它将按照声明的顺序返回模型中的所有字段。

(7) distinct()：去重查询，消除查询结果中的重复行，返回一个新的 QuerySet。例如：

```
>>> Author.objects.filter().values('name').distinct()        #filter 可省略
<QuerySet [{'name': 'dkey'}, {'name': 'jerry'}]>

>>> Author.objects.filter().distinct().values_list('name')
<QuerySet [('dkey',), ('jerry',)]>
```

做完去重之后，如果想把结果封装成一个 list，代码如下：

```
>>> list = [i[0] for i in Author.objects.distinct().values_list('name')]
>>> list
['dkey', 'jerry']
```

(8) dates()：根据日期获取 QuerySet 查询集，返回一个表示 datetime.date 对象列表的表达式。其语法格式是：

```
dates(field, kind, order ='ASC')
```

- 参数 field：是 DateField 模型的名称。
- 参数 kind：取值是 year、month 或 day，功能是在 datetime.date 结果列表中的每个对象都被给定的 type 类型"截断"。year 表示返回该字段的所有不同年份值的列表；month 表示返回该字段的所有不同"年/月"值的列表；day 表示返回该字段的所有不同年/月/日值的列表。
- 参数 order：表示排列顺序，默认为'ASC'，取值可以是'ASC'或者'DESC'。这指定了如何排序结果。

例如：

```
>>> Entry.objects.dates('pub_date', 'year')
[datetime.date(2005, 1, 1)]
>>> Entry.objects.dates('pub_date', 'month')
[datetime.date(2005, 2, 1), datetime.date(2005, 3, 1)]
>>> Entry.objects.dates('pub_date', 'day')
[datetime.date(2005, 2, 20), datetime.date(2005, 3, 20)]
>>> Entry.objects.dates('pub_date', 'day', order='DESC')
[datetime.date(2005, 3, 20), datetime.date(2005, 2, 20)]
>>> Entry.objects.filter(headline__contains='Lennon').dates('pub_date',
'day')
[datetime.date(2005, 3, 20)]
```

(9) datetimes()：返回一个表达式，该表达式表示 QuerySet 是一个 datetime.datetime 对象列表。其语法格式如下所示：

```
datetimes(field_name,kind,order='ASC',tzinfo = None)
```

- 参数 field_name：是 DateTimeField 模型的名称。
- 参数 kind：是 year，month，day，hour，minute 或 second。datetime.datetime 结果列表中的每个对象都被给定的 type 类型"截断"。
- 参数 order：指定如何排序结果，默认为'ASC'，取值有'ASC'或者'DESC'。
- 参数 tzinfo：定义在截断之前将日期时间转换到的时区。事实上，根据使用的时区，给定的日期时间具有不同的表示形式。该参数必须是一个 datetime.tzinfo 对象。如果是 None，则 Django 使用当前时区。

(10) raw()：在模型查询 API 不够用的情况下，可以使用原始的 SQL 语句。Django 提供了两种方法使用原始 SQL 进行查询：一种是使用 Manager.raw()方法，进行原始查询并返回模型实例；另一种是完全避开模型层，直接执行自定义的 SQL 语句。

方法 raw()接受一个原始的 SQL 查询，执行它并返回一个 django.db.models.query.RawQuerySet 实例。这个 RawQuerySet 实例可以迭代 QuerySet 以提供对象实例，例如：

```
>>> raw = Author.objects.raw('select * from polls_author')
>>> type(raw)
<class 'django.db.models.query.RawQuerySet'>
```

也可以使用传入变量的方式查询数据：

```
raw = Author.objects.raw(
    '''
    select * from polls_author where name = '%s' ORDER BY id desc
    ''' % name)
```

通过 raw()方法查询的结果是一个 RawQuerySet 对象，如果想获取所有的值可以这么做：

```
>>> raw[0].__dict__
{'_state': <django.db.models.base.ModelState object at 0x10868fe48>, 'id': 1,
'name': 'dkey', 'email': 'dkey@ywnds.com'}
```

借助 raw()方法可以序列化这个对象：

```python
def Serialization(_obj: object) -> list:
    '''
    _obj: objext -> list, Python 3.6新加入的特性, 用来标识这个方法接收一个对象并返回一个list
    orm.raw 序列化
    '''
    _list = []
    _get = []
    for i in _obj:
        _list.append(i.__dict__)

    for i in _list:
        del i['_state']
        _get.append(i)
    return _get
```

通过上面代码序列化后的使用方式如下:

```
>>> Serialization(raw)
[{'id': 1, 'name': 'dkey', 'email': 'dkey@ywnds.com'}]
```

(11) 链式查询: 可以在 filter 后跟 exclude 的方式实现链式查询。例如:

```
>>> Author.objects.filter(name="dkey").exclude(name="filter")
<QuerySet [<Author: dkey>]>
```

也可以使用一些高级过滤, 例如过滤以某某开头及某某结尾的名称:

```
>>> Author.objects.filter(name__startswith='d').filter(name__endswith='y')
<QuerySet [<Author: dkey>]>
```

(12) 复杂查询 Q filter(): Q filter()方法中的关键字参数查询都是一起进行 "AND" 运算的, 如果你需要执行更复杂的查询(例如 OR 语句), 可以使用 Q 对象实现。例如:

```
>>> from django.db.models import Q

# OR;
>>> Author.objects.filter(Q(name="dkey")|Q(name="jerry"))
<QuerySet [<Author: dkey>, <Author: jerry>, <Author: jerry>]>

# AND;
>>> Author.objects.filter(Q(name="dkey")&Q(name="jerry"))
<QuerySet []>
```

每个接受关键字参数的查询函数(如 filter()、exclude()、get())都可以传递一个或多个 Q 对象作为位置(数据名字无须和 Filter 中保持一致)参数。如果一个查询函数有多个 Q 对象参数, 这些参数的逻辑关系为 "AND"。查询函数可以混合使用 Q 对象和关键字参数, 所有提供给查询函数的参数(关键字参数或 Q 对象)都将 "AND" 运算在一起。但是如果出现 Q 对象, 它必须位于所有关键字参数的前面。例如:

```
>>> Author.objects.filter(Q(name="dkey"),Q(email="jerry@163.com"))
<QuerySet []>
>>> Author.objects.filter(Q(name="dkey"),Q(email="dkey@163.com"))
<QuerySet [<Author: dkey>]>
```

```
>>> Author.objects.filter(email="jerry@163.com",Q(name="dkey"))
  File "<console>", line 1
SyntaxError: positional argument follows keyword argument
```

(13) 反向查询：如果模型有一个 ForeignKey，那么该 ForeignKey 所指的模型实例可以通过一个管理器返回前一个模型的所有实例。默认情况下，这个管理器的名字为 foo_set，其中 foo 是源模型的小写名称。该管理器返回的查询集可以用在过滤和操作上。例如：

下面的代码分别实现了前向查询和反向查询：

```
#前向查询
>>> e = Entry.objects.get(id=2)
>>> e.blog                    #返回相关的blog对象

#反向查询
>>> b = Blog.objects.get(id=1)
>>> b.entry_set.all()         #返回和blog相关的所有对象
```

(14) 跨关联关系查询，例如：

```
# 跨关系反向查询；
>>> Entry.objects.filter(blog__name='Beatles Blog')
<QuerySet [<Entry: hello>]>

>>> Blog.objects.filter(entry__headline__contains='hello')
<QuerySet [<Blog: Beatles Blog>]>

# 跨关系多层查询；
>>> Blog.objects.filter(entry__authors__name__isnull=True)
<QuerySet []>

# 跨关系多层多过滤查询；
>>> Blog.objects.filter(entry__authors__isnull=False,entry__authors__name__contains='dkey')
<QuerySet [<Blog: Beatles Blog>]>
```

(15) 引用模型字段查询：使用 F 可以获取模型字段值，这经常被用来比较两个模型字段。例如：

```
# 过滤n_comments大于等于n_pingbacks的对象；
>>> Entry.objects.filter(n_comments__gte=F('n_pingbacks'))
<QuerySet [<Entry: hello>]>

# 过滤rating小于n_pingbacks + n_comments的对象；
>>> Entry.objects.filter(rating__lt=F('n_pingbacks') + F('n_comments'))
<QuerySet [<Entry: hello>]>
```

(16) 限制返回对象，例如：

```
# 返回前5个对象；
>>> Entry.objects.all()[:5]

# 返回第3个至第5个对象；
>>> Entry.objects.all()[3:5]
```

4.2.3 不返回 QuerySet 的方法

在 Django 框架中,如下所示的 QuerySet 方法不会返回 QuerySet。这些方法不使用缓存,每次查询都会调用数据库。

(1) get():返回匹配给定查找参数的对象,该参数应采用字段查找中描述的格式。例如:

```
>>> Person.objects.get(id=1)
>>> Person.objects.get(id=1, name='dkey')
```

如果没有找到指定参数的对象,则会引发异常。

(2) create():创建对象并将其全部保存在一个步骤中,从而使操作变得简便。例如下面的两段演示的功能是完全等效的:

```
p = Person.objects.create(first_name="Bruce", last_name="Springsteen")
```

和

```
p = Person(first_name="Bruce", last_name="Springsteen")
p.save(force_insert=True)
```

(3) get_or_create():尝试从数据库中获取基于给定 kwargs 的对象,如果没有找到匹配项则创建一个新的对象。其语法格式如下所示:

```
get_or_create(defaults=None, **kwargs)
```

能够返回一个元组(object, created),其中 object 是检索或创建的对象,created 是一个布尔值,指定是否创建了新对象。这是一种编写代码的简洁方法,例如:

```
try:
    obj = Author.objects.get(name='dkey1')
except Author.DoesNotExist:
    obj = Author(name='dkey1', email='dkey@ywnds.com')
    obj.save()
```

随着模型中字段数量的增加,这种模式变得相当不便。可以使用方法 get_or_create()重写上面的代码:

```
obj, created = Author.objects.get_or_create(
    name='dkey',
    defaults={'email': 'dkey@ywnds.com'},
)
```

传递给 get_or_create()方法的任何关键字参数(除可选的 defaults)都将用于获取 get()调用。如果找到一个对象,get_or_create()将返回该对象的元组。如果找到多个对象,get_or_create 将引发 MultipleObjectsReturned 异常。如果找不到对象,get_or_create() 将实例化并创建一个新对象,返回新对象的元组和 True。例如下面的代码根据上述算法创建了一个新对象:

```
params = {k: v for k, v in kwargs.items() if '__' not in k}
params.update({k: v() if callable(v) else v for k, v in defaults.items()})
```

```
obj = self.model(**params)
obj.save()
```

方法 get_or_create()和方法 create()使用手动指定主键时的错误行为类似，如果需要创建对象并且数据库中已存在唯一的主键，则会引发一个 IntegrityError 异常。

(4) update_or_create()：尝试从数据库中获取基于给定 kwargs 的对象。如果找到匹配项，将更新在默认字典中传递的字段，如果没有匹配项需要更新对象则创建一个新的。其语法格式如下所示：

```
update_or_create()
update_or_create(defaults=None, **kwargs)
```

参数 defaults 用于更新对象的由字段和值构成的字典，默认值可以是 callables。

能够返回一个元组(object, created)，其中 object 是创建或更新的对象，created 是一个布尔值，指定是否创建了新对象。这是编写代码的一种便捷方式，例如：

```
defaults = {'email': 'dkey@ywnds.com'}
try:
    obj = Author.objects.get(name='dkey')
    for key, value in defaults.items():
        setattr(obj, key, value)
    obj.save()
except Author.DoesNotExist:
    new_values = {'name': 'dkey'}
    new_values.update(defaults)
    obj = Author(**new_values)
    obj.save()
```

随着模型中字段数量的增加，这种模式变得相当不便。可以使用方法 update_or_create()重写上面的代码：

```
obj, created = Author.objects.update_or_create(
    name='dkey',
    defaults={'email': 'dkey@ywnds.com'},
)
```

如果在数据库级别不强制执行唯一性，则 get_or_create()方法容易出现竞争状况，这可能导致同时插入多行。如 get_or_create() 和 create()，如果手动指定主键并且需要创建一个对象，但主键在数据库中已存在，则会引发 IntegrityError 异常。

(5) bulk_create()：以高效的方式将提供的对象列表插入数据库(通常只有 1 个查询，不管有多少个对象)，其语法格式如下所示：

```
bulk_create(objs, batch_size=None)
```

例如：

```
Author.objects.bulk_create([Author(name='dkey2', email='dkey@ywnds.com'),
    Author(name='dkey3', email='dkey@ywnds.com'),
    ])
```

读者需要注意，模型的 save()方法不会被调用，它不适用于多表继承方案中的子模型，并且不适用于多对多的关系。

(6) count()：返回一个整数，表示与之匹配的数据库中 QuerySet 的对象数。count()方法从不会引发异常。例如：

```
#返回数据库中的信息数据
Entry.objects.count()

#返回在标题中包含"Lennon"的信息数量
Entry.objects.filter(headline__contains='Lennon').count()
```

(7) latest()：根据给定的字段返回表中的最新对象。例如在下面代码中的 Entry 可以根据 pub_date 字段返回表中的最新内容：

```
Entry.objects.latest('pub_date')
```

(8) first()：返回查询集匹配的第一个对象，或者为 None 表示没有匹配的对象。如果 QuerySet 没有定义顺序，那么查询集由主键自动排序。这可能会影响聚合结果。例如：

```
p = Article.objects.order_by('title', 'pub_date').first()
```

注意，这只是 first()的一种方便方法，下面的代码示例等同于上面的代码。

```
try:
    p = Article.objects.order_by('title', 'pub_date')[0]
except IndexError:
    p = None
```

(9) last()：功能与 first()类似，表示返回查询集中的最后一个对象。

(10) in_bulk()：其语法格式如下所示。

```
in_bulk(id_list = None,field_name ='pk')
```

其功能是获取字段值(id_list)的列表以及 field_name 的值，并返回一个字典，将每个值映射到具有给定字段值的对象的实例。如果参数 id_list 未提供，则返回查询集中的所有对象。参数 field_name 必须是唯一的字段，并且默认为主键。例如：

```
>>> Blog.objects.in_bulk([1])
{1: <Blog: Beatles Blog>}
>>> Blog.objects.in_bulk([1, 2])
{1: <Blog: Beatles Blog>, 2: <Blog: Cheddar Talk>}
>>> Blog.objects.in_bulk([])
{}
>>> Blog.objects.in_bulk()
{1: <Blog: Beatles Blog>, 2: <Blog: Cheddar Talk>, 3: <Blog: Django Weblog>}
>>> Blog.objects.in_bulk(['beatles_blog'], field_name='slug')
{'beatles_blog': <Blog: Beatles Blog>}
```

如果传递的 in_bulk()是一个空列表，则会得到一个空字典。

(11) update()：对指定字段执行 SQL 更新查询，并返回匹配的行数(如果某些行已具有新值，则该行数可能不等于更新的行数)。例如，要关闭 2010 年发布的所有博客条目的评论，可以这样实现：

```
>>> Entry.objects.filter(pub_date__year=2010).update(comments_on=False)
```

你可以更新多个字段,这对多少字段并没有限制。例如,在这里我们更新 comments_on 和 headline 字段:

```
>>> Entry.objects.filter(pub_date__year=2010).update(comments_on=False,
headline='This is old')
```

这样 update()方法立即应用。

(12) delete():执行 SQL 删除操作,并返回删除的对象数和每个对象类型具有删除次数的字典。例如,要删除特定博客中的所有条目:

```
>>> b = Blog.objects.get(pk=1)
>>> Entry.objects.filter(blog=b).delete()
```

(13) aggregate():返回通过计算得到的聚合值(平均值、总和等)的字典。例如:

```
>>> from django.db.models import Avg, Min, Max, Count, Sum
>>> Entry.objects.aggregate(Sum('n_comments'))
{'n_comments__sum': 1}
# 别名
>>> Entry.objects.aggregate(comment_avg=Avg('n_comments'))
{'comment_avg': 1.0}
```

4.2.4 字段查找

字段查找是指定 SQL WHERE 子句的内容,实现精确的信息查询功能。在 QuerySet API 中,字段查找被指定为 QuerySet 的 filter()、exclude()或 get()方法的关键字参数,使用__(双下划线)来查询。

通过使用 exact 可以实现精确匹配,这也是默认方式。为了使用方便,当没有提供查找类型(如 Author.objects.get(id=1))时,查找类型被默认为 exact。例如:

```
>>> Author.objects.get(id__exact=1)
<Author: dkey>
>>> Author.objects.filter(id__exact=1)
<QuerySet [<Author: dkey>]>
>>> Author.objects.exclude(id__exact=1)
<QuerySet [<Author: jerry>, <Author: jerry>]>
```

除了使用 exact 外,在 QuerySet API 中还有如下所示的精确查找类型。

- gt:大于。
- gte:大于等于。
- lt:小于。
- lte:小于等于。
- isnull:为 null。
- contains:包含某某,icontains 表示忽略大小写。
- startswith:以某某开始,istartswith 表示忽略大小写。

- endswith：以某某结束，iendswith 表示忽略大小写。
- regex：正则匹配，区分大小写。
- iregex：正则匹配，忽略大小写。
- in：在给定的可迭代对象中，通常是一个列表、元组或查询集。
- range：在某某范围之内。
- date、year、month、week、day、second、hour、minute：时间日期类型。

4.2.5 实战演练：使用 QuerySet API 操作 SQLite 数据库

在下面的实例中，将创建一个 Models 模型文件，然后根据这个模型文件在 SQLite 数据库中创建对应的数据库表和字段，最后使用 QuerySet API 操作 SQLite 数据库中的数据。

> 源码路径：**daima\4\QuerySetAPI**

(1) 在配置文件 settings.py 中设置使用 SQLite 数据库。代码如下：

```
DATABASES = {
    'default': {
        'ENGINE': 'django.db.backends.sqlite3',
        'NAME': os.path.join(BASE_DIR, 'db.sqlite3'),
    }
}
```

(2) 编写 Models 模型文件 models.py，用三个模型类分别实现 SQLite 数据库中的三个表和对应的字段。代码如下：

```
from django.db import models

class Author(models.Model):
    name = models.CharField(max_length=50)
    qq = models.CharField(max_length=10)
    addr = models.TextField()
    email = models.EmailField()

    def __str__(self):
        return self.name


class Article(models.Model):
    title = models.CharField(max_length=50)
    author = models.ForeignKey(Author, on_delete=models.CASCADE)
    content = models.TextField()
    score = models.IntegerField()  # 文章的打分
    tags = models.ManyToManyField('Tag')

    def __str__(self):
        return self.title


class Tag(models.Model):
    name = models.CharField(max_length=50)
```

```
    def __str__(self):
        return self.name
```

假设在 SQLite 数据库中保存博客文章的信息，通过上述代码实现了数据库的设计工作，假设一篇文章只有一个作者(Author)，一个作者可以有多篇文章(Article)，一篇文章可以有多个标签(Tag)。

(3) 创建 migrations，然后使用 migrate 在数据库中生成相应的表。代码如下：

```
python manage.py makemigrations
python manage.py migrate
```

(4) 编写文件 initdb.py，功能是向数据库中添加信息，分别添加作者的信息、文章标题和文章内容信息。代码如下：

```python
import random
from zqxt.wsgi import *
from blog.models import Author, Article, Tag


author_name_list = ['WeizhongTu', 'twz915', 'dachui', 'zhe', 'zhen']
article_title_list = ['Django 教程', 'Python 教程', 'HTML 教程']


def create_authors():
    for author_name in author_name_list:
        author, created = Author.objects.get_or_create(name=author_name)
        # 随机生成 9 位数的 QQ
        author.qq = ''.join(
            str(random.choice(range(10))) for _ in range(9)
        )
        author.addr = 'addr_%s' % (random.randrange(1, 3))
        author.email = '%s@ziqiangxuetang.com' % (author.addr)
        author.save()


def create_articles_and_tags():
    # 随机生成文章
    for article_title in article_title_list:
        # 从文章标题中得到 tag
        tag_name = article_title.split(' ', 1)[0]
        tag, created = Tag.objects.get_or_create(name=tag_name)

        random_author = random.choice(Author.objects.all())

        for i in range(1, 21):
            title = '%s_%s' % (article_title, i)
            article, created = Article.objects.get_or_create(
                title=title, defaults={
                    'author': random_author,  # 随机分配作者
                    'content': '%s 正文' % title,
                    'score': random.randrange(70, 101),  # 随机给文章一个打分
                }
            )
```

```
            article.tags.add(tag)
def main():
   create_authors()
   create_articles_and_tags()

if __name__ == '__main__':
   main()
   print("Done!")
```

(5) 开始测试 Django QuerySet 命令，例如使用 objects.all()获取 Article 中的所有数据：

```
>>> from blog.models import Article, Author, Tag
>>> Article.objects.all()
<QuerySet [<Article: Django 教程_1>, <Article: Django 教程_2>, <Article: Django
教程_3>, <Article: Django 教程_4>, <Article: Django 教程_5>, <Article: Django 教
程_6>, <Article: Django 教程_7>, <Article: Django 教程_8>, <Article: Django 教
程_9>, <Article: Django 教程_10>, <Article: Django 教程_11>, <Article: Django 教
程_12>, <Article: Django 教程_13>, <Article: Django 教程_14>, <Article: Django
教程_15>, <Article: Django 教程_16>, <Article: Django 教程_17>, <Article: Django
教程_18>, <Article: Django 教程_19>, <Article: Django 教程_20>, '...(remaining
elements truncated)...']>
```

(6) 使用如下代码查看 Django QuerySet 执行的 SQL：

```
>>> print(str(Author.objects.all().query))
SELECT "blog_author"."id", "blog_author"."name", "blog_author"."qq",
"blog_author"."addr", "blog_author"."email" FROM "blog_author"
>>> print(str(Author.objects.filter(name="WeizhongTu").query))
SELECT "blog_author"."id", "blog_author"."name", "blog_author"."qq",
"blog_author"."addr", "blog_author"."email" FROM "blog_author" WHERE
"blog_author"."name" = WeizhongTu
```

(7) 我们可以使用 values_list 获取元组形式结果，例如使用如下代码获取作者的 name 信息和 qq 信息：

```
>>> authors = Author.objects.values_list('name', 'qq')
>>> authors
<QuerySet [('WeizhongTu', '867480512'), ('twz915', '800566598'), ('dachui',
'707837909'), ('zhe', '754529421'), ('zhen', '396458386')]>
```

使用 list()获取数据库中作者的信息：

```
>>> list(authors)
[('WeizhongTu', '867480512'), ('twz915', '800566598'), ('dachui', '707837909'),
('zhe', '754529421'), ('zhen', '396458386')]
```

如果只需要获取 1 个字段的信息，可以设置属性 flat 的值为 True：

```
>>> Author.objects.values_list('name', flat=True)
<QuerySet ['WeizhongTu', 'twz915', 'dachui', 'zhe', 'zhen']>
>>> list(Author.objects.values_list('name', flat=True))
['WeizhongTu', 'twz915', 'dachui', 'zhe', 'zhen']
```

(8) 可以使用 filter()过滤信息，例如查询作者是 twz915 的文章的标题信息：

```
>>> Article.objects.filter(author__name='twz915').values_list('title', flat=True)
<QuerySet ['Django 教程_1', 'Django 教程_2', 'Django 教程_3', 'Django 教程_4', 'Django 教程_5', 'Django 教程_6', 'Django 教程_7', 'Django 教程_8', 'Django 教程_9', 'Django 教程_10', 'Django 教程_11', 'Django 教程_12', 'Django 教程_13', 'Django 教程_14', 'Django 教程_15', 'Django 教程_16', 'Django 教程_17', 'Django 教程_18', 'Django 教程_19', 'Django 教程_20']>
```

(9) 我们可以使用 values 获取字典形式的结果，例如获取作者的 name 信息和 qq 信息：

```
>>> Article.objects.filter(author__name='twz915').values_list('title', flat=True)
<QuerySet ['Django 教程_1', 'Django 教程_2', 'Django 教程_3', 'Django 教程_4', 'Django 教程_5', 'Django 教程_6', 'Django 教程_7', 'Django 教程_8', 'Django 教程_9', 'Django 教程_10', 'Django 教程_11', 'Django 教程_12', 'Django 教程_13', 'Django 教程_14', 'Django 教程_15', 'Django 教程_16', 'Django 教程_17', 'Django 教程_18', 'Django 教程_19', 'Django 教程_20']>
Author.objects.values('name', 'qq')
<QuerySet [{'name': 'WeizhongTu', 'qq': '867480512'}, {'name': 'twz915', 'qq': '800566598'}, {'name': 'dachui', 'qq': '707837909'}, {'name': 'zhe', 'qq': '754529421'}, {'name': 'zhen', 'qq': '396458386'}]>
```

或者是：

```
>>> list(Author.objects.values('name', 'qq'))
[{'name': 'WeizhongTu', 'qq': '867480512'}, {'name': 'twz915', 'qq': '800566598'}, {'name': 'dachui', 'qq': '707837909'}, {'name': 'zhe', 'qq': '754529421'}, {'name': 'zhen', 'qq': '396458386'}]
```

同样也可以查询作者是 twz915 的文章的标题：

```
Article.objects.filter(author__name='twz915').values('title')
<QuerySet [{'title': 'Django 教程_1'}, {'title': 'Django 教程_2'}, {'title': 'Django 教程_3'}, {'title': 'Django 教程_4'}, {'title': 'Django 教程_5'}, {'title': 'Django 教程_6'}, {'title': 'Django 教程_7'}, {'title': 'Django 教程_8'}, {'title': 'Django 教程_9'}, {'title': 'Django 教程_10'}, {'title': 'Django 教程_11'}, {'title': 'Django 教程_12'}, {'title': 'Django 教程_13'}, {'title': 'Django 教程_14'}, {'title': 'Django 教程_15'}, {'title': 'Django 教程_16'}, {'title': 'Django 教程_17'}, {'title': 'Django 教程_18'}, {'title': 'Django 教程_19'}, {'title': 'Django 教程_20'}]>
```

(10) 我们可以实现统计和聚合操作，例如计算每一个作者发布的文章总数是多少：

```
>>> from django.db.models import Count
>>>
Article.objects.all().values('author_id').annotate(count=Count('author')).values('author_id', 'count')
<QuerySet [{'author_id': 1, 'count': 20}, {'author_id': 2, 'count': 20}, {'author_id': 3, 'count': 20}]>
```

上述统计功能对应的 SQL 原型是什么呢？可以通过如下命令获得：

```
>>>
Article.objects.all().values('author_id').annotate(count=Count('author')).values('author_id', 'count').query.__str__()
```

```
'SELECT "blog_article"."author_id", COUNT("blog_article"."author_id") AS
"count" FROM "blog_article" GROUP BY "blog_article"."author_id"'
```

上述 SQL 命令可以简化为：

```
SELECT author_id, COUNT(author_id) AS count FROM blog_article GROUP BY author_id
```

同理，我们也可以统计作者的数量和作者的文章数：

```
Article.objects.all().values('author__name').annotate(count=Count('author')
).values('author__name', 'count')
<QuerySet [{'author__name': 'WeizhongTu', 'count': 20}, {'author__name':
'dachui', 'count': 20}, {'author__name': 'twz915', 'count': 20}]>
```

(11) 我们可以实现计算平均值操作，例如计算一个作者的所有文章的得分(score)平均值：

```
>>>
Article.objects.values('author_id').annotate(avg_score=Avg('score')).values
('author_id', 'avg_score')
<QuerySet [{'author_id': 1, 'avg_score': 86.0}, {'author_id': 2, 'avg_score':
85.7}, {'author_id': 3, 'avg_score': 85.0}]>
```

上述统计功能对应的 SQL 原型是什么呢？可以通过如下命令获得：

```
>>>
Article.objects.values('author_id').annotate(avg_score=Avg('score')).values
('author_id', 'avg_score').query.__str__()
'SELECT "blog_article"."author_id", AVG("blog_article"."score") AS
"avg_score" FROM "blog_article" GROUP BY "blog_article"."author_id"'
```

(12) 我们可以实现计算总分操作，例如计算一个作者的所有文章的总分：

```
>>> Article.objects.values('author__name').annotate(sum_score=Sum('score')).
values('author__name', 'sum_score')
<QuerySet [{'author__name': 'WeizhongTu', 'sum_score': 1720}, {'author__name':
'dachui', 'sum_score': 1700}, {'author__name': 'twz915', 'sum_score': 1714}]>
```

上述统计功能对应的 SQL 原型是什么呢？可以通过如下命令获得：

```
>>> Article.objects.values('author__name').annotate(sum_score=Sum('score')).
values('author__name', 'sum_score').query.__str__()
'SELECT "blog_author"."name", SUM("blog_article"."score") AS "sum_score" FROM
"blog_article" INNER JOIN "blog_author" ON ("blog_article"."author_id" =
"blog_author"."id") GROUP BY "blog_author"."name"'
```

(13) 我们也可以实现"多对一"操作，例如从众多信息中提取第一条信息。例如查询 10 篇文章的信息，并且显示第一篇文章的标题、编号和作者名信息，必须按照下面的过程实现：

```
>>> articles = Article.objects.all()[:10]
>>> a1 = articles[0]   # 取第一篇
>>> a1.title
'Django 教程_1'
>>> a1.author_id
2
>>> a1.author.name
'twz915'
```

能不能只查询一次就把作者的信息也查出来呢？当然可以，这时就要用到 select_related。在本实例的数据库中，设计一篇文章只能有一个作者，一个作者可以有多篇文章。假设现在要求在查询文章的时候连同作者一起查询出来，"文章"和"作者"的关系就是多对一。也就是说，一篇文章只可能有一个作者。此时的实现过程如下：

```
>>> articles = Article.objects.all().select_related('author')[:10]
>>> a1 = articles[0]
>>> a1.title
'Django 教程_1'
>>> a1.author.name
'twz915'
```

4.3 实战演练：使用 QuerySet API 操作 MySQL 数据库

MySQL 是 Web 开发中最为常用的数据库之一，本实例讲解使用 QuerySet API 操作 MySQL 数据库的方法，包括连接数据库、添加数据、修改数据和删除数据等操作。

扫码观看本节视频讲解

源码路径：**daima\4\QuerySetAPI**

(1) 首先新建一个名为"QuerySetAPI"的项目，然后进入到 QuerySetAPI 文件夹，新建一个名为"people"的 app。代码如下：

```
django-admin.py startproject QuerySetAPI        # 新建一个项目
cd QuerySetAPI  # 进入到该项目的文件夹
django-admin.py startapp people                  # 新建一个 people 应用(app)
```

(2) 将新建的应用(people)添加到文件 settings.py 中的 INSTALLED_APPS 中，也就是告诉 Django 有这么一个应用。

```
INSTALLED_APPS = (
    'django.contrib.admin',
    'django.contrib.auth',
    'django.contrib.contenttypes',
    'django.contrib.sessions',
    'django.contrib.messages',
    'django.contrib.staticfiles',
    'people',
)
```

(3) 打开文件 people/models.py，新建一个继承自类 models.Model 的子类 aaa，类名 aaa 表示在数据库中创建一个名字为 aaa 的数据库表。类 aaa 里面的属性 name 表示数据库表 aaa 中的一个字段名字是 name。具体实现代码如下所示。

```
from django.db import models

class aaa(models.Model):
    name = models.CharField(max_length=20)
```

在上述代码中,字段 name 的数据类型是 CharField,这相当于 MySQL 数据库中的 varchar 类型,max_length 表示字段 name 的限定长度。

(4) 在设置文件 settings.py 中找到 INSTALLED_APPS 这一项,将上面创建的 APP 程序 people 添加到项目中,对应代码如下所示。

```
INSTALLED_APPS = [
    'django.contrib.admin',
    'django.contrib.auth',
    'django.contrib.contenttypes',
    'django.contrib.sessions',
    'django.contrib.messages',
    'django.contrib.staticfiles',
    'people',
]
```

(5) 在项目的设置文件 settings.py 中找到 DATABASES 配置项,设置想要连接的 MySQL 数据库的参数,其中包括数据库名称、用户名、用户密码、连接端口、服务器地址。

```
DATABASES = {
'default': {
    'ENGINE': 'django.db.backends.mysql',
    'NAME': 'qapi',                          //要连接的数据库名是 qapi
    'USER': 'root',
    'PASSWORD': '66688888',
    'HOST':'127.0.0.1',
    'PORT':'3306',
  }
}
```

(6) 使用如下所示的命令创建数据库表。

```
python manage.py migrate     # 创建表结构
python manage.py makemigrations people  # 让 Django 知道我们的模型有一些变更
python manage.py migrate aaa    # 创建表结构
```

通过上述命令会在连接的 MySQL 数据库中根据模型文件 models.py 创建数据库表,表名的组成结构为:

```
app 名_类名
```

例如本实例的 app 名为 people,类名为 aaa,所以在 MySQL 数据库中的表名显示为 people_aaa,如图 4-2 所示。

图 4-2 成功在 MySQL 数据库中创建的表

注 意

虽然我们没有在模型文件 models.py 中为表设置主键，但是 Django 会自动添加一个 id 作为主键，正如图 4-2 所示。

（7）在系统根目录 QuerySetAPI 中添加 Python 文件 testdb.py，功能是将一条 name 值为 welcome 的数据添加到 MySQL 数据库中。

```
from django.http import HttpResponse

from people.models import aaa

# 数据库操作
def testdb(request):
    test1 = aaa(name='welcome')
    test1.save()
    return HttpResponse("<p>数据添加成功!</p>")
```

（8）编写路径导航文件 urls.py，将上面添加的数据库文件 testdb.py 添加到项目中。文件 urls.py 的具体实现代码如下所示。

```
from django.contrib import admin
from django.conf.urls import include
from django.urls import path
from people import views as learn_views
from testdb import testdb
urlpatterns = [
    path('', learn_views.home, name='home'),
    path('testdb/', testdb, name='testdb'),
]
```

使用如下命令运行项目，在浏览器中输入 http://127.0.0.1:8000/testdb/ 后，一条 name 值为 welcome 的数据添加到 MySQL 数据库中，并且在浏览器中显示 "数据添加成功！" 的提示，如图 4-3 所示。

```
python manage.py runserver
```

成功添加一条数据

浏览器中的执行效果

图 4-3 添加数据执行效果

读者需要注意的是，每当重新浏览或刷新一次 http://127.0.0.1:8000/testdb/，就会在数据库中添加一条数据，例如浏览 10 次会在数据库中添加 10 条数据，如图 4-4 所示。

图 4-4　添加了 10 条数据

(9) 同样道理，我们可以编写程序文件 huoqu.py，使用 QuerySet API 中的方法获取数据库中的数据，具体实现代码如下所示。

```
from django.http import HttpResponse
from people.models import aaa

# 数据库操作
def testdb(request):
    # 初始化
    response = ""
    response1 = ""

    # 通过 objects 这个模型管理器的 all() 获得所有数据行，相当于 SQL 中的 SELECT * FROM
    list = aaa.objects.all()

    # filter 相当于 SQL 中的 WHERE，可设置条件过滤结果
    response2 = aaa.objects.filter(id=1)

    # 获取单个对象
    response3 = aaa.objects.get(id=1)

    # 限制返回的数据，相当于 SQL 中的 OFFSET 0 LIMIT 2;
    aaa.objects.order_by('name')[0:2]

    # 数据排序
    aaa.objects.order_by("id")

    # 上面的方法可以连锁使用
    aaa.objects.filter(name="welcome").order_by("id")

    # 输出所有数据
    for var in list:
        response1 += var.name + " "
    response = response1
    return HttpResponse("<p>" + response + "</p>")
```

执行效果如图 4-5 所示。

```
← → C  🗋 127.0.0.1:8000/huoqu/
welcome welcome welcome welcome welcome welcome welcome welcome welcome welcome
```

图 4-5 执行效果

(10) 同样道理，我们可以编写程序文件 xiu.py，使用 QuerySet API 中的方法 save()或 update()修改数据库中已经存在的数据，具体实现代码如下所示。

```python
from django.http import HttpResponse
from people.models import aaa
# 数据库操作
def testdb(request):
    # 修改其中一个 id=1 的 name 字段，再保存，相当于 SQL 中的 UPDATE
    test1 = aaa.objects.get(id=1)
    test1.name = 'WelcomeToWuhan'
    test1.save()

    #修改其中一个 id=1 的 name 字段，这是另外一种方式
    # Test.objects.filter(id=1).update(name='WelcomeToWuhan')

    # 修改所有的列
    # Test.objects.all().update(name='WelcomeToWuhan')

    return HttpResponse("<p>修改成功</p>")
```

在上述代码中实现了如下三种修改操作。
- 设置表中 id 为 1 的 name 值为"WelcomeToWuhan"，然后使用方法 save()保存，这样只会修改一条数据。
- 使用方法 update()将表中 id 为 1 的 name 值修改为"WelcomeToWuhan"，这样也只会修改一条数据。
- 使用方法 all()和 update()将表中所有的 name 值修改为"WelcomeToWuhan"，这样会修改 n 条数据。

(11) 同样道理，我们可以编写程序文件 shan.py，使用 QuerySet API 中的方法 delete()删除数据库中已经存在的数据，具体实现代码如下所示。

```python
from django.http import HttpResponse
from people.models import aaa

# 数据库操作
def testdb(request):
    # 删除 id=1 的数据
    test1 = aaa.objects.get(id=1)
    test1.delete()
    # 另外一种方式
    # Test.objects.filter(id=1).delete()
    # 删除所有数据
    # Test.objects.all().delete()
    return HttpResponse("<p>删除成功</p>")
```

在上述代码中演示了如下三种删除操作。
- 使用 get() 方法获取表中 id 为 1 的数据，然后使用方法 delete() 删除这条数据，这样只会删除一条数据。
- 使用 filter() 方法获取表中 id 为 1 的数据，然后使用方法 delete() 删除这条数据，这样只会删除一条数据。
- 使用方法 all() 和 delete() 将表中所有数据删除，这样会删除 n 条数据。

第 5 章

使用模板

　　每一个 Web 框架都需要一种很便利的方法用于动态生成 HTML 页面，最常见的做法是使用模板。模板包含所需 HTML 页面的静态部分，以及一些特殊的模板语法，用于将动态内容插入静态部分。究其实质，模板层就是向 HTML 文件中填入动态内容的系统。在本章的内容中，将详细讲解在 Django 中使用模板的知识。

5.1 模板基础

在 Django 框架中提供了模板功能，通过使用里面的模板文件可分离文档的表现形式和具体内容。Django 可以配置一个或多个模板引擎(语言)，也可以不用引擎。Django 自带了一个名为 DTL(Django Template Language)的模板语言，以及另外一种流行的 Jinja2 语言(需要提前安装，pip install Jinja2)。Django 为加载和渲染模板定义了一套标准的 API，与具体的后台无关。加载指的是，根据给定的模板名称调用的对应的代码进行预处理，通常会将它编译好放在内存中。渲染则表示，使用 Context 数据对模板插值并返回生成的字符串。

扫码观看本节视频讲解

5.1.1 配置引擎

模板引擎通过配置文件 settings.py 中的 TEMPLATES 参数来设置模板配置信息。这是一个列表，与引擎一一对应，每个元素都是一个引擎配置字典。当使用 startproject 命令或 PyCharm 创建一个 Django 工程时，会在自动生成的配置文件 settings.py 中自动生成下面的代码。

```
TEMPLATES = [
    {
        'BACKEND': 'django.template.backends.django.DjangoTemplates',
        'DIRS': [],
        'APP_DIRS': True,
        'OPTIONS': {
            # ... some options here ...
        },
    },
]
```

具体说明如下。

- BACKEND：表示后端，内置的后端有 django.template.backends.django.DjangoTemplates 和 django.template.backends.jinja2.Jinja2。
- OPTIONS：里面包含具体的后端设置。
- DIRS：定义了一个目录列表，模板引擎按列表顺序搜索这些目录以查找模板源文件。
- APP_DIRS：告诉模板引擎是否应该进入每个已安装的应用中查找模板。通常将该选项保持为 True。

每种模板引擎后端都定义了一个惯用的名称作为应用内部存放模板的子目录名称。例如 Django 为它自己的模板引擎指定的是"templates"，为 jinja2 指定的名字是"jinja2"。尤其是，django 允许有多个模板引擎后端实例，且每个实例有不同的配置选项。在这种情况下，必须为每个配置指定一个唯一的名字。

在 DTL 引擎的 OPTIONS 配置项中接收以下参数。
- 'autoescape'：一个布尔值，用于控制是否启用 HTML 自动转义功能。默认为 True。
- context_processors：以 "." 为分隔符的 Python 调用路径的列表。默认是个空列表。
- 'debug'：打开/关闭模板调试模式的布尔值。默认和文件 setting.py 中的 DEBUG 有相同的值。
- 'loaders'：模板加载器类的虚拟 Python 路径列表。默认值取决于 DIRS 和 APP_DIRS 的值。
- string_if_invalid：非法变量时输出的字符串。默认为空字符串。
- file_charset：用于读取磁盘上的模板文件的字符集编码。默认为 FILE_CHARSET 的值。
- 'libraries'：用于注册模板引擎。这可以用于添加新的库或为现有库添加备用标签。
- 'builtins'：以圆点分隔的 Python 路径的列表。

5.1.2 Django 模板的基础用法

在 Django 框架中，模板通常是静态的 HTML、CSS 和 JS 文件，用于定义一个页面的显示样式或外观。为了实现样式与业务逻辑的分离，在调用这些静态模板文件时，需要借助于 View 视图传递过来的变量(Variable)或内容对象(Context Object)才能被渲染成一个完整的页面。

1. Template 的工作过程

我们继续使用一个博客类网站进行举例，假设当用户访问/blog/articles/21/的时候，URL 路由器会调用视图文件 views.py 中的方法 articles_detail，方法 articles_detail 的功能是提取数据库中 id 号为 21 的这篇文章，然后通过方法 render 将提取到的数据传递到模板文件/blog/articles_detail.html。假设 URL 文件 blog/urls.py 的实现代码如下所示。

```
from django.urls import path

from . import views

urlpatterns = [
  path('blog/articles/<int:id>/', views.articles_detail,
name='articles_detail'),
]
```

假设视图文件 blog/views.py 的实现代码如下所示。

```
from django.shortcuts import render, get_object_or_404
from .models import Articles

def articles_detail(request, id):
   articles = get_object_or_404(Articles, pk=id)
   return render(request, 'blog/articles_detail.html', {"articles":
articles})
```

这样在模板文件中可以通过双括号{{ articles }}的方式显示上面定义的变量或内容对象，也可以通过点号"."来直接访问变量的属性。例如，下面是模板文件 blog/articles_detail.html 的实现代码。

```
{% block content %}
{{ article.title }}
{{ article.pub_date }}
{{ article.body }}
{% endblock %}
```

2. 模板(Template)文件的位置

读者在开发 Django Web 项目时，建议将 HTML 格式的模板文件放在 app/templates/app/ 目录中，而不是简单地放在 app/templates/ 目录中。当我们多加一层 app 目录后，Django 只会查找 app 文件夹中的模板文件，这样会提高项目的安全性。另外，在视图文件 views.py 中也建议通过 app/template_name.html 调用 template，这样做的好处是防止与其他同名 template 发生冲突。

5.1.3 实战演练：使用简易模板

在下面的实例代码中，演示了在 Django 框架中使用模板的过程。

> 源码路径：**daima\5\zqxt_tmpl**

（1）分别创建一个名为"zqxt_tmpl"的项目和一个名称为"learn"的应用。
（2）将"learn"应用加入 settings.INSTALLED_APPS 中，具体实现代码如下所示。

```
INSTALLED_APPS = (
    'django.contrib.admin',
    'django.contrib.auth',
    'django.contrib.contenttypes',
    'django.contrib.sessions',
    'django.contrib.messages',
    'django.contrib.staticfiles',
    'learn',
)
```

（3）打开文件 learn/views.py 编写一个首页的视图，具体实现代码如下所示。

```
from django.shortcuts import render
def home(request):
    return render(request, 'home.html')
```

（4）在"learn"目录下新建一个"templates"文件夹用于保存模板文件，然后在里面新建一个 home.html 文件作为模板。文件 home.html 的具体实现代码如下所示。

```
<!DOCTYPE html>
<html>
<head>
    <title>欢迎光临</title>
```

```
</head>
<body>
欢迎选择浪潮产品!
</body>
</html>
```

(5) 为了将视图函数对应到网址,对文件 zqxt_tmpl/urls.py 的代码进行如下所示的修改。

```
from django.urls import path
from django.contrib import admin
from learn import views as learn_views

urlpatterns = [
   path(r'', learn_views.home, name='home'),
   path(r'admin/', admin.site.urls),
]
```

(6) 输入如下命令启动服务器:

```
python manage.py runserver
```

执行后将显示模板的内容,执行效果如图 5-1 所示。

图 5-1　执行效果(一)

5.2　模板标签 Tags

在 Django 的模板中,变量被存放在双大括号{{ }}中,代码存放在 {% tag_name %}标签中。Django 框架中有很多自带标签,可以满足绝大部分程序的开发需求。

5.2.1　常用的模板标签

扫码观看本节视频讲解

(1) autoescape:控制当前自动转义的行为,有 on 和 off 两个选项,例如:

```
{% autoescape on %}
   {{ body }}
{% endautoescape %}
```

(2) block:定义一个子模板可以覆盖的块。

(3) comment:注释,例如{% comment %} 和 {% endcomment %}之间的内容被解释为注释。

(4) crsf_token:一个防止 CSRF 攻击(跨站点请求伪造)的标签。

(5) cycle:循环给出的字符串或者变量,可以混用。例如:

```
{% for o in some_list %}
    <tr class="{% cycle 'row1' rowvalue2 'row3' %}">
        ...
    </tr>
{% endfor %}
```

值得注意的是，这里的变量的值默认不是自动转义的，为了提高代码安全性，可以使用强制转义的方法进行转义，例如：

```
{% for o in some_list %}
    <tr class="{% filter force_escape %}{% cycle rowvalue1 rowvalue2 %}{% endfilter %}">
        ...
    </tr>
{% endfor %}
```

在某些情况下，可能想在循环外部引用循环中的下一个值，这时需要用 as 给 cycle 标签设置一个名字，这个名字代表的是当前循环的值。但是在 cycle 标签里面，可以用这个变量来获得循环中的下一个值。例如：

```
<tr>
    <td class="{% cycle 'row1' 'row2' as rowcolors %}">...</td>
    <td class="{{ rowcolors }}">...</td>
</tr>
<tr>
    <td class="{% cycle rowcolors %}">...</td>
    <td class="{{ rowcolors }}">...</td>
</tr>
```

对应的渲染的结果是：

```
<tr>
    <td class="row1">...</td>
    <td class="row1">...</td>
</tr>
<tr>
    <td class="row2">...</td>
    <td class="row2">...</td>
</tr>
```

但是一旦定义了 cycle 标签，默认就会使用循环中的第一个值。当你仅仅是想定义一个循环，而不想打印循环的值的时候(比如在父模板定义变量以方便继承)，可以用 cycle 的 silent 参数(必须保证 silent 是 cycle 的最后一个参数)，并且 silent 也具有继承的特点。例如下面的代码：

```
{% cycle 'row1' 'row2' as rowcolors silent %}
{% cycle rowcolors %}
```

尽管上面第 2 行中的 cycle 没有 silent 参数，但是由于 rowcoclors 是前面定义的且包含 silent 参数，所以第 2 个 cycle 也具有 silent 循环的特点。

(6) debug：输出所有的调试信息，包括当前上下文和导入的模块。

(7) extends：表示当前模板继承了一个父模板，接收一个包含父模板名字的变量或者字符串常量。

(8) filter：通过可用的过滤器过滤内容，过滤器之间还可以相互调用。例如：

```
{% filter force_escape|lower %}
    This text will be HTML-escaped, and will appear in all lowercase.
{% endfilter %}
```

(9) firstof：返回列表中第一个可用(非 False)的变量或者字符串，注意 firstof 中的变量不是自动转义的。例如：

```
{% firstof var1 var2 var3 "fallback value" %}
```

(10) for：for 循环，可以在后面加入 reversed 参数遍历逆序的列表。例如：

```
{% for obj in list reversed %}
```

还可以根据列表的数据来写 for 语句，例如下面是对于字典类型数据的 for 循环。

```
{% for key, value in data.items %}
    {{ key }}: {{ value }}
{% endfor %}
```

另外，在 for 循环中还有一系列有用的变量，具体说明如表 5-1 所示。

表 5-1 for 循环中的变量

变　　量	描　　述
forloop.counter	当前循环的索引，从 1 开始
forloop.counter0	当前循环的索引，从 0 开始
forloop.revcounter	当前循环的索引(从后面算起)，从 1 开始
forloop.revcounter0	当前循环的索引(从后面算起)，从 0 开始
forloop.first	如果这是第一次循环返回真
forloop.last	如果这是最后一次循环返回真
forloop.parentloop	如果是嵌套循环，指的是外一层循环

(11) for...empty：如果 for 循环中的参数列表为空，则执行 empty 里面的内容。例如：

```
<ul>
{% for athlete in athlete_list %}
    <li>{{ athlete.name }}</li>
{% empty %}
    <li>Sorry, no athlete in this list!</li>
{% endfor %}
<ul>
```

(12) if：这是一个条件语句，例如：

```
{% if athlete_list %}
    Number of athletes: {{ athlete_list|length }}
{% elif athlete_in_locker_room_list %}
    Athletes should be out of the locker room soon!
{% else %}
    No athletes.
{% endif %}
```

(13) 布尔操作符：在 if 标签中只可以使用 and、or 和 not 三个布尔操作符，还有==、!=、<、>、<=、>=、in、not in 等操作符。在 if 标签里面，通过这些操作符可以开发出复杂的表达式。

(14) ifchanged：检测一个值在循环的最后有没有改变。这个标签是在循环里面使用的，有如下所示的两个用法。

- 当没有接收参数时，比较的是 ifchanged 标签里面的内容相比以前是否有变化，有变化时生效。
- 当接收一个或一个以上的参数的时候，如果有一个或者一个以上的参数发生变化，则生效。

在 ifchanged 中可以有 else 标签，例如：

```
{% for match in matches %}
    <div style="background-color:
        {% ifchanged match.ballot_id %}
            {% cycle "red" "blue" %}
        {% else %}
            grey
        {% endifchanged %}
    ">{{ match }}</div>
{% endfor %}
```

(15) ifequal：仅当两个参数相等的时候输出块中的内容，可以配合 else 输出。例如：

```
{% ifequal user.username "adrian" %}
...
{% endifequal %}
```

(16) ifnotequal：功能和用法与 ifequal 标签类似。

(17) include：用于加载一个模板并包含引用当前模板的上下文渲染它，接收一个变量或者字符串参数，也可以在 include 的时候传递一些参数进来。例如：

```
{% include "name_snippet.html" with person="Jane" greeting="Hello" %}
```

如果只想接收传递的参数，不接收当前模板的上下文时，可以使用 only 参数。例如：

```
{% include "name_snippet.html" with greeting="Hi" only %}
```

(18) load：加载一个自定义的模板标签集合。

(19) now：显示当前的时间日期，接收格式化字符串的参数。例如：

```
It is {% now "jS F Y H:i" %}
```

在现实中已经定义好了一些格式化字符串参数，例如：

- DATE_FORMAT(月日年)。
- DATETIME_FORMAT(月日年时)。
- SHORT_DATE_FORMAT(月/日/年)。
- SHORT_DATETIME_FORMAT(月/日/年/时)。

(20) regroup：通过共同的属性对一个列表的相似对象重新分组，假如存在如下一个城市(city)的列表。

```
cities = [
    {'name': 'Mumbai', 'population': '19,000,000', 'country': 'India'},
    {'name': 'Calcutta', 'population': '15,000,000', 'country': 'India'},
    {'name': 'New York', 'population': '20,000,000', 'country': 'USA'},
    {'name': 'Chicago', 'population': '7,000,000', 'country': 'USA'},
    {'name': 'Tokyo', 'population': '33,000,000', 'country': 'Japan'},
]
```

如果想按照国家 country 这个属性来重新分组，目的是得到下面的分组结果。

```
India
Mumbai: 19,000,000
Calcutta: 15,000,000
USA
New York: 20,000,000
Chicago: 7,000,000
Japan
Tokyo: 33,000,000
```

则可以通过如下代码实现上述要求的分组功能。

```
{% regroup cities by country as country_list %}
<ul>
{% for country in country_list %}
    <li>{{ country.grouper }}
    <ul>
        {% for item in country.list %}
          <li>{{ item.name }}: {{ item.population }}</li>
        {% endfor %}
    </ul>
    </li>
{% endfor %}
</ul>
```

值得注意的是，regroup 并不会重新排序，所以必须确保 city 在 regroup 之前已经按 country 排好序，否则将得不到预期想要的结果。如果不确定 city 在 regroup 之前已经按 country 排好序，可以用 dictsort 进行过滤器排序。例如：

```
{% regroup cities|dictsort:"country" by country as country_list %}
```

(21) spaceless：移除 html 标签之间的空格，需要注意的是标签之间的空格，标签与内容之间的空格不会被删除。例如：

```
{% spaceless %}
    <p>
        <a href="foo/">Foo</a>
    </p>
{% endspaceless %}
```

运行结果是：

```
<p><a href="foo/">Foo</a></p>
```

(22) ssi：在页面上输出给定文件的内容，例如：

```
{% ssi /home/html/ljworld.com/includes/right_generic.html %}
```

使用参数 parsed 可以将输入的内容作为一个模板,从而可以使用当前模板的上下文。例如下面的演示代码:

```
{% ssi /home/html/ljworld.com/includes/right_generic.html parsed %}
```

(23) url:返回一个绝对路径的引用(没有域名的 url),接收的第一个参数是一个视图函数的名字,然后从 urls 配置文件里面找到那个视图函数对应的 url。

(24) widthratio:计算给定值与最大值的比率,然后把这个比率与一个常数相乘,返回最终的结果。例如:

```
<img src="bar.gif" height="10" width="{% widthratio this_value max_value 100 %}" />
```

(25) with:用更简单的变量名缓存复杂的变量名。例如:

```
{% with total=business.employees.count %}
   {{ total }} employee{{ total|pluralize }}
{% endwith %}
```

5.2.2 实战演练:在模板中使用 for 循环显示列表内容

在下面的实例中,演示了在 Django 模板中使用 for 循环显示列表内容的过程。

源码路径:**daima\5\biaoqian**

(1) 通过如下命令分别创建一个名为"biaoqian"的项目和一个名称为"learn"的应用程序。

```
django-admin.py startproject biaoqian
cd biaoqian
python manage.py startapp learn
```

(2) 将上面创建的"learn"应用加入 settings.INSTALLED_APPS 中。

```
INSTALLED_APPS = (
   'django.contrib.admin',
   'django.contrib.auth',
   'django.contrib.contenttypes',
   'django.contrib.sessions',
   'django.contrib.messages',
   'django.contrib.staticfiles',
   'learn',
)
```

(3) 为了将视图函数对应到网址,对文件 urls.py 的代码进行如下所示的修改。

```
from django.contrib import admin
from django.conf.urls import include, url
from learn import views as learn_views
urlpatterns = [
   url(r'^$', learn_views.home, name='home'),
```

```
    url(r'^admin/', admin.site.urls),
]
```

(4) 在视图文件 view.py 中传递一个 List 列表到模板 home.html，在列表中包含 5 门编程语言的名字。文件 view.py 的具体代码如下所示。

```
from django.shortcuts import render

def home(request):
    TutorialList = ["Java", "C", "C++", "Python", "C#"]
    return render(request, 'home.html', {'TutorialList': TutorialList})
```

(5) 在模板文件 home.html 中使用 for 循环遍历列表中的内容，在 for 循环最后要有一个结束标记 "endfor"。文件 home.html 的具体实现代码如下所示。

```
5大最流行的编程语言列表:
{% for i in TutorialList %}
{{ i }}
{% endfor %}
```

(6) 输入如下命令启动服务器运行 Web 程序。

```
python manage.py runserver
```

执行后会在网页中显示在列表中存储的 5 门编程语言的名字，执行效果如图 5-2 所示。

图 5-2　执行效果(二)

5.3　模板过滤器 Filter

在 Django 框架中，可以通过使用过滤器来改变变量在模板中的显示，例如{{ article.title | lower }}中的过滤器 lower 可以将文章的标题转换为小写形式。

5.3.1　常用的内置过滤器

扫码观看本节视频讲解

为模板过滤器提供参数的方式是：过滤器后加一个冒号，然后在后面紧跟参数，中间不能有空格。在目前的 Django 版本中，只能为过滤器最多提供一个参数。

(1) add：把 add 后的参数加给 value。例如：

```
{{ value|add:"2" }}
```

过滤器首先会强制把两个值转换成 Int 类型。 如果强制转换失败，它会试图使用各种方式把两个值相加。例如，如果 value 为 4，则会输出 6。对于下面的例子：

```
{{ first|add:second }}
```

如果 first 是[1, 2, 3]，second 是[4, 5, 6]，将会输出[1, 2, 3, 4, 5, 6]。

(2) addslashes：在引号前面加上斜杠，常用于在 CSV 中转义字符串。例如：

`{{ value|addslashes }}`

如果 value 是"I'm using Django"，则输出将会变成 "I\'m using Django"。

(3) capfirst：大写变量的第一个字母。如果第一个字符不是字母，该过滤器将不会生效。例如：

`{{ value|capfirst }}`

如果 value 是"django"，输出将变成"Django"。

(4) center：在给定的宽度范围内居中。例如：

`"{{ value|center:"15" }}"`

如果 value 值是"django"，则上述代码会打印输出"Django"。

(5) cut：移除 value 中所有的与给定参数相同的字符串。例如：

`{{ value|cut:" " }}`

如果 value 为"String with spaces"，则会输出"Stringwithspaces"。

(6) date：根据给定格式对一个日期变量进行格式化，可用的格式字符串如表 5-2 所示。

表 5-2 可用的格式字符串

格式化字符	描 述	示例输出
a	'a.m.'或'p.m.'	'a.m.'
A	'AM'或'PM'	'AM'
b	月份，文字形式，3 个字母，小写	jan
B	未实现	
c	ISO 8601 格式	2008-01-02T10:30:00.000123+02:00
d	月的日子，带前导零的 2 位数字	'01'到'31'
D	周几的文字表述形式，3 个字母	'Fri'
e	时区名称	' ', 'GMT', '-500', 'US/Eastern'等
E	月份，分地区	
f	时间	'1', '1:30'
F	月，文字形式。	'January'
g	12 小时格式，无前导零	'1'到'12'
G	24 小时格式，无前导零	'0'到'23'
h	12 小时格式	'01'到'12'
H	24 小时格式	'00'到'23'
i	分钟	'00'到'59'
I	夏令时间，无论是否生效	'1'或'0'
j	没有前导零的月份的日子	'1'到'31'

续表

格式化字符	描述	示例输出
l	星期几，完整英文名	'Friday'
L	布尔值是否是一个闰年	True 或 False
m	月，2 位数字带前导零	'01'到'12'
M	月，文字，3 个字母	Jan
n	月，无前导零	'1'到'12'
N	美联社风格的月份缩写	'Jan.', 'Feb.', 'March', 'May'
o	ISO-8601 周编号	'1999'
O	与格林威治时间的差，单位为小时	'+0200'
P	时间为 12 小时	'1 am', '1:30 pm', 'midnight', 'noon', '12:30 pm'>
r	RFC 5322 格式化日期	'Thu, 21 Dec 2000 16:01:07 +0200'
s	秒，带前导零的 2 位数字	'00'到'59'
S	一个月的英文序数后缀，2 个字符	'st', 'nd', 'rd'或'th'
t	给定月份的天数	28 to 31
T	本机的时区	'EST', 'MDT'
u	微秒	000000 to 999999
U	自 Unix Epoch 以来的秒数(1970 年 1 月 1 日 00:00:00 UTC)	
w	星期几，数字无前导零。	'0'(星期日)至'6'(星期六)
W	ISO-8601 周数，周数从星期一开始	1，53
y	年份，2 位数字	'99'
Y	年，4 位数	'1999'
z	一年中的日子	0 到 365
Z	时区偏移量，单位为秒	-43200 到 43200

例如：

`{{ value|date:"D d M Y" }}`

如果 value 是一个 datetime 对象，比如 datetime.datetime.now()，输出将是字符串'Wed 09 Jan 2008'。可以将 date 与 time 过滤器结合使用，以呈现 datetime 值的完整表示形式。例如：

`{{ value|date:"D d M Y" }} {{ value|time:"H:i" }}`

(7) default：为变量提供一个默认值。例如：

`{{ value|default:"nothing" }}`

(8) default_if_none：如果(且仅当)value 为 None，则使用给定的默认值。例如：

`{{ value|default_if_none:"nothing" }}`

(9) dictsort：接收一个包含字典元素的列表，并返回按参数中给出的键排序的列表。例如：

```
{{ value|dictsort:"name" }}
```

如果 value 为：

```
[
    {'name': 'zed', 'age': 19},
    {'name': 'amy', 'age': 22},
    {'name': 'joe', 'age': 31},
]
```

那么输出将是：

```
[
    {'name': 'amy', 'age': 22},
    {'name': 'joe', 'age': 31},
    {'name': 'zed', 'age': 19},
]
```

还可以做更复杂的事情，例如：

```
{% for book in books|dictsort:"author.age" %}
    * {{ book.title }} ({{ book.author.name }})
{% endfor %}
```

如果 books 是：

```
[
    {'title': '1984', 'author': {'name': 'George', 'age': 45}},
    {'title': 'Timequake', 'author': {'name': 'Kurt', 'age': 75}},
    {'title': 'Alice', 'author': {'name': 'Lewis', 'age': 33}},
]
```

那么输出将是：

```
* Alice (Lewis)
* 1984 (George)
* Timequake (Kurt)
```

另外，dictsort 也可以按指定索引对多维列表进行排序。例如：

```
{{ value|dictsort:0 }}
```

如果 value 为：

```
[
    ('a', '42'),
    ('c', 'string'),
    ('b', 'foo'),
]
```

那么输出将是：

```
[
    ('a', '42'),
    ('b', 'foo'),
    ('c', 'string'),
]
```

在使用时必须提供整数索引，不能是字符串。例如下面的形式会产生空输出：

```
{{ values|dictsort:"0" }}
10. dictsortreversed
```

(10) dictsortreversed：如果 value 的值是一个字典，那么返回值是按照关键字排序的结果的反序。

(11) divisibleby：如果 value 可以被参数整除，则返回 True。例如：

```
{{ value|divisibleby:"3" }}
```

如果 value 是 21，则输出 True。

(12) escape：转义字符串的 HTML。转义仅在字符串输出时应用，因此在链接的过滤器序列中 escape 的位置无关紧要，就像它是最后一个过滤器。 如果要立即应用转义，则使用 force_escape 过滤器。

(13) escapejs：将转义用于 JavaScript 字符串的字符，确保在使用模板生成 JavaScript / JSON 时避免语法错误。例如：

```
{{ value|escapejs }}
```

如果 value 为 testing\r\njavascript \'string" escaping，输出将为 testing\\u000D\\u000Ajavascript\\u0027string\\u0022\\u003Cb\\u003Eescaping\\u003C/b\\u003E。

(14) filesizeformat：格式化为直观的文件大小形式(即'13 KB', '4.1 MB', '102 bytes'等)。例如：

```
{{ value|filesizeformat }}
```

如果 value 为 123456789，输出将是 117.7 MB。

(15) first：返回列表中的第一项。例如：

```
{{ value|first }}
```

如果 value 是列表['a', 'b', 'c']，则输出为'a'。

(16) floatformat

当不使用参数时，将浮点数舍入到小数点后一位，但前提是要显示小数部分。例如：

```
value              模板语法                    输出
34.23234    {{ value | floatformat }}       34.2
34.00000    {{ value | floatformat }}       34
34.26000    {{ value | floatformat }}       34.3
```

如果与数字整数参数一起使用，将数字四舍五入为小数位数。例如：

```
value              模板语法                       输出
34.23234    {{ value | floatformat: 3 }}      34.232
34.00000    {{ value | floatformat: 3 }}      34.000
34.26000    {{ value | floatformat: 3 }}      34.260
```

特别有用的是传递 0(零)作为参数，它将使 float 类型浮动到最接近的整数。例如：

```
value              模板语法                        输出
34.23234    {{ value | floatformat: "0" }}      34
```

34.00000	{{ value | floatformat: "0" }}	34
39.56000	{{ value | floatformat: "0" }}	40

如果传递给 floatformat 的参数为负，则会将一个数字四舍五入到小数点后的位置，但前提是要显示一个小数部分。例如：

value	模板语法	输出
34.23234	{{ value | floatformat: " - 3" }}	34.232
34.00000	{{ value | floatformat: " - 3" }}	34
34.26000	{{ value | floatformat: " - 3" }}	34.260

(17) force_escape：立即转义 HTML 字符串。

(18) get_digit：给定一个整数，返回所请求的数字，1 表示最右边的数字，2 表示右边第二个数字，以此类推。例如：

```
{{ value|get_digit:"2" }}
```

如果 value 为 123456789，则输出 8。

(19) iriencode：将 IRI(国际化资源标识符)转换为适合包含在 URL 中的字符串。例如：

```
{{ value|iriencode }}
```

如果 value 是?test=1&me=2，则输出是?test=1&me=2。

(20) join：使用字符串连接列表，类似于 Python 的 str.join(list)。例如：

```
{{ value|join:" // " }}
```

如果 value 是列表['a', 'b', 'c']，输出为 a // b // c。

(21) last：返回列表中的最后一个项目，类似于 first 过滤器。例如：

```
{{ value|last }}
```

(22) length：返回对象的长度。这适用于字符串和列表。例如：

```
{{ value|length }}
```

如果 value 是['a', 'b', 'c', 'd']或"abcd"，输出将为 4。对于未定义的变量，过滤器返回 0。

(23) length_is：如果对象的长度等于参数值，则返回 True，否则返回 False。例如：

```
{{ value|length_is:"4" }}
```

如果 value 是['a', 'b', 'c', 'd']或"abcd"，输出将为 True。

(24) linebreaks：替换纯文本中的换行符为<p>标签。例如：

```
{{ value|linebreaks }}
```

如果 value 是 Joel\nis a slug，输出将为<p>Joel
is a slug</p>。

(25) linebreaksbr：替换纯文本中的换行符为
标签。例如：

```
{{ value|linebreaksbr }}
```

如果 value 是 Joel\nis a slug，输出将为 Joel
is a slug。

(26) linenumbers：显示带行号的文本。例如：

```
{{ value|linenumbers }}
```

如果 value 为：

```
one
two
three
```

则输出将是：

```
1. one
2. two
3. three
27. ljust
```

(27) ljust：设置在指定的宽度下左对齐，例如在下面的代码中，如果 value 的值为 Django，则会打印输出 Django。

```
"{{ value|ljust:"10" }}"
```

(28) lower：将字符串全部转换为小写。例如：

```
{{ value|lower }}
```

(29) make_list：将对象转换为字符的列表。对于字符串，直接拆分为单个字符的列表。对于整数，在创建列表之前将参数强制转换为 Unicode 字符串。

```
{{ value|make_list }}
```

如果 value 是字符串"Joel"，输出将是列表['J', 'o', 'e', 'l']。如果 value 为 123，输出为列表['1', '2', '3']。

(30) phone2numeric：将电话号码(可能包含字母)转换为其等效数字。例如：

```
{{ value|phone2numeric }}
```

如果 value 为 800-COLLECT，输出将为 800-2655328。

(31) pluralize：如果值不是 1，则返回一个复数形式，通常在后面添加"s"表示。例如：

```
You have {{ num_messages }} message{{ num_messages|pluralize }}
```

如果 num_messages 是 1，则输出为 You have 1 message；如果 num_messages 是 2，输出为 You have 2 messages。

另外，如果需要的不是"s"后缀的话，可以提供一个备选的参数给过滤器：

```
You have {{ num_walruses }} walrus{{ num_walruses|pluralize:"es" }}
```

对于非一般形式的复数，可以同时指定单复数形式，用逗号隔开。例如：

```
You have {{ num_cherries }} cherr{{ num_cherries|pluralize:"y,ies" }}
```

(32) pprint：用于调试的过滤器。

(33) random：返回给定列表中的随机项。例如：

```
{{ value|random }}
```

(34) rjust：右对齐给定宽度字段中的值，例如在下面的代码中，如果 value 的值为 Django，则会打印输出右对齐的 Django。

```
"{{ value|rjust:"10" }}"
```

(35) safe：将字符串标记为安全，不需要转义。

(36) safeseq：将 safe 过滤器应用于序列的每个元素，此标签与其他对序列进行过滤操作(例如 join)的标签一起使用时非常有用。例如：

```
{{ some_list|safeseq|join:", " }}
```

在这种情况下，不能直接使用 safe 过滤器，因为它首先将变量转换为字符串，而不是使用序列的各个元素。

(37) slice：返回列表的一部分。也就是切片，与 Python 的列表切片的语法相同。例如：

```
{{ some_list|slice:":2" }}
```

如果 some_list 是['a'，'b'，'c']，输出将为['a'，'b']。

(38) slugify：转换为 ASCII 码。将空格转换为连字符。删除不是字母、数字、下划线或连字符的字符。转换为小写。还会去除前导空格和尾随空格。例如：

```
{{ value|slugify }}
```

如果 value 是 Joel is a slug，输出为 joel-is-a-slug。

(39) stringformat：根据参数，格式化变量。例如：

```
{{ value|stringformat:"E" }}
```

如果 value 为 10，输出将为 1.000000E+01。

(40) striptags：尽可能去除 HTML 中的标签。例如：

```
{{ value|striptags }}
```

如果 value 是Joel <button>is</button> a slug，则输出 Joel is a slug。

(41) time：根据给定的格式参数显示对应格式的时间。给定格式可以是预定义的 TIME_FORMAT，也可以是与 date 过滤器相同的自定义格式。例如：

```
{{ value|time:"H:i" }}
```

如果 value 等于 datetime.datetime.now()，则输出字符串 01:23。time 过滤器只接收格式字符串中与时间相关的参数，而不是日期。如果需要格式化 date 值，则改用 date 过滤器。

(42) timesince：将日期格式设为自该日期起的时间(例如，"4 天，6 小时")。采用一个可选参数，这是一个包含用作比较点的日期的变量。例如，如果 blog_date 表示 2006 年 6 月 1 日午夜的日期实例，并且 comment_date 是 2006 年 6 月 1 日 08:00，则以下内容将返回"8 hours"。

```
{{ blog_date|timesince:comment_date }}
```

(43) timeuntil：类似于 timesince，能够测量从现在开始直到给定日期或日期时间的时间。例如，如果今天是 2006 年 6 月 1 日，而 conference_date 是 2006 年 6 月 29 日，则

`{{ conference_date | timeuntil }}`将返回"4 weeks"。可选参数是一个包含用作比较点的日期变量。如果 from_date 为 2006 年 6 月 22 日，则以下内容将返回"1 weeks"。

```
{{ conference_date|timeuntil:from_date }}
```

(44) title：将所有单词的首字母大写，其他字母小写。例如：

```
{{ value|title }}
```

如果 value 为"my FIRST post"，则输出为"My First Post"。

(45) truncatechars：如果字符串包含的字符总个数多于指定的字符数量，那么会被截掉后面的部分。截断的字符串将以"..."结尾。例如：

```
{{ value|truncatechars:9 }}
```

如果 value 是 Joel is a slug，则输出为 Joel i...。

(46) truncatechars_html：类似于 truncatechars，但是会保留 HTML 标记。例如：

```
{{ value|truncatechars_html:9 }}
```

如果 value 是<p>Joel is a slug</p>，则输出<p>Joel i...</p>。

(47) truncatewords：在一定数量的字符后截断字符串。与 truncatechars 不同的是，这个以字的个数计数，而不是以字符计数。例如：

```
{{ value|truncatewords:2 }}
```

如果 value 是 Joel is a slug，输出为 Joel is ...，字符串中的换行符将被删除。

(48) truncatewords_html：类似于 truncatewords，但是保留 HTML 标记。例如：

```
{{ value|truncatewords_html:2 }}
```

HTML 内容中的换行符将被保留。

(49) unordered_list：接收一个嵌套的列表，返回一个 HTML 的无序列表，但不包含开始和结束的<ul>标签。例如，如果 var 包含['States', ['Kansas', ['Lawrence', 'Topeka'], 'Illinois']]，那么`{{ var|unordered_list }}`将返回：

```
<li>States
<ul>
        <li>Kansas
        <ul>
                <li>Lawrence</li>
                <li>Topeka</li>
        </ul>
        </li>
        <li>Illinois</li>
</ul>
</li>
```

(50) upper：将字符串全部转换为大写的形式。例如：

```
{{ value|upper }}
```

(51) urlencode：转义要在 URL 中使用的值。例如：

```
{{ value|urlencode }}
```

如果 value 为 https://www.example.org/foo?a=b&c=d，则输出 https%3A//www.example.org/foo%3Fa%3Db%26c%3Dd。

(52) urlize：将文字中的网址和电子邮件地址转换为可单击的链接。该模板标签适用于前缀为 http://、https:// 的链接，或者 www。由 urlize 生成的链接会向其中添加 rel="nofollow" 属性，例如：

`{{ value|urlize }}`

如果 value 是 Check out www.djangoproject.com，则输出 Check out <a href="http://www.djangoproject.com" rel="nofollow">www.djangoproject.com</a>。

除了超级链接之外，urlize 也会将电子邮件地址转换为邮件地址链接。如果 value 是 Send questions to foo@example.com，则输出 Send questions to <a href="mailto:foo@example.com">foo@example.com</a>。

(53) urlizetrunc：将网址和电子邮件地址转换为可单击的链接，就像 urlize，但截断长度超过给定字符数限制的网址。例如：

`{{ value|urlizetrunc:15 }}`

如果 value 是 Check out www.djangoproject.com，则输出 Check out <a href="http://www.djangoproject.com" rel="nofollow">www.djangopr...</a>'。

> **注意**
> 与 urlize 一样，此过滤器应仅应用于纯文本。

(54) wordcount：返回单词的个数。例如：

`{{ value|wordcount }}`

如果 value 是 Joel is a slug，则输出 4。

(55) wordwrap：以指定的行长度，换行单词。例如：

`{{ value|wordwrap:5 }}`

如果 value 是 Joel is a slug，输出为：

```
Joel
is a
slug
```

(56) yesno：将 True、False 和 None，映射成字符串 yeah，no，maybe。例如：

`{{ value|yesno:"yeah,no,maybe" }}`

5.3.2 国际化标签和过滤器

在 Django 中还提供了一些模板标签和过滤器，用以控制模板中的国际化功能。它们允许对翻译、格式化和时区转换进行粒度控制。

(1) i18n：此标签允许在模板中指定可翻译文本。要启用它，需要将 USE_I18N 设置为

True，然后加载{% load i18n %}。

(2) l10n：此标签提供对模板的本地化控制，只需要使用{% load l10n %}。通常将 USE_L10N 设置为 True，以便本地化默认处于活动状态。

(3) tz：此标签对模板中的时区进行控制。像 l10n，只需要使用{% load tz %}，但通常还会将 USE_TZ 设置为 True，以便默认情况下转换为本地时间。

5.3.3 其他标签和过滤器库

除了前面介绍的模板标签外，在 Django 中还附带了其他模板标签，必须在 INSTALLED_APPS 设置中显式启用，并在模板中启用{% load %}标记。

(1) django.contrib.humanize：一组 Django 模板过滤器，用于向数据添加更加可读的信息。

(2) static：用于链接保存在 STATIC_ROOT 中的静态文件。例如：

```
{% load static %}
<img src="{% static "images/hi.jpg" %}" alt="Hi!" />
```

还可以使用变量：

```
{% load static %}
<link rel="stylesheet" href="{% static user_stylesheet %}" type="text/css" media="screen" />
```

还可以像下面这样使用：

```
{% load static %}
{% static "images/hi.jpg" as myphoto %}
<img src="{{ myphoto }}"></img>
```

5.3.4 实战演练：使用过滤器提取列表和字典中的内容

在下面的实例中，演示了使用如下模板过滤器(filter)的方法。

- add：给变量加上相应的值。
- addslashes：给变量中的引号前加上斜线。
- capfirst：首字母大写。
- cut：从字符串中移除指定的字符。
- date：格式化日期字符串。
- default：如果值是 False，就替换成设置的默认值，否则就使用本来的值。
- default_if_none：如果值是 None，就替换成设置的默认值，否则就使用本来的值。

源码路径：daima\5\py-filter

(1) 编写 URL 导航文件 urls.py，设置 View 视图对应的 URL。代码如下：

```
from django.contrib import admin
from django.urls import path
```

```
from tem import views
urlpatterns = [
    path('query/',views.query),
]
```

(2) 编写视图文件 views.py，在视图函数 query()中分别创建列表、字典和类对象。代码如下：

```
from django.shortcuts import render
import datetime
class Animal(object,):
    def __init__(self,name,sex):
        self.name=name
        self.sex=sex

def query(request):
    l=["存","正","参"]
    d={'name':'见','age':12,'sex':'M'}
    c=Animal('alex','M')
    test='world'
    test1='hello kitty'
    t=datetime.datetime.now()
    e=[]
    a='<a href="">click</a>'
    return render(request,'index.html',locals())
```

(3) 编写视图文件 index.html，使用模板过滤器处理视图文件 views.py 中的列表、字典和类对象。代码如下：

```
<h1>hello {{ l.1 }}</h1>
<h1>hello {{ l.2 }}</h1>
<h1>hello {{ d.name }}</h1>
<h1>hello {{ d.age }}</h1>
<h1>hello {{ d.sex }}</h1>
<h1>hello {{ c.name }}</h1>
<h1>hello {{ c.sex }}</h1>
<h1>{{ d.name }}的真实年龄:{{ d.age|add:12 }}</h1>
<h1>hello {{ test|capfirst }}</h1>
<h1>{{ test1 }}</h1> <!-- helo kitty-->
<h1>{{ test1|cut:' ' }}</h1> <!-- 去除空格 helokitty-->
<h1>{{ t }}</h1>
<h1>{{ t|date:'Y-m-d' }}</h1>
<h1>{{ e }}</h1>
<h1>{{e|default:'空列表'}}</h1><!-- 若e为空列表,则过滤default用后面值来代替 -->
<h1>{{ e|default_if_none:'空列表' }}</h1> <!-- 若e不为空列表,则过滤default用后面值来代替 -->
<h1>{{ a }}</h1>
<h1>{{ a|safe }}</h1>
<!-- 与上面<h1>{{ a|safe }}</h1>等价 -->
{% autoescape off %}
<h1>{{ a }}</h1>
{% endautoescape %}
```

输入 http://127.0.0.1:8000/query/后的执行效果如图 5-3 所示。

图 5-3　执行效果(三)

5.4　模板继承

在 Django Web 程序中，模板可以用继承的方式来实现复用，此功能需要使用 extends 标签实现。在本节的内容中，将详细讲解 Django 模板继承的知识。

5.4.1　模板继承介绍

扫码观看本节视频讲解

Django 使用了"模板继承"的概念：这就是{% extends "base.html" %}所做的事。这意味着"首先载入名为'base'的模板中的内容到当前模板，然后再处理本模板中的其余内容。"总之，模板继承可以在模板间大大减少冗余内容：每一个模板只需要定义它独特的部分即可。例如，在下面演示模板继承代码中，模板文件 template.html 中的 content 模块会替换掉模板文件 base.html 中的 content 模块。同时模板文件 template.html 继承了模板文件 base.html 的 sidebar 和 footer 模块。

```
# base.html
{% block sidebar %}
{% endblock %}

{% block content %}
{% endblock %}

{% block footer %}
{% endblock %}

# template.html
{% extends "base.html" %}
{% block content %}
    {{ some code }}
```

如果不使用模板继承功能，在项目中有一个页面需要修改还好，而如果多处并且多个页面都需要修改，那么就大大增加了工作量。当多个页面中都存在相同的部分时，使用模板继承就大大减少了开发人员和维护人员的压力。在模板继承中，父模板中放置大部分子模板共用的且不变的内容，在每一个子模板中可以重写父模板中的内容。

模板继承的典型应用就是在网站开头或者网站结尾处，其主要特性如下。

- 如果在模板中使用 extends 标签，它必须是模板中的第一个标签。
- 不能在一个模板中定义多个相同名字的 block 标签。
- 子模板不必定义全部父模板中的 blocks，如果子模板没有定义 block，则使用父模板中的默认值。
- 如果发现在模板中有大量的复制内容，则应该把内容移动到父模板中。
- 可以获取父模板中 block 的内容。
- 为了具有更好的可读性，可以给 endblock 标签设置一个名字。

下面演示使用模板继承的具体流程。

(1) 创建根级模板。

创建模板文件 base.html，用于存放整个站点共用的内容。代码如下：

```
<!DOCTYPE html>
<html>
<head>

    <title>{% block title %}{% endblock title %} </title>

    <link rel="stylesheet" type="text/css" href="/static/css/style.css">

    {% block custom_css %}{% endblock custom_css %}

</head>
<body>

{% block left %}{% endblock left%}
{% block content %}{% endblock content %}

</body>
<script src="/static/js/jquery.min.js" type="text/javascript"></script>
{% block cutsom_js %}{% endblock custom_js %}
</html>
```

另外，为了保证浏览器的页面加载进度尽可能不影响用户的浏览体验，通常将 JavaScript 文件放在 dom 树的最后分支中。

(2) 创建分支模板。

创建模板文件 home.html，设置此文件继承于根级模板文件 base.html。代码如下：

```
{% extend 'base.html' %}

{% block titlt %}生鲜超市{% endblock title %}

{% block custom_css %}
<link rel="stylesheet" type="text/css" href=" /static /css/reset.css">
{% endblock custom_css %}

{% block left %}
<p>左侧菜单栏</p>
```

```
{% endblock left%}

{% block content %}
<p>中央内容</p>
{% endblock content %}

{% block cutsom_js %}
<script src="/static/js/jquery-migrate-1.2.1.min.js"
type="text/javascript"></script>
{% endblock custom_js %}
```

所以最后的模板文件 home.html 为：

```
<!DOCTYPE html>
<html>
<head>

    <title>生鲜超市</title>

<link rel="stylesheet" type="text/css" href="/static/css/style.css">
<link rel="stylesheet" type="text/css" href="/static /css/reset.css">

</head>
<body>

<p>左侧菜单栏</p>
<p>中央内容</p>

</body>
<script src="/static/js/jquery.min.js" type="text/javascript"></script>
<script src="/static/js/jquery-migrate-1.2.1.min.js"
type="text/javascript"></script>
</html>
```

5.4.2 实战演练：使用模板继承

在下面的实例代码中，演示了在 Django 模板中使用继承模板显示内容的过程。

源码路径：**daima\5\HelloWorld**

（1）通过如下命令分别创建一个名为"HelloWorld"的项目和一个名为"jicheng"的应用程序。

```
django-admin.py startproject HelloWorld
cd HelloWorld
python manage.py startapp jicheng
```

（2）修改视图文件 view.py，在里面增加一个新的对象，用于向模板中提交数据。

```
from django.shortcuts import render

from django.http import HttpResponse

def hello(request):
```

```
    context = {}
    context['hello'] = 'Hello World!'
    return render(request, 'hello.html', context)
```

(3) 接下来需要向 Django 说明模板文件的路径，在前面我们使用的是将应用加入设置文件 settings.py 的 settings.INSTALLED_APPS 中，现在我们使用一种新的设置方法。打开设置文件 settings.py，修改 TEMPLATES 中的 DIRS 为[BASE_DIR+"/jicheng/templates",]，具体代码如下所示。

```
TEMPLATES = [
    {
        'BACKEND': 'django.template.backends.django.DjangoTemplates',
        'DIRS': [BASE_DIR+"/jicheng/templates",],
        'APP_DIRS': True,
        'OPTIONS': {
            'context_processors': [
                'django.template.context_processors.debug',
                'django.template.context_processors.request',
                'django.contrib.auth.context_processors.auth',
                'django.contrib.messages.context_processors.messages',
            ],
    },
```

(4) 在子目录"jicheng"下面创建目录"templates"，然后在里面创建模板文件 base.html 和 hello.html。其中文件 base.html 的具体实现代码如下所示。

```
<html>
  <head>
    <title>Hello World!</title>
  </head>

  <body>
    <h1>Hello World!</h1>
    {% block mainbody %}
      <p>original</p>
    {% endblock %}
  </body>
</html>
```

在上述代码中，名字为 mainbody 的 block 标签可以被继承者替换。通过所用的{% block %}标签告诉模板引擎，子模板可以重载这些部分。

模板文件 hello.html 继承 base.html，并替换特定的 block，文件 hello.html 的具体实现代码如下所示。

```
{% extends "base.html" %}

{% block mainbody %}
<p>继承了 base.html 文件</p>
{% endblock %}
```

在上述代码中，第一行代码说明模板文件 hello.html 继承自文件 base.html。使用相同名字的 block 标签可以替换文件 base.html 中相应的 block。

(5) 输入如下命令启动服务器运行 Web 程序：

```
python manage.py runserver
```

执行后的效果如图 5-4 所示。

图 5-4　执行效果(四)

5.5　自定义模板标签和过滤器

虽然 Django 模板包含的内置 tags 和 filters 可以满足大多数应用程序的需求，但是在一些情况下，可能会发现需要的功能未被核心模板集覆盖。这时候可以通过 Python 代码自定义 tags 和 filters 扩展集成模板引擎，然后通过 {% load %} 标签使用这些自定义模板的标签和过滤器。

扫码观看本节视频讲解

5.5.1　基本方法

在 Django 中自定义标签和过滤器之前，首先需要实现代码布局和工程文件的组织。我们可以为自定义标签和过滤器新建一个 app，也可以在原有的某个 app 中添加。不管怎么样，自定义标签和过滤器的第一步是在 app 中新建一个包 templatetags(名字是固定的，不能改变)，此包和 views.py、models.py 等文件处于同一级别目录下。读者要切记，这是一个包，不要忘记创建 __init__.py 文件以使得该目录可以作为 Python 的包。

在添加 templatetags 包后，需要重新启动服务器，才能在模板中使用标签或过滤器。然后将自定义的标签和过滤器放在 templatetags 包下的一个模块里。这个模块的名字是后面载入标签时使用的标签名，所以要谨慎地选择名字以防与其他应用下的自定义标签和过滤器名字冲突，当然更不能与 Django 内置的模块名冲突。

假设你自定义的标签/过滤器在一个名为 poll_extras.py 的文件中，那么你的 app 目录结构看起来应该是这样的：

```
polls/
    __init__.py
    models.py
    templatetags/
        __init__.py
        poll_extras.py
    views.py
```

为了让{% load xxx %}标签正常工作，包含自定义标签的app必须在INSTALLED_APPS中注册，然后就可以在模板中像下面这样使用：

```
{% load poll_extras %}
```

在包templatetags中放多少个模块没有限制，只需要记住{% load xxx %}将会载入给定模块名中的标签/过滤器，而不是app中所有的标签和过滤器。

要在模块内自定义标签，首先，这个模块必须包含一个名为 register 的变量，它是template.Library 的一个实例，所有的标签和过滤器都是在其中注册的，所以需要把如下的内容放在模块的顶部。

```
from django import template
register = template.Library()
```

5.5.2 自定义模板过滤器

1. 编写过滤器

自定义过滤器就是一个带有如下一个或两个参数的Python函数。

- 变量的值：不一定是字符串形式。
- 参数的值：可以有一个初始值，或者完全不要这个参数。

注意，这个Python函数的第一个参数是要过滤的对象，第二个参数才是自定义的参数。而且最多只能有两个参数，所以只能自定义一个参数，这是过滤器的先天限制。例如，在{{ var|foo:"bar" }}中，foo过滤器应当传入变量var和参数"bar"。

由于模板语言没有提供异常处理，任何从过滤器中抛出的异常都将会显示为服务器错误。例如，下面是一个定义过滤器的例子：

```
def cut(value, arg):
    """将value中的所有arg部分切除掉"""
    return value.replace(arg, '')
```

下面是这个过滤器的使用方法：

```
{{ somevariable|cut:"0" }}
```

大多数过滤器没有参数，在这种情况下，我们自定义的过滤器函数不带额外的参数即可，但基本的value参数是必带的。例如：

```
def lower(value): # Only one argument.
    """Converts a string into all lowercase"""
    return value.lower()
```

2. 注册过滤器

一旦写好了过滤器函数，接下来需要注册它，注册方法是调用register.filter()，比如：

```
register.filter('cut', cut)
register.filter('lower', lower)
```

方法 Library.filter()需要如下两个参数。

- 过滤器的名称：一个字符串对象。
- 编译的函数：刚才写的过滤器函数。

还可以把方法 register.filter()用作装饰器，例如以下面的方式注册过滤器：

```
@register.filter(name='cut')
def cut(value, arg):
    return value.replace(arg, '')

@register.filter
def lower(value):
    return value.lower()
```

上面第二个例子没有声明 name 参数，Django 将使用函数名作为过滤器的名字。由此可见，使用自定义过滤器的方法非常简单，使用方法和普通的过滤器没有什么区别。

5.5.3 自定义模板标签

标签比过滤器更复杂，因为标签可以做任何事情。Django 提供了大量的快捷方式，使得编写标签工作比较容易。对于我们一般的自定义标签来说，simple_tag 是最重要的，它可以帮助你将 Python 函数注册为一个简单的模板标签。

1. simple_tag

原型：django.template.Library.simple_tag()

为了简单化模板标签的创建，Django 提供了一个辅助函数 simple_tag，这个函数是 django.template.Library 的一个方法。例如想编写一个返回当前时间的模板标签，那么可以编写如下代码实现函数 current_time()：

```
import datetime
from django import template
register = template.Library()

@register.simple_tag
def current_time(format_string):
    return datetime.datetime.now().strftime(format_string)
```

在使用函数 simple_tag()时需要注意：

- 如果不需要额外的转义，可以使用 mark_safe()让输出不进行转义，前提是要确保代码中不包含 XSS 漏洞。如果要创建小型 HTML 片段，强烈建议使用 format_html() 而不是 mark_safe()。
- 如果模板标签需要访问当前上下文，可以在注册标签时使用 takes_context 参数，例如：

```
@register.simple_tag(takes_context=True)
def current_time(context, format_string):
    timezone = context['timezone']
    return your_get_current_time_method(timezone, format_string)
```

注意，第一个参数必须为 context！

- 如果需要重命名标签，那么可以给它提供自定义的名称。例如：

```
register.simple_tag(lambda x: x - 1, name='minusone')

@register.simple_tag(name='minustwo')
def some_function(value):
    return value - 2
```

- 函数 simple_tag() 可以接收任意数量的位置参数和关键字参数，例如：

```
@register.simple_tag
def my_tag(a, b, *args, **kwargs):
    warning = kwargs['warning']
    profile = kwargs['profile']
    ...
    return ...
```

然后在模板中，可以将任意数量的由空格分隔的参数传递给模板标签。像在 Python 中一样，关键字参数的值使用等号(=)赋予，并且必须在位置参数之后提供。例如：

```
{% my_tag 123 "abcd" book.title warning=message|lower profile=user.profile %}
```

可以将标签结果存储在模板变量中，而不是直接输出。这是通过使用 as 参数后跟变量名来实现的。例如：

```
{% current_time "%Y-%m-%d %I:%M %p" as the_time %}
<p>The time is {{ the_time }}.</p>
```

2. inclusion_tag()

原型：django.template.Library.inclusion_tag()

另一种常见类型的模板标签是通过渲染一个模板来显示一些数据。例如，Django 的 Admin 界面使用自定义模板标签显示"添加/更改"表单页面底部的按钮。这些按钮看起来总是相同，但链接的目标却是根据正在编辑的对象变化的。这种类型的标签被称为"Inclusion 标签"。

下面展示一个根据给定的 tutorials 中创建的 Poll 对象输出一个选项列表的自定义 Inclusion 标签，在模板中它是这么调用的：

```
{% show_results poll %}
```

而输出是这样的：

```
<ul>
  <li>First choice</li>
  <li>Second choice</li>
  <li>Third choice</li>
</ul>
```

具体的编写方法如下。

- 首先，编写 Python 函数：

```
def show_results(poll):
    choices = poll.choice_set.all()
    return {'choices': choices}
```

- 然后创建用于标签渲染的模板 results.html：

```
<ul>
{% for choice in choices %}
    <li> {{ choice }} </li>
{% endfor %}
</ul>
```

- 最后，通过调用 Library 对象的 inclusion_tag()装饰器方法创建并注册 Inclusion 标签：

```
@register.inclusion_tag('results.html')
def show_results(poll):
    ...
```

或者使用 django.template.Template 实例注册 Inclusion 标签：

```
from django.template.loader import get_template
t = get_template('results.html')
register.inclusion_tag(t)(show_results)
```

函数 inclusion_tag()可以接收任意数量的位置参数和关键字参数，例如：

```
@register.inclusion_tag('my_template.html')
def my_tag(a, b, *args, **kwargs):
    warning = kwargs['warning']
    profile = kwargs['profile']
    ...
    return ...
```

然后在模板中，可以将任意数量的由空格分隔的参数传递给模板标签。像在 Python 中一样，关键字参数的值的设置使用等号(=)，并且必须在位置参数之后提供。例如：

```
{% my_tag 123 "abcd" book.title warning=message|lower profile=user.profile %}
```

可以在标签中传递上下文中的参数，例如，想要将上下文 context 中的 home_link 和 home_title 这两个变量传递给模板，代码如下所示：

```
@register.inclusion_tag('link.html', takes_context=True)
def jump_link(context):
    return {
        'link': context['home_link'],
        'title': context['home_title'],
    }
```

此处需要注意，函数的第一个参数必须叫作 context，context 必须是一个字典类型。在 register.inclusion_tag()这一行代码中指定了 takes_context=True 和模板的名字。模板 link.html 很简单，代码如下所示：

```
Jump directly to <a href="{{ link }}">{{ title }}</a>.
```

然后，当任何时候想调用这个自定义的标签，只需要加载它本身，不需要添加任何参数，{{ link }}和{{ title }}会自动从标签中获取参数值。例如：

```
{% jump_link %}
```

通过使用 takes_context=True，表示不需要传递参数给这个模板标签，它会自己去获取上下文。

5.5.4 实战演练：创建自定义模板过滤器

在下面的内容中，将通过一个具体实例的实现过程，详细讲解创建自定义模板过滤器的方法。

源码路径：**daima\5\Django-Template**

(1) 创建一个名字为 basic_app 的 app，然后在里面创建包 templatetags。

(2) 在包 templatetags 中创建 Python 文件 my_extras.py，这是自定义模板过滤器的实现文件，在里面编写自定义过滤器函数 cut()，此函数的功能是将 value 中的所有 arg 部分切除掉。代码如下：

```python
from django import template

register = template.Library()

@register.filter(name='cut')          #使用装饰器注册
def cut(value,arg):
    """
    这会从字符串中删除"arg"的所有值！
    """
    return value.replace(arg,'')
```

在上述代码中，使用方法 register.filter()注册了自定义过滤器 cut。

(3) 在视图文件 views.py 中为不同的 URL 创建对应的视图函数。代码如下：

```python
from django.shortcuts import render,HttpResponse

def index(request):
    context_dict ={
                'got':'winter is coming','num':624,
            }
    return render(request,'basic_app/index.html',context_dict)

def base(request):
    return render(request,'basic_app/base.html')

def relative(request):
    return render(request,'basic_app/realtive_url_templates.html')

def other(request):
    return render(request,'basic_app/other.html')
```

(4) 在模板文件 index.html 中使用上面自定义的过滤器 cut，代码如下：

```html
<!DOCTYPE html>
   {% extends 'basic_app/base.html' %}
     {% block body_block %}
       <h1>Welcome to index.</h1>
       <h1>This is index.html page showing!</h1>

       <!--接下来的两个标签使用内置的Django模板过滤器-->
       <!--<h2> {{ got|title }} </h2>-->
       <!--<h2> {{ num|add:"1" }} </h2>-->

       <!-- 下面使用自定义的模板过滤器-->
       <h1>{{ got|cut:"is" }}</h1> <!-- 使用注册的 -->
       <h1> {{ got|cut:"coming" }}</h1> <!-- 使用装饰器的-->
     {% endblock %}
```

执行程序后会发现使用自定义过滤器 cut 处理了在视图文件 views.py 中定义的字符串"winter is coming"，分别删除了字符串 "winter is coming"中的"is"和"coming"。执行效果如图 5-5 所示。

图 5-5　执行效果(五)

第 6 章

表　单

在动态 Web 应用程序中，表单是实现动态网页效果的核心。在 Web 程序中可以经常见到表单的身影，例如通过表单可以让用户提交数据或上传文件，也可以让用户输入注册信息和登录用户名、密码。在 Django 框架中提供了表单类 Forms，其功能是把用户输入的数据转化成 Python 对象格式，便于后续的操作(例如存储、修改和删除)。在本章的内容中，将详细讲解在 Django 中使用表单的知识。

6.1 表单介绍

在 Web 开发领域中,表单是开发动态 Web 的基础,几乎大部分常见的 Web 交互功能都离不开表单。在本书前面介绍的 Django 实例中,曾经用到过表单的知识。在本节的内容中,将简要介绍表单的基础知识。

扫码观看本节视频讲解

6.1.1 HTML 表单介绍

Django 开发的是动态 Web 服务,而非单纯提供静态页面。动态服务的本质在于和用户进行互动,接收用户的输入,根据输入内容的不同返回不同的内容给用户。返回数据是服务器后端做的,而接收用户输入就需要靠 HTML 表单。表单<form>...</form>可以收集其内部标签中的用户输入,然后将数据发送到服务端。

作为一个 HTML 表单,必须设置如下两个内容。

- 提交目的地:用户数据发送的目的 URL。
- 提交方式:发送数据所使用的 HTTP 方法。

例如,Django Admin 的登录页面就是一个表单,如图 6-1 所示。在这个登录表单包含几个<input>元素:type="text"用于用户名,type="password"表示密码,type="submit"表示"登录"按钮。另外还包含一些用户看不到的隐藏的文本字段,Django 使用它们来提高安全性和决定下一步的行为。还需要告诉浏览器表单数据应该发往<form>的 action 属性指定的 URL:/admin/,而且应该使用 method 属性指定的 HTTP post 方法发送数据。当单击<input type="submit" value="登录">元素时,数据将发送给/admin/。

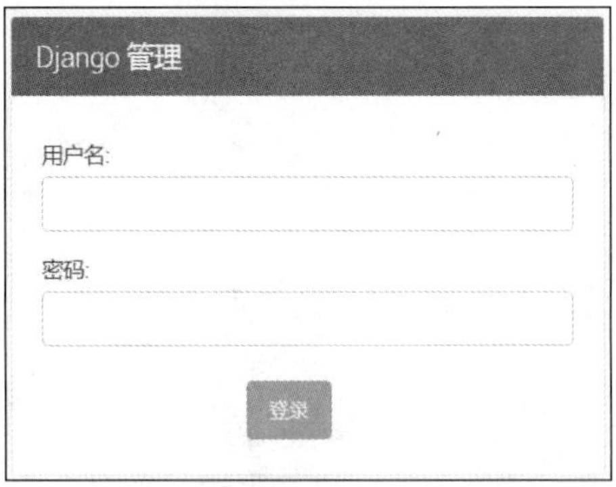

图 6-1 Django Admin 的登录页面

Django Admin 的登录页面的 HTML 源码如下。

```
<form action="/admin/login/?next=/admin/" method="post" id="login-form">

    <input type='hidden' name='csrfmiddlewaretoken'
value='NNHZaDVJGduajNMECXygKZkAt8vyEcw9HS2qm2Vdf7brDZrA0qK1R0I7M2p3TKcs' />
```

```html
    <div class="form-row">
        <label class="required" for="id_username">用户名:</label>
        <input type="text" name="username" autofocus maxlength="254" required id="id_username" />
    </div>

    <div class="form-row">
        <label class="required" for="id_password">密码:</label>
        <input type="password" name="password" required id="id_password" />
        <input type="hidden" name="next" value="/admin/" />
    </div>

    <div class="submit-row">
        <label> </label><input type="submit" value="登录" />
    </div>
</form>
```

在 Web 应用中，有两种提交处理表单数据的方式——POST 和 GET，具体说明如下。

- GET：将用户数据以"键=值"的形式，用"&"符号组合在一起成为一个整体字符串，最后添加前缀"？"，将字符串拼接到 url 内，生成一个类似 https://docs.djangoproject.com/search/?q=forms&release=1 的 URL。
- POST：浏览器会组合表单数据、对它们进行编码，然后打包将它们发送到服务器，数据不会出现在 URL 中。

GET 方法通常用来请求数据，不适合密码表单这一类保密信息的发送，也不适合数据量大的表单和二进制数据应用。对于这些类型的数据，应该使用 POST 方法。但是，GET 特别适用于网页搜索的表单，因为这种表示一个 GET 请求的 URL 可以很容易地设置书签、分享和重新提交。

6.1.2　Django 中的表单

在动态 Web 应用中，处理表单是一件挺复杂的事情。例如在 Django 的 admin 后台管理模块中，许多不同类型的数据可能需要在一张表单中显示，渲染成 HTML，然后在表单中编辑数据，将数据传到服务器，验证和清理数据，然后保存或跳过进行下一步处理。

在 Django 中实现表单功能时，在通常情况下，需要自己手动在 HTML 页面中编写 form 标签和其他元素。但是这样会费时费力，而且有可能写得不太恰当，数据验证也比较麻烦。为了提高开发效率，Django 在内部集成了一个表单模块 form，专门帮助我们快速处理表单相关的内容。Django 的表单模块给开发者提供了如下三个主要功能。

- 准备和重构数据用于页面渲染。
- 为数据创建 HTML 表单元素。
- 接收和处理用户从表单发送过来的数据。

在 Django 中编写 form 表单的方法，和我们在模型系统里编写一个模型的方法类似。在模型中，一个字段代表数据表的一列，而 form 表单中的一个字段代表<form>中的一个<input>元素。

6.2 使用表单

Django 表单系统的核心组件是 Form 类，它与 Django 模型描述对象的逻辑结构、行为以及它呈现给我们内容的形式的方式大致相同，Form 类描述一张表单并决定它如何工作及呈现。

6.2.1 使用表单类 Form 的方法

扫码观看本节视频讲解

类似于模型类的字段映射到数据库字段的方式，表单类的字段会映射到 HTML 表单的 \<input\>元素。ModelForm 通过 Form 映射模型类的字段到 HTML 表单的 \<input\> 元素，Django 的 admin 系统就是基于这个模式实现的。表单字段本身也是类，它们管理表单数据并在提交表单时执行验证。DateField 和 FileField 处理的数据类型差别很大，所以必须用来处理不同的字段。在浏览器中，表单字段以 HTML "控件"(用户界面的一个片段)的形式展现给我们。每个字段类型都有与之相匹配的控件类，但是在必要时可以覆盖。

在接下来的内容中，将用一个提交"用户名"数据的表单为例，介绍在 Django Web 中使用表单类 Form 的方法。

(1) 创建一个提交"用户名"的表单。

假设想从表单中接收用户名数据，在一般情况下，需要在 HTML 中手动编写一个如下所示的表单元素。

```html
<form action="/your-name/" method="post">
   <label for="your_name">Your name: </label>
   <input id="your_name" type="text" name="your_name" value="{{ current_name }}">
   <input type="submit" value="OK">
</form>
```

在上述代码中，\<form action="/your-name/" method="post"\>这一行代码定义了发送表单数据目的地/your-name/和提交表单数据的方法 POST。另外，在 form 元素内部还定义了一个说明标签\<label\>和一个发送按钮"submit"，以及最关键的接收用户输入的\<input\>元素。

(2) 编写表单类。

通过 Django 提供的 Form 类自动生成上面的表单代码，不需要开发者手动编写上述 HTML 代码。首先，在当前 Django 工程的 app 内新建一个 forms.py 文件(这个格式是 Django 的惯用手法，就像 views.py、models.py 等)，然后输入下面的内容：

```python
from django import forms

class NameForm(forms.Form):
    your_name = forms.CharField(label='Your name', max_length=100)
```

对上述代码的具体说明如下。

- 提前导入 forms 模块。

- 所有的表单类都要继承 forms.Form 类。
- 每个表单字段都有自己的字段类型，例如 CharField，它们分别对应一种 HTML 语言中的<form>元素中的表单元素，这一点和 Django 模型系统的设计非常相似。
- label 用于设置说明标签。
- max_length 用于限制表单的最大长度为 100，它同时起到两个作用：一是在浏览器页面限制用户输入不可超过 100 个字符；二是在后端服务器验证用户输入的长度不可超过 100 个字符。

> **注意**
> 由于浏览器页面是可以被篡改、伪造、禁用、跳过的，所有的 HTML 手段的数据验证只能防止意外而不能防止恶意行为，是没有安全保证的，破坏分子完全可以跳过浏览器的防御手段伪造发送请求！所以，在服务器后端，必须将前端当作"裸机"来对待，再次进行完全彻底的数据验证和安全防护！

每个 Django 表单的实例都有一个内置的方法 is_valid()，用来验证接收的数据是否合法。如果所有数据都合法，那么该方法将返回 True，并将所有的表单数据转存到它的属性 cleaned_data 中，该属性是一个字典类型数据。

当将上面的表单渲染成真正的 HTML 元素时，其内容变为：

```html
<label for="your_name">Your name: </label>
<input id="your_name" type="text" name="your_name" maxlength="100" required />
```

一定要注意，渲染后的 HTML 元素不包含<form>标签本身以及提交按钮，为什么要这样呢？为了方便我们控制表单动作、CSS、JS 以及其他类似 bootstrap 框架的嵌入。

(3) 视图处理。

接下来需要在视图中实例化编写好的表单类，例如视图文件 views.py 的代码如下。

```python
from django.shortcuts import render
from django.http import HttpResponseRedirect

from .forms import NameForm

def get_name(request):
    # 如果 form 通过 POST 方法发送数据
    if request.method == 'POST':
        # 接收 request.POST 参数构造 form 类的实例
        form = NameForm(request.POST)
        # 验证数据是否合法
        if form.is_valid():
            # 处理 form.cleaned_data 中的数据
            # ...
            # 重定向到一个新的 URL
            return HttpResponseRedirect('/thanks/')

    # 如果是通过 GET 方法请求数据，返回一个空的表单
    else:
        form = NameForm()

    return render(request, 'name.html', {'form': form})
```

对上述代码的具体说明如下。
- 对于 GET 方法请求页面时，返回空的表单，让用户可以填入数据。
- 对于 POST 方法，接收表单数据，并验证。
- 如果数据合法，按照正常业务逻辑继续执行下去。
- 如果不合法，返回一个包含先前数据的表单给前端页面，方便用户修改。
- 通过表单的 is_bound 属性可以获知一个表单已经绑定了数据，还是一个空表。

(4) 模板处理。

在 Django 的模板中，只需要按下面的处理就可以得到完整的 HTML 页面。

```
<form action="/your-name/" method="post">
    {% csrf_token %}
    {{ form }}
    <input type="submit" value="Submit" />
</form>
```

对上述代码的具体说明如下。
- `<form>...</form>` 标签要自己写。
- 在使用 POST 方法提交数据时，必须添加 `{% csrf_token %}` 标签，用于处理 csrf 安全机制。
- `{{ form }}` 代表 Django 为你生成其他所有的 form 标签元素，也就是我们上面做的事情。
- 需要手动添加提交按钮。

在默认情况下，Django 支持 HTML 5 的表单验证功能，比如邮箱地址验证、必填项目验证等。

(5) 创建多个表单。

在上面的例子中只用到了一个用户名输入框，这实在是太简单了，在实际项目中可能会有更多的表单元素。例如下面的例子：

```
from django import forms

class ContactForm(forms.Form):
    subject = forms.CharField(max_length=100)
    message = forms.CharField(widget=forms.Textarea)
    sender = forms.EmailField()
    cc_myself = forms.BooleanField(required=False)
```

在上述代码中创建了 4 个表单框。实际上，Django 内置的表单模块为开发者内置了许多表单字段，例如 BooleanField、CharField、ChoiceField、TypedChoiceField、DateField 和 DateTimeField 等。每一个表单字段类型都对应一种 Widget 类，每一种 Widget 类都对应 HMTL 语言中的一种 input 元素类型，比如`<input type="text">`。在 HTML 中实际使用什么类型的 input，就需要在 Django 的表单字段中选择相应的 field。比如要使用一个`<input type="text">`，可以选择一个 CharField。

一旦我们的表单接收数据并通过了验证，那么就可以从 form.cleaned_data 字典中读取所有的表单数据，例如下面是一个 views.py 视图文件及处理表单数据的例子。

```python
from django.core.mail import send_mail

if form.is_valid():
    subject = form.cleaned_data['subject']
    message = form.cleaned_data['message']
    sender = form.cleaned_data['sender']
    cc_myself = form.cleaned_data['cc_myself']

    recipients = ['info@example.com']
    if cc_myself:
        recipients.append(sender)

    send_mail(subject, message, sender, recipients)
    return HttpResponseRedirect('/thanks/')
```

6.2.2 实战演练：第一个表单程序

在下面的实例代码中，演示了在 Django Web 中使用表单计算两个数字的和的过程。

源码路径：**daima\6\zqxt_form2**

（1）首先新建一个名为 zqxt_form2 的项目，然后进入 zqxt_form2 文件夹新建一个名为 tools 的 app。

```
django-admin startproject zqxt_form2
python manage.py startapp tools
```

（2）在 tools 文件夹中新建文件 forms.py，具体实现代码如下所示。

```python
from django import forms
class AddForm(forms.Form):
    a = forms.IntegerField()
    b = forms.IntegerField()
```

（3）编写视图文件 views.py，实现两个数字的求和处理，具体实现代码如下所示。

```python
# coding:utf-8
from django.shortcuts import render
from django.http import HttpResponse

# 引入我们创建的表单类
from .forms import AddForm

def index(request):
    if request.method == 'POST':# 当提交表单时

        form = AddForm(request.POST) # form 包含提交的数据

        if form.is_valid():# 如果提交的数据合法
            a = form.cleaned_data['a']
            b = form.cleaned_data['b']
            return HttpResponse(str(int(a) + int(b)))

    else:# 当正常访问时
```

```
       form = AddForm()
   return render(request, 'index.html', {'form': form})
```

(4) 编写模板文件 index.html，实现一个简单的表单效果，具体实现代码如下所示。

```
<form method='post'>
{% csrf_token %}
{{ form }}
<input type="submit" value="提交">
</form>
```

(5) 在文件 urls.py 中设置将视图函数对应到网址，具体实现代码如下所示。

```
from django.urls import path
from django.contrib import admin
from tools import views as tools_views

urlpatterns = [
    path('', tools_views.index, name='home'),
    path('admin/', admin.site.urls),
]
```

在浏览器中运行后会显示一个表单效果，在表单中可以输入两个数字，执行效果如图 6-2 所示。

图 6-2　表单效果

单击"提交"按钮后会计算这两个数字的和，并显示求和结果，如图 6-3 所示。

图 6-3　显示求和结果

6.3　表单的典型应用

在本书前面的内容中曾经讲解过使用表单的知识，表单是实现动态 Web 的重要媒介之一。在本节的内容中，将进一步讲解在 Django 框架中开发表单程序的知识。

6.3.1　表单 forms 的设计与使用

扫码观看本节视频讲解

在 Django Web 程序中，表单是由各种字段组成的。开发人员可以自定义表单(forms.Form)，也可以使用模型 Models 创建(forms.ModelForm)表单。读者需要注意的是，在模型中使用属性 verbose_name 来描述一个字段，而在表单中使用 label 来描述一个字段。

1. 两个表单

下面举两个联系人表单的例子，其中第一个表单类 ContactForm1 由自定义方式实现，第二个表单类 ContactForm2 在模型 Model 中创建实现。

```
from django import forms
from .models import Contact

class ContactForm1(forms.Form):

    name = forms.CharField(label="用户名", max_length=255)
    email = forms.EmailField(label="邮箱地址")


class ContactForm2(forms.ModelForm):

    class Meta:
        model = Contact
        fields = ('name', 'email',)
```

在 Django Web 程序中，惯用做法是在 app 文件夹下创建一个表单文件 forms.py，在里面保存 app 中的所有表单，这样做的好处是便于集中管理表单。在使用上述表单时，在视图文件 views.py 中使用 import 命令导入即可。

2. 表单实例化

在 Django Web 程序中，只有实例化后的表单才有意义。例如通过下面的代码可以实例化一个空表单，空表单中没有任何数据，可以通过 {{ form }}在模板中渲染这个表单。

```
form = ContactForm()
```

用户提交的数据可以通过以下代码与表单继续结合，生成与数据结合过的表单(Bound forms)。Django 只能对 Bound forms 进行验证。

```
form = ContactForm(data=request.POST, files=request.FILES)
```

3. 在模板文件中使用 form

在模板文件中可以使用{{ form.as_p }}、{{ form.as_li }}和{{ form.as_table }}等标签来渲染表单。如果想详细控制每个 field 的格式，可以使用下面的方式实现。

```
{% block content %}
<div class="form-wrapper">
  <form method="post" action="" enctype="multipart/form-data">
    {% csrf_token %}
    {% for field in form %}
      <div class="fieldWrapper">
    {{ field.errors }}
    {{ field.label_tag }} {{ field }}
```

```
        {% if field.help_text %}
          <p class="help">{{ field.help_text|safe }}</p>
        {% endif %}
        </div>
      {% endfor %}
      <div class="button-wrapper submit">
        <input type="submit" value="Submit" />
      </div>
  </form>
</div>
{% endblock %}
```

4. 表单的使用案例

(1) 假设现在需要设计一个表单让用户实现会员注册功能，需要先在 app 目录下新建表单文件 forms.py，然后创建一个 RegistrationForm 类。演示代码如下：

```python
from django import forms
from django.contrib.auth.models import User

class RegistrationForm(forms.Form):
    username = forms.CharField(label='Username', max_length=50)
    email = forms.EmailField(label='Email',)
    password1 = forms.CharField(label='Password', widget=forms.Password)
    password2 = forms.CharField(label='Password Confirmation', widget=forms.Password)
```

当然也可以不用新建 forms.py 文件，而是直接在 HTML 模板文件中实现表单。但是笔者建议使用表单文件 forms.py 实现，这样做的好处是所有的表单在一个文件中，便于后期维护。

(2) 再看视图文件 views.py，下面是使用 RegistrationForm 后的 views.py 文件的演示代码。

```python
from django.shortcuts import render, get_object_or_404
from django.contrib.auth.models import User
from .forms import RegistrationForm
from django.http import HttpResponseRedirect


def register(request):
    if request.method == 'POST':

        form = RegistrationForm(request.POST)
        if form.is_valid():
            username = form.cleaned_data['username']
            email = form.cleaned_data['email']
            password = form.cleaned_data['password2']
            # 使用内置User自带create_user方法创建用户，不用使用save()
            user = User.objects.create_user(username=username, password=password, email=email)
            # 如果直接使用objects.create()方法后不需要使用save()
            return HttpResponseRedirect("/accounts/login/")
```

```
    else:
        form = RegistrationForm()
    return render(request, 'users/registration.html', {'form': form})
```

（3）最后看模板文件，下面是模板文件 registration.html 的演示代码。如果需要通过表单实现图片或文件上传功能，则需要给 form 元素添加 enctype="multipart/form-data"属性。

```
<form action="." method="POST">
{{ form.as_p }}
</form>
```

综上所述，我们总结 RegistrationForm 的运行流程如下所示。

- 当用户通过 post 方法提交表单中的数据时，将提交的数据与 RegistrationForm 相互结合，然后验证表单 RegistrationForm 中的数据是否合法。
- 如果表单中的数据合法，则先用 Django User 模型自带的方法 create_user 创建一个 user 对象，然后再创建 user_profile(用户配置)。用户提交表单后，提交的数据被分别存储在两张数据库表中。
- 如果新用户注册成功，通过 HttpResponseRedirect 方法跳转到用户登录页面。
- 如果用户没有提交表单或不是通过 post 方法提交表单，则会跳转到用户注册页面，生成一张空的 RegistrationForm。

5. 表单的验证

在 Django Web 程序中，每个 forms 类可以使用 clean 方法自定义表单验证。如果只是想验证某些字段，可以通过 "clean_字段名" 的方式自定义实现表单验证。如果用户提交的数据没有通过验证，则会返回 ValidationError 错误。如果用户提交的数据有效，则会将提交的数据存储在数据库中。

例如在前面的用户注册演示里，可以在 RegistrationForm 中使用 clean 方法添加验证功能，包括用户名验证、邮箱格式验证和密码验证功能。例如下面的演示代码：

```
from django import forms
from django.contrib.auth.models import User
import re


def email_check(email):
    pattern = re.compile(r"\"?([-a-zA-Z0-9.`?{}]+@\w+\.\w+)\"?")
    return re.match(pattern, email)


class RegistrationForm(forms.Form):

    username = forms.CharField(label='Username', max_length=50)
    email = forms.EmailField(label='Email',)
    password1 = forms.CharField(label='Password', widget=forms.Password)
    password2 = forms.CharField(label='Password Confirmation',
widget=forms.Password)
```

```python
    def clean_username(self):
        username = self.cleaned_data.get('username')

        if len(username) < 6:
            raise forms.ValidationError("用户名的长度必须大于6.")
        elif len(username) > 50:
            raise forms.ValidationError("用户名长度太长.")
        else:
            filter_result = User.objects.filter(username__exact=username)
            if len(filter_result) > 0:
                raise forms.ValidationError("用户名已经存在.")

        return username

    def clean_email(self):
        email = self.cleaned_data.get('email')

        if email_check(email):
            filter_result = User.objects.filter(email__exact=email)
            if len(filter_result) > 0:
                raise forms.ValidationError("邮箱已经存在.")
        else:
            raise forms.ValidationError("请输入一个合法的邮箱.")

        return email

    def clean_password1(self):
        password1 = self.cleaned_data.get('password1')

        if len(password1) < 6:
            raise forms.ValidationError("密码长度太短.")
        elif len(password1) > 20:
            raise forms.ValidationError("密码长度太长.")

        return password1

    def clean_password2(self):
        password1 = self.cleaned_data.get('password1')
        password2 = self.cleaned_data.get('password2')

        if password1 and password2 and password1 != password2:
            raise forms.ValidationError("两次输入的密码不一致.")

        return password2
```

6. 在通用视图中使用表单

在 Django Web 程序中定义表单类后就可以在基于类的视图(Class Based View)中使用表单。例如，下面是一个创建一篇新博客文章的演示代码。

```python
from django.views.generic.edit import CreateView
from .models import Articles
```

```
from .forms import ArticlesForm

class ArticlesCreateView(CreateView):
    model = Articles
    form_class = ArticlesForm
    template_name = 'blog/A_create_form'
```

7. 自定义表单组件

在 Django Web 程序中，在 forms 的每个字段中都可以设置输入组件，例如多选框、复选框和文本框等，并且还可以定义每个组件的 CSS 属性。如果不指定组件，Django 会使用默认的、外观不够美观的组件。例如在下面的演示代码中，使用 Textarea 作为表单字段的输入控件，并且还设置了 Textarea 的 CSS 样式。

```
from django import forms

class ContactForm(forms.Form):
    name = forms.CharField(
        max_length=255,
        widget=forms.Textarea(
            attrs={'class': 'custom'},
        ),
    )
```

为了提高用户的选择输入体验，我们可美化表单组件的样式。例如在下面的演示代码中，使用 SelectDateWidget 组件实现了选择年份时间的功能，在复选框 CheckboxSelectMultiple 中设置了修饰颜色。

```
from django import forms

BIRTH_YEAR_CHOICES = ('1990', '1991', '1992')
COLORS_CHOICES = (
    ('red', 'Red'),
    ('green', 'Green'),
    ('black', 'Black'),
)

class SimpleForm(forms.Form):
    birth_year = 
forms.DateField(widget=forms.SelectDateWidget(years=BIRTH_YEAR_CHOICES))
    favorite_colors = forms.MultipleChoiceField(
        required=False,
        widget=forms.CheckboxSelectMultiple,
        choices=COLORS_CHOICES,
    )
```

8. 表单数据初始化

在 Django Web 程序中，有时需要在表单中设置一些提示信息作为初始数据，此功能可以通过方法 initial 实现。例如下面的演示代码：

```
form = ContactForm(
    initial={
```

```
            'name': '请输入姓名',
        },)
```

这一功能在修改信息时比较常见,此时通常在表单中显示原来的数据信息。例如下面的方法仅适用于由模型创建的 ModelForm,而不适用于自定义的表单。

```
contact = Contact.objects.get(id=1)
form = ContactForm(instance = contact)
```

对于自定义的表单,可以设置 default_data 属性实现表单数据的初始化。

```
default_data = {'name': 'John', 'email': 'someone@hotmail.com', }
form = ContactForm(default_data)
```

9. 使用 Formset

在 Django Web 程序中,有时需要在一个页面上使用多个表单,例如一次性添加多本书的信息,此时可以使用表单集合 formset 实现。例如,下面的演示代码创建了一个表单集合 FormSet。

```
from django import forms


class BookForm(forms.Form):
    name = forms.CharField(max_length=100)
    title = forms.CharField()
    pub_date = forms.DateField(required=False)


# forms.py - 图书表单集合

from django.forms import formset_factory
from .forms import BookForm

# extra: 额外的空表单数量
# max_num: 包含表单数量(不含空表单)

BookFormSet = formset_factory(BookForm, extra=2, max_num=1)
```

接下来在视图文件 views.py 中,可以像使用 form 一样使用 formset,例如下面的演示代码:

```
from .forms import BookFormSet
from django.shortcuts import render

def manage_books(request):
    if request.method == 'POST':
        formset = BookFormSet(request.POST, request.FILES)
        if formset.is_valid():
            pass
    else:
        formset = BookFormSet()
    return render(request, 'manage_books.html', {'formset': formset})
```

最后在模板文件中可以使用 formset，例如下面的演示代码：

```
<form action="." method="POST">
{{ formset }}
</form>
```

10. 自定义字段属性和错误信息

在 Django 表单中，我们可以设置每个字段是否为必需字段，并且可以设置字段的最大长度和最小长度，并且还可以针对每个属性自定义错误信息。例如下面的演示代码：

```python
from django import forms

class LoginForm(forms.Form):
    username = forms.CharField(
        required=True,
        max_length=20,
        min_length=6,
        error_messages={
            'required': '用户名不能为空',
            'max_length': '用户名长度不得超过20个字符',
            'min_length': '用户名长度不得少于6个字符',
        }
    )
    password = forms.CharField(
        required=True,
        max_length=20,
        min_length=6,
        error_messages={
            'required': '密码不能为空',
            'max_length': '密码长度不得超过20个字符',
            'min_length': '密码长度不得少于6个字符',
        }
    )
```

对于继承于类 ModelForm 的表单，可以在 Meta 选项下的组件中自定义错误信息。例如下面的演示代码：

```python
from django.forms import ModelForm, Textarea
from myapp.models import Author

class AuthorForm(ModelForm):
    class Meta:
        model = Author
        fields = ('name', 'title', 'birth_date')
        widgets = {
            'name': Textarea(attrs={'cols': 80, 'rows': 20}),  # 关键是这一行
        }
        labels = {
            'name': _('Author'),
        }
        help_texts = {
            'name': _('Some useful help text.'),
        }
```

```
    error_messages = {
        'name': {
            'max_length': _("This writer's name is too long."),
        },
    }
```

6.3.2 实战演练：简易用户登录验证系统

在下面的实例代码中，演示了使用 Django 框架开发一个简易用户登录验证系统的过程。

源码路径：daima\6\\denglu\

（1）新建一个名称为 biaodan1 的项目，然后进入 biaodan1 文件夹新建一个名为 people 的 app。

```
django-admin startproject biaodan1
cd biaodan1
python manage.py startapp people
```

（2）在设置文件 settings.py 的 INSTALLED_APPS 中将上面创建的 people 添加进去：

```
INSTALLED_APPS = [
    'django.contrib.admin',
    'django.contrib.auth',
    'django.contrib.contenttypes',
    'django.contrib.sessions',
    'django.contrib.messages',
    'django.contrib.staticfiles',
    'people',
]
```

（3）编写视图文件 views.py 的代码，具体实现代码如下所示。

```
from people.models import User
from functools import wraps
# 说明：这个装饰器的作用，就是在每个视图函数被调用时，都验证下有没有登录，
# 如果登录，则可以执行新的视图函数，否则自动跳转到登录页面
def check_login(f):
    @wraps(f)
    def inner(request,*arg,**kwargs):
        if request.session.get('is_login')=='1':
            return f(request,*arg,**kwargs)
        else:
            return redirect('/login/')
    return inner

def login(request):
    # 如果是 POST 请求，则说明是单击登录按扭 FORM 表单跳转到此的，那么就要验证密码，并保存 session
    if request.method=="POST":
        username=request.POST.get('username')
        password=request.POST.get('password')

        user=User.objects.filter(username=username,password=password)
```

```
        print(user)
        if user:
            #登录成功
            # 1,生成特殊字符串
            # 2,这个字符串当成key,此key在数据库的session表(在数据库中一个表名是session
的表)中对应一个value
            # 3,在响应中,用cookies保存这个key (即向浏览器写一个cookie,此cookie的值
即是这个key特殊字符)
            request.session['is_login']='1'  # 这个session用于后面访问每个页面(调用
每个视图函数时要用到,即判断是否已经登录)
            request.session['username']=username  # 这个要存储的session是用于后面,
每个页面上要显示出来,登录状态的用户名用。
            # 说明:如果需要在页面上显示出来的用户信息太多(有时还有积分、姓名、年龄等信息),
可以只用session保存user_id
            request.session['user_id']=user[0].id
            return redirect('/index/')
    # 如果是GET请求,就说明用户刚开始登录,使用URL直接进入登录页面
    return render(request,'login.html')

@check_login
def index(request):
    user_id1=request.session.get('user_id')
    # 使用user_id去数据库中找到对应的user信息
    userobj=User.objects.filter(id=user_id1)
    print(userobj)
    if userobj:
        return render(request,'index.html',{"user":userobj[0]})
    else:
        return render(request,'index.html',{'user','匿名用户'})
```

(4) 编写模型文件 models.py 的代码,具体实现代码如下所示。

```
from django.db import models
class User(models.Model):
    username=models.CharField(max_length=16)
    password=models.CharField(max_length=32)
```

(5) 编写 URL 导航文件 urls.py 的代码,具体实现代码如下所示。

```
from django.urls import path
from django.contrib import admin
from people import views
urlpatterns = [
    path('admin/', admin.site.urls),
    path('login/', views.login),
    path('index/', views.index),
]
```

(6) 在 people 目录下创建 templates 子目录,在里面新建两个模板文件。其中第一个模板文件 index.html 的代码如下所示。

```
<body>
  <h1>这是一个 index 页面</h1>
  <p>欢迎:{{user.username}}--{{user.password}}</p>
</body>
```

第二个模板文件 login.html 的代码如下所示。

```html
<body>
<h1>欢迎登录！</h1>
<form action="/login/" method="post">
    {% csrf_token %}
    <p>
        用户名：
        <input type="text" name="username">
    </p>
    <p>
        密码：
        <input type="text" name="password">
    </p>
    <p>
        <input type="submit" value="登录">
    </p>
    <hr>
</form>
</body>
```

（7）将控制命令定位到项目根目录 biaodan1，通过如下命令根据模型文件 models.py 创建数据库表。

```
python manage.py makemigrations
python manage.py migrate
```

（8）通过如下命令创建一个合法用户数据，下面的命令创建的用户名是 admin，密码是数字 123。

```
python manage.py shell

>>> from people.models import User
>>> User.objects.create(username="admin", password="123")
```

开始测试程序，如果没有登录，输入 http://localhost:8000/index/后会自动跳转到 login 登录表单页面，如图 6-4 所示。输入用户名 admin 和密码 123 后会来到登录成功页面 index.html，在页面显示用户名和密码，如图 6-5 所示。

图 6-4　登录表单界面

图 6-5　登录成功界面

在浏览器的 Cookie 界面中会显示用 Session 保存的登录信息，如图 6-6 所示。

图 6-6　浏览器中存储的信息

6.3.3　实战演练：文件上传系统

在下面的实例代码中，演示了使用 Django 框架开发一个简易文件上传系统的过程。

源码路径：daima\6\biaodan2

(1) 使用如下命令创建一个名为 biaodan2 的项目。

```
django-admin.py startproject biaodan2  # 新建一个项目
```

(2) 定位到 biaodan2 的根目录，然后创建一个名为 pic_upload 的 app 程序。

```
cd biaodan2  # 进入到该项目的文件夹
django-admin.py startapp pic_upload  # 新建一个应用(app)
```

(3) 本项目的模型文件是 pic_upload/models.py，具体实现代码如下所示。

```python
from django.db import models
from datetime import date
from django.urls import reverse
from uuid import uuid4
import os
def path_and_rename(instance, filename):
    upload_to = 'mypictures'
    ext = filename.split('.')[-1]
    filename = '{}.{}'.format(uuid4().hex, ext)
    # 返回文件的整个路径
    return os.path.join(path_and_rename, filename)
# 在这里创建模型
class Picture(models.Model):
    title = models.CharField("标题", max_length=100, blank=True, default='')
    image = models.ImageField("图片", upload_to="mypictures", blank=True)
    date = models.DateField(default=date.today)

    def __str__(self):
        return self.title
```

```python
# 对于使用 Django 自带的通用视图非常重要
def get_absolute_url(self):
    return reverse('pic_upload:pic_detail', args=[str(self.id)])
```

- 当使用 ImageField 或 FileField 时必须定义 upload_to 选项，这是文件上传后所在的文件夹。数据库本身并不存储文件，而只是存储文件所在的链接路径。如果项目的图片文件根目录是 media，那么在本实例中上传图片后会被存储在 media/mypictures/文件夹中。
- Django 自带通用视图，在完成对象编辑或创建后需要一个返回页面，所以必须在模型中定义一个返回链接 get_absolute_url。在本项目中，图片上传成功后会跳转到图片详情页面。
- 自定义方法 path_and_rename，使所有上传的图片都会以随机的 uuid 字符串命名。这一点非常重要，如果我们没有对上传图片的名字做任何修改，例如上传了两张同样名字的图片，后面那张将覆盖前面那张，这显然不是我们想要的结果，所以我们实际在操作用户上传的图片过程中有必要动态地定义文件上传的路径，并分配一个随机的名字。

(4) 因为在项目中设置了如下所示的 3 个页面。
- 查看图片列表页面。
- 查看图片详情页面。
- 图片上传表单页面。

所以需要编写 3 个对应的 URLs，在 pic_upload 文件夹中创建程序文件 urls.py，具体实现代码如下所示。

```python
app_name = 'pic_upload'

urlpatterns = [

    # 展示所有图片
    path('', views.PicList.as_view(), name='pic_list'),

    # 上传图片
    re_path(r'^pic/upload/$',
        views.PicUpload.as_view(), name='pic_upload'),

    # 展示图片
    re_path(r'^pic/(?P<pk>\d+)/$',
        views.PicDetail.as_view(), name='pic_detail'),

]
```

(5) 编写 3 个视图来处理 URL 发来的三种请求，使用 Django 中的 ListView 展示已经上传图片的列表页面，使用 DetailView 展示图片详情页面，使用 CreateView 展示图片上传表单页面。实例文件 pic_upload/views.py 的具体实现代码如下所示。

```python
class PicList(ListView):
    queryset = Picture.objects.all().order_by('-date')
```

```python
    # ListView 默认 Context_object_name 是 object_list
    context_object_name = 'latest_picture_list'

    # 默认 template_name = 'pic_upload/picture_list.html'


class PicDetail(DetailView):
    model = Picture
    # DetailView 默认 Context_object_name 是 picture

    # 下面是 DetailView 默认模板,可以换成自己的模板
    # template_name = 'pic_upload/picture_detail.html'


class PicUpload(CreateView):
    model = Picture

    # 可以通过 fields 选项自定义需要显示的表单
    fields = ['title', 'image']

    # CreateView 默认 context_object_name 是 form

    # 下面是 CreateView 默认模板,可以换成自己的模板
    # template_name = 'pic_upload/picture_form.html'
```

当使用 Django 的通用视图时,Django 会使用默认内容的对象名字 context_object_name 和模板名字 template_name。

(6) 在 pic_upload 目录中创建一个 templates 文件夹,然后在 templates 目录中再创建一个 pic_upload 文件夹,把所有模板文件放在这个文件夹里面。模板文件 picture_list.html 的具体实现代码如下。

```
{% block content %}
<h3>图片列表</h3>
<ul>
{% if latest_picture_list %}
   {% for picture in latest_picture_list %}
   <li><a href="{% url 'pic_upload:pic_detail' picture.id %}">{{ picture.title }}</a>
      - {{ picture.date| date:"Y-m-j" }}</li>
   {% endfor %}
{% else %}
   <li>还没有上传新图片</li>
{% endif %}
</ul>
<p><a href="{% url 'pic_upload:pic_upload' %}">上传新图片</a></p>
{% endblock %}
```

模板文件 picture_detail.html 的具体实现代码如下所示。

```
{% block content %}
<h3>{{ picture.title }}</h3>

{% if picture.image %}
```

```
        <p><img src="{{ picture.image.url }}"/></p>
{% endif %}

<p>上传日期: {{ picture.date | date:"Y-m-j" }} </p>
<p><a href="{% url 'pic_upload:pic_list' %}">查看所有图片</a> |
    <a href="{% url 'pic_upload:pic_upload' %}">上传新图片</a></p>
{% endblock %}
```

模板文件 picture_form.html 的具体实现代码如下所示。

```
{% block content %}
<h3>上传新图片</h3>
<form method="post" enctype="multipart/form-data">{% csrf_token %}
    {{ form.as_p }}
    <input type="submit" value="确定" />
</form>
{% endblock %}
```

- 在上传图片的表单 form 中必须加上 enctype="multipart/form-data"，这表示图片等文件格式，否则无法成功上传图片。这相当于告诉表单，有数据或文件通过表单上传。
- 在模板中使用图片的方式是<img src="{{ picture.image.url }}">，而不是<img src="{{ picture.image }}">，否则图片无法显示。

(7) 在系统文件 settings.py 中同时设置 MEDIA_ROOT 和 MEDIA_URL，并且需要在 biaodan2 根目录下新建一个 media 文件夹。这样图片就会被上传到 biaodan2/media/mypictures/ 目录下。

```
INSTALLED_APPS = [
    'django.contrib.admin',
    'django.contrib.auth',
    'django.contrib.contenttypes',
    'django.contrib.sessions',
    'django.contrib.messages',
    'django.contrib.staticfiles',
    'pic_upload',
]

# 设置媒体文件夹，对于图片和文件上传很重要
MEDIA_ROOT = os.path.join(BASE_DIR, 'media')
MEDIA_URL = '/media/'
```

(8) 在 biaodan2 根目录下的文件 urls.py 中，不仅要把 app 的 urls 添加进去，而且还要在最后通过 static 方法把 MEDIA_URL 添加进去，只有这样才能在模板中正确显示图片，因为图片属于静态文件。文件 biaodan2/urls.py 的具体实现代码如下所示。

```
from django.contrib import admin
from django.conf.urls import url, include

# 对于显示静态文件非常重要
from django.conf import settings
from django.conf.urls.static import static
```

```
urlpatterns = [
            url('admin/', admin.site.urls),
            url('', include('pic_upload.urls')),
] + static(settings.MEDIA_URL, document_root=settings.MEDIA_ROOT)
```

开始调试程序，将命令定位到 biaodan2 根目录，使用如下命令根据模型文件创建数据库表。

```
python manage.py makemigrations
python manage.py migrate
```

输入如下命令开始运行程序。

```
python manage.py runserver
```

上传图片列表界面效果如图 6-7 所示，单击"上传新图片"链接进入上传表单界面，如图 6-8 所示。

图 6-7　上传图片列表界面　　　　图 6-8　上传表单界面

将同一幅本地图片"西客站 2.jpg"上传 3 次，上传后会自动命名为不同的名字，这样可以避免漏传，如图 6-9 所示。

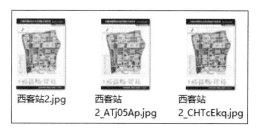

图 6-9　自动命名重复的上传图片

第 7 章

站点配置和管理

在动态 Web 应用程序中，需要对整个站点的整体结构和管理后台进行架构设计，以确保整个 Web 项目的正确运行。在本章的内容中，将详细讲解在 Django Web 项目中实现站点管理的基本知识，包括系统配置、静态文件和后台管理等知识，为读者步入本书后面知识的学习打下基础。

7.1 系统配置文件

我们创建好一个 Django Web 工程后，会自动生成一个名为 settings.py 的文件，通过这个文件可以配置和管理 Django 项目的管理运维信息。在本书前面的内容中，已经多次用到了这个文件。在本节的内容中，将详细讲解配置文件 settings.py 的知识。

扫码观看本节视频讲解

7.1.1 配置文件的特性

在大多数情况下，我们不需要设置 settings.py 文件中的大多数内容，因为这些内容系统已经默认设置好了，我们只需要根据项目的实际情况(例如我们的数据库名、app 名)修改其中的几个选项即可。文件 settings.py 中的默认设置来自于文件 django/conf/global_settings.py。在编译 Django 框架时，会先载入文件 global_settings.py 中的默认配置信息，然后加载工程中的 settings.py 文件，并重写改变的用户自己的设置。

在 Django Web 工程中，工程配置文件 settings.py 的主要特性如下所示。

(1) 配置项。

配置文件 settings.py 中的所有配置项都是大写的。

(2) 默认值。

创建好一个 Django Web 工程后，在配置文件 settings.py 中会初始化一些默认的配置，这些默认配置设置了最基础的工程信息。

(3) 查看配置区别。

可以从文件 global_settings.py 中导入全局配置，在一般情况下这是不必要的。在配置过程中，可以随时通过如下命令来查看当前 settings 文件和默认设置的不同之处。

```
python manage.py diffsettings
```

(4) 安全性。

在配置文件 settings.py 中包含的信息十分重要，因为里面涉及账户信息，所以需要严格控制 settings.py 文件的访问权限。

7.1.2 基本配置

(1) INSTALLED_APPS：一个字符串 tuple(元组)，内容是本 Django 安装中的所有应用。每个字符串应该是一个包含 Django 应用程序的 Python 包的路径全称。在创建一个 Django 工程和一个 app 后，我们需要在里面添加 app 项目的名字，否则会找不到执行的项目。例如在下面的代码中，app1 就是一个我们新建的 app 名字。

```
INSTALLED_APPS = [
 'django.contrib.admin',
```

```
'django.contrib.auth',
'django.contrib.contenttypes',
'django.contrib.sessions',
'django.contrib.messages',
'django.contrib.staticfiles',
'app1',
# 默认已有，如果没有只要添加 app 名称即可，例如：'app1'
# 新建的应用都要在这里添加
]
```

(2) MIDDLEWARE：中间件配置信息。中间件就是一个类，在请求到来和结束后，Django 会根据自己的规则在合适的时机执行中间件中相应的方法。例如创建项目时默认添加的 django.contrib.sessions.middleware.SessionMiddleware 就是一个中间件，这个中间件的功能是启用 Session 验证功能。

(3) ROOT_URLCONF：一个字符串，表示你的根 URL 路径导航文件的模块名，在新建工程后会自动生成。

(4) TEMPLATES：设置模板文件 HTML 的位置，在 Django 2.0 以上的版本中，无须开发者填写 templates 的路径，Django 会自动检索当前工程下的 templates 目录。

(5) SECRET_KEY：工程密钥，在创建工程后会自动生成，也可以从系统环境中、配置文件中或硬编码的配置中得到密钥。

(6) DEBUG：一个开关，默认值是 False，设置是否打开调试模式。

(7) ALLOWED_HOSTS：设置允许访问 Web 的 host 列表，只有在列表中的 host 才能被访问。后面所跟的属性值是一个字符串列表值，这个字符串列表值表示当下这个 Django 站点可以提供的 host/domain(主机/域名)。这是一种安全措施，通过使用伪造的 HTTP 主机标头提交请求来防止攻击者中毒缓存并触发带有恶意主机链接的密码重置电子邮件，即使在许多看似安全的 Web 服务器配置下也是如此。

(8) DATABASES：设置数据库连接参数，默认为 SQLite3 数据库，此时开发者不用作任何修改。如果是其他数据库，则需要开发者自定义设置。例如是 MySQL 数据库，则可以通过如下代码继续设置。

```
DATABASES = {
    'default': {
        'ENGINE': 'django.db.backends.mysql',
        'NAME': 'blog',            #这里的 blog 是数据库名
        'USER': 'root',            #用户名
        'PASSWORD': '123',         #密码
        'HOST':'localhost',        #默认 localhost
        'PORT':'3306'              #默认端口 3306
    }
}
```

(9) AUTH_PASSWORD_VALIDATORS：密码验证规则，比如当运行如下命令创建超级管理员时，输入的密码必须满足里面的验证条件。

```
mange.py createsuperuser
```

在创建一个 Django 工程后，在 AUTH_PASSWORD_VALIDATORS 中会默认启用如下

所示的验证规则。

```
AUTH_PASSWORD_VALIDATORS = [
    {#用户属性相似性验证
        'NAME':
'django.contrib.auth.password_validation.UserAttributeSimilarityValidator',
    },
    {#最小长度验证
        'NAME':
'django.contrib.auth.password_validation.MinimumLengthValidator',
    },
    {#通用密码验证
        'NAME':
'django.contrib.auth.password_validation.CommonPasswordValidator',
    },
    {#数字密码验证
        'NAME':
'django.contrib.auth.password_validation.NumericPasswordValidator',
    },
]
```

(10) LANGUAGE_CODE：一个表示默认语言的字符串，默认值是 'en-us'。

(11) TIME_ZONE：一个表示时区的字符串，国内用户的默认值是'UTC'。

(12) USE_TZ：设置时区，通常和上面的 TIME_ZONE 选项一起使用，默认值为 True。当将 USE_TZ 设置为 True 时，Django 会使用系统默认设置的时区，即 America/Chicago，此时的 TIME_ZONE 不管有没有设置都不起作用。如果 USE_TZ 设置为 False，而 TIME_ZONE 设置为 None，则 Django 还是会使用默认的 America/Chicago 时间。若 TIME_ZONE 设置为其他时区，则还要分情况，如果是 Windows 系统，则 TIME_ZONE 设置是没用的，Django 会使用本机的时间。如果为其他系统，则使用该时区的时间，如果设置 USE_TZ = False, TIME_ZONE = 'Asia/Shanghai'，则使用上海的 UTC 时间。

(13) USE_I18N：设置当前的 Django 工程是否开启国际化功能，默认是开启的。如果不需要国际化支持，那么可以将此项的值设置为 False，此时 Django 会进行一些优化，不会加载国际化支持机制。

(14) USE_L10N：设置当前的 Django 工程是否开启本地化功能，默认值是 True，表示是开启的。使一个程序本地化，意味着需要提供根据 I18N 抽取出来的块进行翻译和格式本地化。USE_L10N 需要和上面的 USE_I18N 选项一起使用。

(15) STATIC_URL：设置当前工程的静态文件目录，默认值为'/static/'。这个 static 是在 Django 工程的具体 app 下建立的 static 目录，用来存放静态资源。而 STATICFILES_DIRS 一般用来设置通用的静态资源，对应的目录不放在 app 下，而是放在工程根目录中。

7.2 静态文件

在开发 Web 程序的过程中，除了前面介绍的模板文件外，还会经常使用其他类型的文件，例如图片、脚本和 CSS 文件。在 Django 框

扫码观看本节视频讲解

架中，将这些文件统称为"静态文件"。在本节的内容中，将详细讲解 Django 静态文件的知识。

7.2.1 静态文件介绍

对于小型 Web 项目来说，对图片、脚本和 CSS 等静态文件的保存要求可能会很低。但是在大型 Web 项目中，特别是由好几个应用组成的大型 Web 项目中，如果静态文件保存得杂乱无章，会直接影响到项目的开发效率和后期维护。

在 Django 框架中，使用 django.contrib.staticfiles 将各个应用程序的静态文件(和一些开发者指明的目录里的文件)统一收集起来，这样在开发项目的过程中，这些文件就会被集中在一个便于分发的地方。

在 Django 的 STATICFILES_FINDERS 设置中包含一系列的查找器，它们知道去哪里找到 static 文件。AppDirectoriesFinder 是默认查找器中的一个，它会在每个 INSTALLED_APPS 中指定的应用的子文件中寻找名称为 static 的特定文件。在后台管理模块中，采用和前台模块相同的方式管理静态文件。假设现在有一个名为 polls 的 app 项目，那么我们应该首先在 polls 目录下创建一个名为 static 的目录。Django 会自动在该目录下查找静态文件，这种方式和 Diango 自动在"polls/templates/"目录下查找模板目录 template 的方式类似。

假设在上面刚创建的 static 文件夹中创建一个名为 polls 的文件夹，然后在 polls 文件夹中创建一个名为 style.css 的样式文件。也就是说，样式表文件的路径是 polls/static/polls/style.css。因为 AppDirectoriesFinder 的存在，我们可以在 Django 中使用 polls/style.css 的形式引用此文件，具体方法类似于引用模板路径的方式。

> **注 意**
> 虽然我们可以像管理模板文件一样，把静态文件直接放入"polls/static"目录，而不是创建另一个名为 polls 的子文件夹，但是这实际上是不被建议的。Django 只会使用第一个找到的静态文件。如果在其他应用中有一个相同名字的静态文件，Django 将无法区分它们。这时需要设置 Django 选择正确的静态文件，其中最简单的方式就是把它们放入各自的命名空间，也就是把这些静态文件放入另一个与应用名相同的目录中。

假设样式文件 polls/static/polls/style.css 的代码如下所示。

```
li a {
    color: green;
}
```

在模板文件 polls/templates/polls/index.html 中可以通过如下代码使用上面的静态样式文件。

```
{% load static %}
<link rel="stylesheet" type="text/css" href="{% static 'polls/style.css' %}" />
```

在上述代码中，{% static %}模板标签会生成静态文件的绝对路径。

接下来创建一个用于存放图像文件的目录，假设在"polls/static/polls"目录下创建一个名为 images 的文件夹，然后在这个目录中放一张名为 background.gif 的图片。换言之，这张静态图片的路径是：polls/static/polls/images/background.gif。然后在静态样式文件 polls/static/polls/style.css 中添加如下代码。

```
body {
    background: white url("images/background.gif") no-repeat;
}
```

此时在浏览中执行模板文件会显示上面设置的图片。

> **注意**
>
> {% static %}模板标签在静态文件(例如样式表)中是不可用的，因为它们不是由 Django 生成的。我们仍然需要使用"相对路径"的方式在静态文件之间互相引用。这样就可以任意改变 STATIC_URL，而无须修改大量的静态文件。

7.2.2 实战演练：在登录表单中使用静态文件

在下面的实例文件中，演示了在 Django Web 登录程序中使用静态文件的过程。

源码路径：daima\7\jingtai

（1）使用如下命令创建一个名为"jingtai"的工程，定位到工程根目录，然后创建一个名为 blog 的 app。

```
django-admin.py startproject jingtai
cd jingtai
python manage.py startapp blog
```

（2）在 blog 目录下新建 static 和 templates 文件夹，分别用于保存静态文件和模板文件。将样式文件 other.min.css 和 amazeui.css 放在 static/css 文件里。

（3）在模板文件中使用静态文件，首先创建模板文件 home.html，通过代码 "/static/css/amazeui.css" 使用静态 CSS 文件 amazeui.css。读者需要注意，在 static 前面必须加 "/"。模板文件 home.html 的具体实现代码如下所示。

```
<!DOCTYPE html>
<html lang="en">
<head>
  <meta charset="UTF-8">
  <meta name="viewport" content="width=device-width, initial-scale=1.0">
  <meta http-equiv="X-UA-Compatible" content="ie=edge">
  <title>Document</title>
  <link rel="stylesheet" href="/static/css/amazeui.css" />
  <link rel="stylesheet" href="/static/css/other.min.css" />
</head>
<body class="login-container">
  <div class="login-box">
    <div class="logo-img">
      <img src="/static/img/logo2_03.png" alt="" />
    </div>
    <form action="" class="am-form" data-am-validator>
```

```html
      <button class="am-btn am-btn-secondary" type="submit">
<a href="login.html">跳转到登录界面</a></button>
    </form>
  </div>
</body>
</html>
```

在模板文件 login.html 中也使用了静态 CSS 文件，主要实现代码如下所示。

```html
  <link rel="stylesheet" href="/static/css/amazeui.css" />
  <link rel="stylesheet" href="/static/css/other.min.css" />
</head>
<body class="login-container">
  <div class="login-box">
    <div class="logo-img">
      <img src="/static/img/logo2_03.png" alt="" />
    </div>
    <form action="" class="am-form" data-am-validator>
      <div class="am-form-group">
        <label for="doc-vld-name-2"><i class="am-icon-user"></i></label>
        <input type="text" id="doc-vld-name-2" minlength="3" placeholder="输入用户名(至少 3 个字符)" required/>
      </div>

      <div class="am-form-group">
        <label for="doc-vld-email-2"><i class="am-icon-key"></i></label>
        <input type="email" id="doc-vld-email-2" placeholder="输入邮箱" required/>
      </div>
      <button class="am-btn am-btn-secondary" type="submit">登录</button>
    </form>
```

(4) 在路径导航文件 urls.py 中设置了项目的 URL 链接，主要实现代码如下所示。

```python
urlpatterns = [
    path('admin/', admin.site.urls),
    path(r'', views.home),
    path(r'login.html', views.login),
]
```

在浏览器中输入 http://127.0.0.1:8000/ 会跳转到主页，如图 7-1 所示。单击"跳转到登录界面"按钮进入登录表单界面，如图 7-2 所示。

图 7-1　系统主页

图 7-2　登录表单界面

7.3　Django Admin 管理

在一个动态 Web 程序中，最主流的做法是实现前台和后台的分离。在 Django 框架中，为开发者提供了 Admin 管理模块，可以帮助开发者快速搭建一个功能强大的后台管理系统。在本节的内容中，将详细讲解 Django Admin 管理模块的基本知识。

扫码观看本节视频讲解

7.3.1　Django Admin 基础

在 Django Web 程序中，通过使用 Admin 模块，可以高效地对数据库表实现增加、删除、查询和修改功能。如果 Django 没有提供 Admin 模块，开发者不但需要自己手动开发后台管理系统，并且需要手动编写增加、删除、查询和修改数据库功能的代码，这样不但会带来更大的开发工作量，而且不利于系统的后期维护工作和后台管理工作。通过使用 Admin 模块的功能，将所有需要管理的模型(数据表)集中在一个平台，不仅可以选择性地管理模型(数据表)，而且还可以快速设置数据条目查询、过滤和搜索条件。

1. 创建超级用户 superuser

使用 Django Admin 的第一步是创建超级用户(superuser)，使用如下命令并根据指示分别输入用户名和密码即可创建超级管理员。

```
python manage.py createsuperuser
```

此时在浏览器访问 http://127.0.0.1:8000/admin/ 就可以看到后台登录界面，如图 7-3 所示。

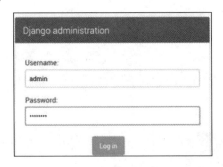

图 7-3　后台登录界面

2. 注册模型(数据表)

假设有一个名字为 blog 的 app，里面包含一个名字为 Articles(文章)的模型。如果想对 Articles 进行管理，我们只需打开 blog 目录下的文件 admin.py，然后使用 admin.site.register 方法注册 Articles 模型即可。演示代码如下所示。

```
from django.contrib import admin
from .models import Articles
```

```
#注册模型
admin.site.register(Articles)
```

此时登录后台后会看到 Articles 数据表中的信息，单击标题即可对文章进行修改。在这个列表中只会显示 Title 字段，并不会显示作者和发布日期等相关信息，也没有分页和过滤条件。

3. 自定义数据表显示选项

在现实应用中，我们需要自定义显示数据表中的哪些字段，也需要设置可以编辑修改哪些字段，并可以对数据库表中的信息进行排序，同时可以设置查询指定的选项内容。在 Admin 模块中，内置了 list_display、list_filter、list_per_page、list_editable、date 和 ordering 等选项，通过这些选项可以轻松实现上面要求的自定义功能。

要想自定义显示数据表中的某些字段，只需对前面的演示文件 blog/admin.py 进行如下改进即可。我们可以先定义 ArticlesAdmin 类，然后使用 admin.site.register(Articles, ArticlesAdmin)方法注册即可实现。

```python
from django.contrib import admin
from .models import Articles,

# Register your models here.

class ArticlesAdmin(admin.ModelAdmin):

    '''设置列表可显示的字段'''
    list_display = ('title', 'author', 'status', 'mod_date',)

    '''设置过滤选项'''
    list_filter = ('status', 'pub_date',)

    '''每页显示条目数'''
    list_per_page = 10

    '''设置可编辑字段'''
    list_editable = ('status',)

    '''按日期月份筛选'''
    date = 'pub_date'

    '''按发布日期排序'''
    ordering = ('-mod_date',)

admin.site.register(Articles, ArticlesAdmin)
```

此时登录后台会看到 Articles 数据表中的展示内容，例如 Articles 标题、作者、状态、修改时间和分页信息，效果类似于图 7-4。

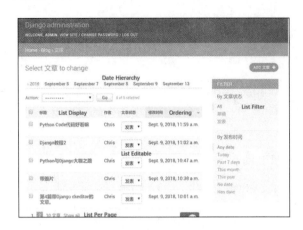

图 7-4　展示数据库中的 Articles

4. 使用 raw_id_fields 选项实现单对多关系

我们知道，新闻网站中的文章往往属于不同的类型，例如体育新闻、娱乐新闻等。假设创建了一个名为 Fenlei 的模型类，用于表示 Articles 的所属类型。在里面有一个父类(ForeignKey)，一个父类可能有多个子类。模型类 Fenlei 的具体实现代码如下所示。

```
class Fenlei(models.Model):
    """文章分类"""
    name = models.CharField('分类名', max_length=30, unique=True)
    slug = models.SlugField('slug', max_length=40)
    parent_Fenlei = models.ForeignKey('self', verbose_name="父级分类", blank=True, null=True, on_delete=models.CASCADE)
```

现在把模型类 Fenlei 添加到 admin 中，因为我们需要根据类别名(name)生成 slug，所以还需要在文件 blog/admin.py 中使用 prepopulated_fields 选项。

```
class FenleiAdmin(admin.ModelAdmin):
    prepopulated_fields = {'slug': ('name',)}

admin.site.register(Fenlei, FenleiAdmin)
```

在 Django Admin 模块中，下拉菜单是默认的单对多关系的选择器，如图 7-5 所示。如果 ForeignKey 非常多，那么下拉菜单将会非常长。此时可以设置 ForeignKey 使用 raw_id_fields 选项，如图 7-6 所示。

图 7-5　默认使用下拉菜单　　　　　图 7-6　使用 raw_id_fields

5. 使用 filter_horizontal 选项实现多对多关系

在 Django Admin 模块中，列表框是默认多对多关系(Many To Many)的选择器。我们可以使用 filter_horizontal 或 filter_vertical 选项设置不同的选择器样式，如图 7-7 所示。

图 7-7　左侧是默认复选框样式，右侧是使用 filter_horizontal 设置的效果

6. 使用类 InlineModelAdmin 在同一页面中显示多个数据表数据

在现实应用中，在一个文章类别下通常会包含多篇文章。如果希望在查看或编辑某个文章类别信息时，同时显示并编辑同属该类别的所有文章信息，可以先定义 ArticlesList 类，然后把其添加到 FenleiAdmin 中。这样就可以在同一页面上编辑修改文章类别信息，也可以修改所属文章信息。例如文件 blog/admin.py 的演示代码如下所示。

```python
from django.contrib import admin
from .models import Articles, Fenlei, Tag

class ArticlesList(admin.TabularInline):
    model = Articles
    '''设置列表可显示的字段'''
    fields = ('title', 'author', 'status', 'mod_date',)

class FenleiAdmin(admin.ModelAdmin):
    prepopulated_fields = {'slug': ('name',)}
    raw_id_fields = ("parent_Fenlei", )
    inlines = [ArticlesList, ]

admin.site.register(Fenlei, FenleiAdmin)
```

7.3.2　实战演练：使用 Django Admin 系统

在下面的实例代码中，演示了使用 Django 框架开发一个博客系统的过程。

源码路径：daima\7\zqxt_admin

(1) 新建一个名称为 zqxt_admin 的项目，然后进入 zqxt_admin 文件夹，新建一个名为 blog 的 app。代码如下：

```
django-admin startproject zqxt_admin
cd zqxt_admin
```

```
# 创建 blog 这个 app
python manage.py startapp blog
```

(2) 修改 blog 文件夹中的文件 models.py，具体实现代码如下所示。

```
# -*- coding: utf-8 -*-
from __future__ import unicode_literals

from django.db import models
from django.utils.encoding import python_2_unicode_compatible
@python_2_unicode_compatible
class Article(models.Model):
    title = models.CharField('标题', max_length=256)
    content = models.TextField('内容')
    pub_date = models.DateTimeField('发表时间', auto_now_add=True, editable=True)
    update_time = models.DateTimeField('更新时间', auto_now=True, null=True)
    def __str__(self):
        return self.title
class Person(models.Model):
    first_name = models.CharField(max_length=50)
    last_name = models.CharField(max_length=50)
    def my_property(self):
        return self.first_name + ' ' + self.last_name
    my_property.short_description = "Full name of the person"
    full_name = property(my_property)
```

(3) 将 blog 加入 settings.py 文件中的 INSTALLED_APPS 中，具体实现代码如下所示。

```
INSTALLED_APPS = (
    'django.contrib.admin',
    'django.contrib.auth',
    'django.contrib.contenttypes',
    'django.contrib.sessions',
    'django.contrib.messages',
    'django.contrib.staticfiles',
    'blog',
)
```

(4) 通过如下所示的命令同步所有的数据库表。

```
# 进入包含有 manage.py 的文件夹
python manage.py makemigrations
python manage.py migrate
```

(5) 进入文件夹 blog，修改里面的文件 admin.py(如果没有则新建一个)，具体实现代码如下所示。

```
from django.contrib import admin
from .models import Article, Person
class ArticleAdmin(admin.ModelAdmin):
    list_display = ('title', 'pub_date', 'update_time',)
class PersonAdmin(admin.ModelAdmin):
    list_display = ('full_name',)
admin.site.register(Article, ArticleAdmin)
admin.site.register(Person, PersonAdmin)
```

输入下面的命令启动服务器。

```
python manage.py runserver
```

在浏览器中输入"http://localhost:8000/admin"后会显示一个用户登录表单界面，如图 7-8 所示。

图 7-8　用户登录表单界面

我们可以创建一个超级管理员用户，使用 CMD 命令进入包含 manage.py 的文件夹 zqxt_admin。然后输入如下命令创建一个超级账号，根据提示分别输入账号、邮箱地址和密码。

```
python manage.py createsuperuser
```

此时可以使用超级账号登录后台管理系统，登录成功后的界面效果如图 7-9 所示。

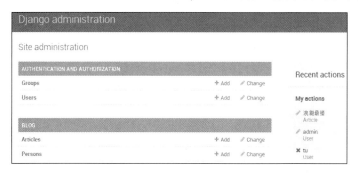

图 7-9　登录成功后的界面效果

管理员可以修改、删除或添加账号信息，如图 7-10 所示。

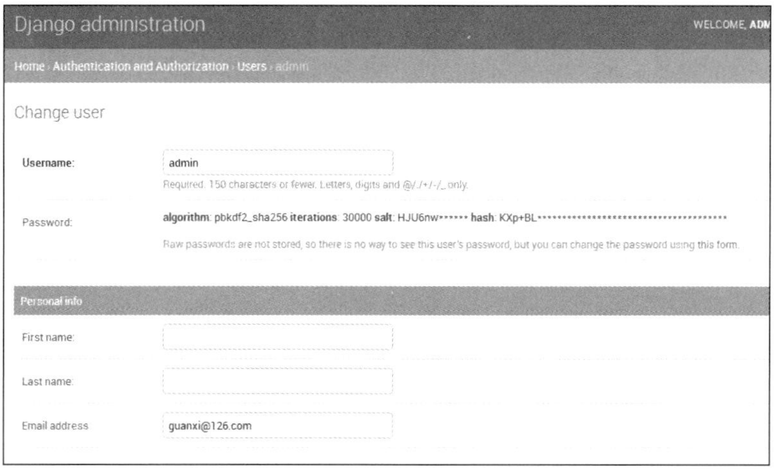

图 7-10　账号管理

也可以对系统内已经发布的博客信息进行管理维护，如图 7-11 所示。

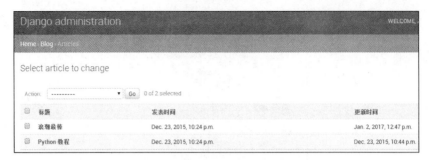

图 7-11　博客信息管理

也可以直接修改用户账号信息的密码，如图 7-12 所示。

图 7-12　修改用户账号信息的密码

第 8 章

站点的安全性

作为一个在网络中运行的 Web 站点来说，安全性问题十分重要。Django 是一款著名的 Web 框架，提供了强大的工具和文档来防止导致安全问题的常见错误，帮助开发者快速搭建一个安全的 Web 站点。在本章的内容中，将详细讲解在 Django Web 项目中实现站点安全管理的知识，为读者步入本书后面知识的学习打下基础。

8.1 Django 安全概述

在 Web 网站中经常面对各种各样的攻击,为了提高网站的安全性,Django 针对常见的攻击提供了内置的模块系统来提高网站的安全性。在本节的内容中,将简要介绍 Django 安全方面的知识。

扫码观看本节视频讲解

8.1.1 跨站脚本(XSS)防护

XSS 攻击通常指的是通过利用网页开发时留下的漏洞,通过巧妙的方法注入恶意指令代码到网页,使用户加载并执行攻击者恶意制造的网页程序。这些恶意网页程序通常是 JavaScript,但实际上也可以包括 Java、VBScript、ActiveX、Flash 甚至是普通的 HTML。常见的 XSS 攻击是将恶意脚本存储在数据库中,然后检索并盗取里面的数据,或者通过让用户单击一个链接,攻击者的 JavaScript 被用户的浏览器执行。然而,XSS 攻击可以来自任何不受信任的源数据,如 Cookie 或 Web 服务,任何没有经过充分处理就包含在网页中的数据。

使用 Django 内置的模板可以保护我们的 Web 免受大多数 XSS 攻击,在 Django 模板中会编码特殊字符,这些字符在 HTML 中都是特别危险的。虽然这可以防止大多数恶意输入的用户,但它不能完全保证万无一失。 例如,它不会防护以下内容:

```
<style class=>...</style>
```

如果将 var 设置为 'class1 onmouseover=javascript:func()',这可能会导致在未经授权的 JavaScript 的执行,这取决于浏览器如何呈现不完整的 HTML。如果我们使用模板系统输出了 HTML 以外的内容,可能会有完全不同的字符和单词需要编码。另外,在数据库中存储 HTML 的时候需要非常小心,尤其是当 HTML 被检索然后展示出来的时候。

8.1.2 跨站请求伪造(CSRF)防护

CSRF 攻击是指恶意用户在另一个用户不知情或者未同意的情况下,冒充他的身份来执行操作。

Django 对大多数类型的 CSRF 攻击提供了内置的保护措施,在适当情况下,我们可以开启并使用这些保护措施。但是,对于任何解决技术都有它的局限性。例如,通过使用 CSRF 模块,可以在全局范围内或设置在某个视图中启用身份验证,开发者只要知道在做什么的情况下才启用这个操作即可保护自己的 Web。

CSRF 防护是通过检查每个 POST 请求的一个随机数(nonce)来实现的,这确保了恶意用户不能简单"回放"你网站上面表单的 POST,以及让另一个登录的用户无意中提交表单。恶意用户必须知道这个随机数,这个随机数是用户特定的(被存放在 Cookie 里)值,具有极强的安全性。

当使用 HTTPS 部署 Web 时，CsrfViewMiddleware 会检查 HTTP referer 协议头是否设置为同源的 URL(包括子域和端口)。因为 HTTPS 提供了附加的安全保护，转发不安全的连接请求时，必须确保链接使用 HTTPS，并使用 HSTS 支持的浏览器。

8.1.3　SQL 注入保护

SQL 注入是一种常见的黑客攻击类型，恶意用户可以在系统数据库中执行任意 SQL 代码。这可能会导致记录删除或者数据泄露。

通过使用 Django 的查询集，生成的 SQL 会被底层数据库驱动正确地转义，提高安全性。然而，Django 也允许开发者编写原始查询或者执行自定义 SQL。这些功能应该谨慎使用，并且我们应该时刻小心正确转义任何用户可以控制的参数。另外，在使用 extra() 的时候应该谨慎行事。

8.1.4　点击劫持保护

点击劫持是指在一个 Web 页面下隐藏了一个透明的 iframe(opacity：0)，用外层假页面诱导用户点击，实际上是在隐藏的 frame 上触发了点击事件进行一些用户不知情的操作。Django 在 X-Frame-Options 中间件的表单中提供了点击劫持保护机制，它在支持的浏览器中可以保护站点免于在 Frame 中渲染。也可以在每个视图中禁止这一保护，或者配置要发送的额外的协议头。

如果不需要在三方站点的 Frame 中展示页面内容，或者只需要在 Frame 中展示一部分内容的站点时，都强烈推荐大家启用 X-Frame-Options 中间件。

8.1.5　SSL/HTTPS

我们将 Web 站点部署在 HTTPS 下总是更安全的，尽管这不是 100%的安全。如果不这样，恶意的网络用户可能会嗅探授权证书，或者在其他的客户端和服务端之间传输的信息，或者在一些情况下，活跃的网络攻击者会修改在两边传输的数据。如果想要使用 HTTPS 提供保护，那么需要在服务器上启用 HTTPS，另外可能还需要去做一些额外的操作。

- 如果必要的话，需要设置 SECURE_PROXY_SSL_HEADER，确保我们已经彻底了解风险。
- 也可以使用设置重定向的方式使用 HTTPS 保护，这样通过 HTTP 的请求会重定向到 HTTPS。
- 也可以通过自定义的中间件的方式实现 HTTPS 保护，此时需要注意 SECURE_PROXY_SSL_HEADER 警告。
- 使用"安全的"Cookie：如果浏览器的连接一开始通过 HTTP，这是大多数浏览器的通常情况，已存在的 Cookie 可能会被泄露。因此，应该将 SESSION_COOKIE_SECURE 和 CSRF_COOKIE_SECURE 设置为 True。这样会使浏览器只在 HTTPS 连接中发送这些 Cookie。此时需要注意，这意味着这些会话在

HTTP 下不能工作，并且 CSRF 保护功能会在 HTTP 下阻止接收任何 POST 数据(如果你把所有 HTTP 请求都重定向到 HTTPS 之后就没问题了)。

- 使用 HTTP 强制安全传输 (HSTS)：HSTS 是一个 HTTP 协议头，能够通知浏览器到特定站点的所有链接都一直使用 HTTPS。HTTPS 的安全基础是 SSL，因此加密的详细内容就需要 SSL。HTTPS 存在不同于 HTTP 的默认端口及一个加密/身份验证层(在 HTTP 与 TCP 之间)。这个系统提供了身份验证与加密通讯方法。

8.1.6 Host 协议头验证

在某些情况下，Django 使用客户端提供的 Host 协议头来构造 URL。虽然可以审查这些值以免受到跨站脚本攻击(XSS)，但是一个假的 Host 值可以用于跨站请求伪造(CSRF)、有害的缓存攻击以及 E-mail 中的有害链接。

因为即使表面上看起来安全的 Web 服务器也容易被篡改主机头，所以 Django 再次在方法 django.http.HttpRequest.get_host()中验证在 ALLOWED_HOSTS 中设置 的 IP，通过 ALLOWED_HOSTS 可以只允许列表中的 IP 地址访问我们的 Web。

8.2 使用 Cookie 和 Session

通过使用 Cookie 和 Session，服务器就可以利用它们记录客户端的访问状态，这样用户就不用在每次访问不同页面都需要登录了。并且还可以确保只有是 Web 的会员才能访问某些内容，这样提高了网站的安全性。另外，通过使用 Cookie 和 Session，可以存储用户在京东或天猫等网站的会员信息，这样用户下次无须输入用户名和密码即可登录，并且通过 Cookie 还可以实现购物车功能。

扫码观看本节视频讲解

8.2.1 Django 框架中的 Cookie

在 Django 框架中，通过如下所示的代码设置 Cookie。

`response.set_cookie(key,value,expires)`

- Key：Cookie 的名称。
- Value：保存的 Cookie 的值。
- Expires：保存的时间，以秒为单位。

例如下面设置了一个 Cookie 的例子，设置的数据将被保存到客户端浏览器中。

`response.set_cookie('username','John',60*60*24)`

通常在 Django 的视图中先生成不包含 Cookie 的 response，然后使用 set_cookie 设置一个 Cookie，最后把 response 返回给客户端浏览器。

下面演示了 3 个设置 Cookie 的例子。

```
#例子1.不使用模板
response = HttpResponse("hello world")
response.set_cookie(key,value,expires)
return response
#例子2.使用模板
response = render(request,'xxx.html', context)
response.set_cookie(key,value,expires)
return response
#例子3.重定向
response = HttpResponseRedirect('/login/')
response.set_cookie(key,value,expires)
return response
```

在 Django 框架中,通过如下所示的代码获取用户发来请求中的 Cookie。

```
request.COOKIES['username']
request.COOKIES.get('username')
```

在 Django 框架中,通过如下所示的代码检查 Cookie 是否已经存在。

```
request.COOKIES.has_key('<cookie_name>')
```

在 Django 框架中,通过如下所示的代码删除一个已经存在的 Cookie。

```
response.delete_cookie('username')
```

例如下面演示了使用 Cookie 验证用户是否已登录。

```
# 如果登录成功,设置cookie
def login(request):
    if request.method == 'POST':
        form = LoginForm(request.POST)

        if form.is_valid():
            username = form.cleaned_data['username']
            password = form.cleaned_data['password']

            user = User.objects.filter(username__exact=username,
password__exact=password)

            if user:
                response = HttpResponseRedirect('/index/')
                # 将username写入浏览器cookie,失效时间为3600秒
                response.set_cookie('username', username, 3600)
                return response

            else:
                return HttpResponseRedirect('/login/')

    else:
        form = LoginForm()

    return render(request, 'users/login.html', {'form': form})


# 通过cookie判断用户是否已登录
```

```python
def index(request):

    #提取浏览器中的cookie,如果不为空,表示为已登录账号
    username = request.COOKIES.get('username', '')
    if not username:
        return HttpResponseRedirect('/login/')
    return render(request, 'index.html', {'username': username})
```

在下面的实例代码中,演示了在 Django Web 程序中使用 Cookie 实现用户注册和登录验证功能的过程。

源码路径:**daima\8\mysite5**

(1) 通过如下命令新建一个名为"mysite5"的工程,然后定位到工程根目录,新建一个名为"online"的 app。

```
django-admin.py startproject mysite5
cd mysite5
python manage.py startapp online
```

(2) 在设置文件 settings.py 的 INSTALLED_APPS 中加入 online,并将 MIDDLEWARE 里面的"django.middleware.csrf.CsrfViewMiddlewar"注释掉。最终代码如下所示。

```python
INSTALLED_APPS = [
    'django.contrib.admin',
    'django.contrib.auth',
    'django.contrib.contenttypes',
    'django.contrib.sessions',
    'django.contrib.messages',
    'django.contrib.staticfiles',
    'online',
]

MIDDLEWARE = [
    'django.middleware.security.SecurityMiddleware',
    'django.contrib.sessions.middleware.SessionMiddleware',
    'django.middleware.common.CommonMiddleware',
    #'django.middleware.csrf.CsrfViewMiddleware',
    'django.contrib.auth.middleware.AuthenticationMiddleware',
    'django.contrib.messages.middleware.MessageMiddleware',
    'django.middleware.clickjacking.XFrameOptionsMiddleware',
]
```

(3) 在工程主目录编写路径导航文件 urls.py,主要实现代码如下所示。

```python
urlpatterns = [
    path(r'online/',include("online.urls")),
]
```

(4) 在 app 目录"online"中编写路径导航文件 urls.py,分别设置登录验证表单、注册表单、主页和退出 4 个页面的导航链接,主要实现代码如下所示。

```python
urlpatterns = [
    path(r'', views.login, name='login'),
```

```
path('login/', views.login, name='login'),
path('regist/', views.regist, name='regist'),
path('index/', views.index, name='index'),
path('logout/', views.logout, name='logout'),
]
```

(5) 编写模型文件 online/models.py，设置创建数据库表 User，表中有两个字段 username 和 password，主要实现代码如下所示。

```
from django.db import models
class User(models.Model):
    username = models.CharField(max_length=50)
    password = models.CharField(max_length=50)

    def __unicode__(self):
        return self.username
```

(6) 通过如下命令创建数据库表。

```
python manage.py makemigrations
python manage.py migrate
```

(7) 编写视图文件 online/views.py，分别实现用户注册、登录验证、退出、登录成功的功能，主要实现代码如下所示。

```
#表单
class UserForm(forms.Form):
    username = forms.CharField(label='用户名',max_length=100)
    password = forms.CharField(label='密码',widget=forms.PasswordInput())

#注册
def regist(req):
    if req.method == 'POST':
        uf = UserForm(req.POST)
        if uf.is_valid():
            #获得表单数据
            username = uf.cleaned_data['username']
            password = uf.cleaned_data['password']
            #添加到数据库
            User.objects.create(username= username,password=password)
            return HttpResponse('regist success!!')
    else:
        uf = UserForm()
    return render_to_response('regist.html',{'uf':uf}, RequestContext(req))

#登录
def login(req):
    if req.method == 'POST':
        uf = UserForm(req.POST)
        if uf.is_valid():
            #获取表单用户密码
            username = uf.cleaned_data['username']
            password = uf.cleaned_data['password']
            #获取的表单数据与数据库进行比较
```

```
            user = User.objects.filter(username__exact = 
username,password__exact = password)
            if user:
                #比较成功，跳转到 index
                response = HttpResponseRedirect('/online/index/')
                #将 username 写入浏览器 cookie,失效时间为 3600s
                response.set_cookie('username',username,3600)
                return response
            else:
                #比较失败，还在 login 页面
                return HttpResponseRedirect('/online/login/')
    else:
        uf = UserForm()
    return render_to_response('login.html',{'uf':uf},RequestContext(req))

#登录成功
def index(req):
    username = req.COOKIES.get('username','')
    return render_to_response('index.html' ,{'username':username})

#退出
def logout(req):
    response = HttpResponse('logout !!')
    #清理 cookie 里保存的 username
    response.delete_cookie('username')
    return response
```

省略模板文件的实现过程，输入 http://127.0.0.1:8000/online/login/进入系统登录表单界面，单击"注册"链接进入注册表单页面，登录成功会显示提示信息。执行效果如图 8-1 所示。

(a) 登录表单页面　　　　(b) 注册页面　　　　(c) 登录成功页面

图 8-1　执行效果

8.2.2　Django 框架中的 Session

在 Web 应用程序中，Session 又被称为会话，其功能和使用场景与 Cookie 类似，能够存储少量的数据或信息。但是由于 Session 数据被存储在服务器上，而不是客户端上，因此 Session 比 Cookie 更加安全，所以通常用来保存后台管理员的登录数据。另外，使用 Session 的另一个好处是即使用户关闭了浏览器，Session 仍将保持到会话过期。

Session 工作的流程如下所示。

(1) 当客户端向服务器发送请求时,首先查看本地是否有 Cookie 文件。如果有,则会在 HTTP 的请求头(Request Headers)中包含一行 Cookie 信息。

(2) 当服务器接收到请求后,根据 Cookie 信息得到 sessionId,根据 sessionId 找到对应的 Session,通过这个 Session 可以判断用户是否登录。

在 Django 框架中,使用如下所示的代码设置 Session 的值。

```
request.session['key'] = value
request.session.set_expiry(time)        #设置过期时间,0 表示浏览器关闭则失效
```

- 如果 time 是一个整数,Session 会在整数秒数后失效。
- 如果 time 是一个 datatime 或 timedelta,Session 就会在这个时间后失效。
- 如果 time 是 0,用户关闭浏览器 Session 就会失效。
- 如果 time 是 None,Session 会依赖全局 Session 失效策略。

在 Django 框架中,使用如下所示的代码获取 Session 的值。

```
request.session.get('key', None)
```

在 Django 框架中,使用如下所示的代码删除 Session 的值。

```
del request.session['key']
```

在 Django 框架中,使用如下所示的代码判断是否在 Session 中存储了某个数据。

```
'fav_color' in request.session
```

在 Django 框架中,使用如下所示的代码获取所有 Session 的 key、value 和 item 的值。

```
request.session.keys()
request.session.values()
request.session.items()
```

在 Django 框架中,可以在配置文件 settings.py 中设置项目的 Session。

```
SESSION_COOKIE_AGE = 60 * 30                #表示 30*60 秒,即 30 分钟
SESSION_EXPIRE_AT_BROWSER_CLOSE = True
```

如果设置 SESSION_EXPIRE_AT_BROWSER_CLOSE 的值为 False,表示会话 Session 可以在用户浏览器中保持有效期。如果设置为 True,表示关闭浏览器则 Session 失效。

例如下面是在 Django 中通过使用 Session 来判断用户是否已登录的演示代码。

```
# 如果登录成功,设置 session
def login(request):
    if request.method == 'POST':
        form = LoginForm(request.POST)

        if form.is_valid():
            username = form.cleaned_data['username']
            password = form.cleaned_data['password']

            user = User.objects.filter(username__exact=username,
password__exact=password)
```

```python
        if user:
            # 将 username 写入 session,存入服务器
            request.session['username'] = username
            return HttpResponseRedirect('/index/')
        else:
            return HttpResponseRedirect('/login/')
    else:
        form = LoginForm()

    return render(request, 'users/login.html', {'form': form})


# 通过 session 判断用户是否已登录
def index(request):

    # 获取 session 中的 username
    username = request.session.get('username', '')
    if not username:
        return HttpResponseRedirect('/login/')
    return render(request, 'index.html', {'username': username})
```

下面是通过 Session 控制不让用户连续评论两次的演示代码。在现实应用中，还可以通过 Session 来控制用户登录时间、单位时间内连续输错密码次数等。

```python
from django.http import HttpResponse


def post_comment(request, new_comment):
    if request.session.get('has_commented', False):
        return HttpResponse("You've already commented.")
    c = comments.Comment(comment=new_comment)
    c.save()
    request.session['has_commented'] = True
    return HttpResponse('Thanks for your comment!')
```

在下面的实例代码中，演示了在 Django Web 程序中使用 Session 存储登录数据的过程。

源码路径：daima\8\test_session\

（1）通过如下命令新建一个名为"test_session"的工程，然后定位到工程根目录，新建一个名为"online"的 app。

```
django-admin.py startproject test_session
cd test_session
python manage.py startapp online
```

（2）在设置文件 settings.py 的 INSTALLED_APPS 中加入 online，并将 MIDDLEWARE 里面的"django.middleware.csrf.CsrfViewMiddlewar"注释掉。最终代码如下所示。

```
INSTALLED_APPS = [
    'django.contrib.admin',
    'django.contrib.auth',
```

```
    'django.contrib.contenttypes',
    'django.contrib.sessions',
    'django.contrib.messages',
    'django.contrib.staticfiles',
    'online',
]

MIDDLEWARE = [
    'django.middleware.security.SecurityMiddleware',
    'django.contrib.sessions.middleware.SessionMiddleware',
    'django.middleware.common.CommonMiddleware',
    #'django.middleware.csrf.CsrfViewMiddleware',
    'django.contrib.auth.middleware.AuthenticationMiddleware',
    'django.contrib.messages.middleware.MessageMiddleware',
    'django.middleware.clickjacking.XFrameOptionsMiddleware',
]
```

(3) 编写路径导航文件 urls.py，分别设置登录表单、主页和退出 3 个页面的导航链接，主要实现代码如下所示。

```
from online import views
urlpatterns = [
    path('login/', views.login),
    path('index/', views.index),
    path('logout/', views.logout),
]
```

(4) 视图文件 online/views.py 的主要实现代码如下所示。

```
class UserForm(forms.Form):
    username = forms.CharField()
    password = forms.CharField()


# 用户登录
def login(req):
    if req.method == "POST":
        uf = UserForm(req.POST)
        if uf.is_valid():
            username = uf.cleaned_data['username']
            password = uf.cleaned_data['password']
            # 把获取表单的用户名传递给 session 对象
            req.session['username'] = username
            req.session['password'] = password

            return HttpResponseRedirect('/index/')
    else:
        uf = UserForm()
    return render_to_response('login.html', {'uf': uf})


# 登录之后跳转页
def index(req):
    username = req.session.get('username', 'anybody')
    password = req.session.get('password', '')
```

```
        return render_to_response('index.html', {'username': username})

# 注销动作
def logout(req):
    del req.session['username']    # 删除session
    del req.session['password']
    return HttpResponse('logout ok!')
```

(5) 登录表单页面的模板文件是 login.html，具体实现代码如下所示。

```
<form method = 'post'>
{{uf.as_p}}
<input type="submit" value = "ok"/>
</form>
```

(6) 系统主页的模板文件是 index.html，具体实现代码如下所示。

```
<div>
  <h1>welcome {{username}}</h1>
  <a href="/logout">logout</a>
</div>
```

执行后登录表单界面效果如图 8-2 所示。随便输入用户名和密码，例如分别输入用户名"admin"和密码"admin"，单击 ok 按钮后会进入系统主页，效果如图 8-3 所示。其中 welcome 后面的"admin"是用 Session 存储的。

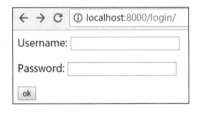

图 8-2　登录表单

图 8-3　系统主页

8.3 点击劫持保护

Django 的点击劫持中间件和装饰器，为开发者提供了简捷易用的、对点击劫持的保护。在本节的内容中，详细讲解 Django 实现点击劫持保护的方法。

扫码观看本节视频讲解

8.3.1 点击劫持的例子

假设现在有一个在线商店页面，已登录的用户可以点击"现在购买"按钮来购买一个商品。会员用户为了方便，可以一直保持在这个商店网站中的登录状态。一个攻击者可能在他们自己的页面上创建一个"我喜欢 Ponies"的按钮，并且在一个透明的 iframe 中加载这个商店的页面，把"现在购买"的按钮隐藏起来覆盖在"我喜欢 Ponies"上。如果用户访问了攻击者的站点，点击"我喜欢 Ponies"按钮会触发对"现在购买"按钮的无意识的点击，不知不觉中购买了商品。

在当前市面的浏览器产品中，基本上都遵循 X-Frame-Options 协议头，它表明一个资源是否允许加载到 frame 或者 iframe 中。如果响应包含值为 SAMEORIGIN 的协议头，浏览器会在 frame 中只加载同源请求的资源。如果协议头设置为 DENY，浏览器会在加载 frame 时屏蔽所有资源，无论请求来自于哪个站点。

Django 提供了一些简单的方法，在我们的站点响应中包含这个协议头：

- 一个简单的中间件，在所有响应中设置协议头。
- 一系列的视图装饰器，可以用于覆盖中间件，或者只用于设置指定视图的协议头。

8.3.2 使用 X-Frame-Options

在 Django 中为了实现点击劫持保护，建议为所有响应设置 X-Frame-Options。

(1) 为所有的响应设置 X-Frame-Options。

要为站点中的所有响应设置相同的 X-Frame-Options 值，需要将'django.middleware.clickjacking.XFrameOptionsMiddleware'设置为 MIDDLEWARE_CLASSES：

```
MIDDLEWARE_CLASSES = (
    ...
    'django.middleware.clickjacking.XFrameOptionsMiddleware',
    ...
)
```

中间件 X-Frame-Options 可以在使用 startproject 生成的设置文件中开启。在通常情况下，这个中间件会为任何开放的 HttpResponse 设置 X-Frame-Options 协议头为 SAMEORIGIN。如果你想用 DENY 来替代它，需要设置 X_FRAME_OPTIONS：

```
X_FRAME_OPTIONS = 'DENY'
```

在使用这个中间件时可能需要配置一些视图，而并不需要为它设置 X-Frame-Options 协议头。对于这些情况，可以使用一个视图装饰器来告诉中间件不必设置协议头：

```
from django.http import HttpResponse
from django.views.decorators.clickjacking import xframe_options_exempt

@xframe_options_exempt
def ok_to_load_in_a_frame(request):
    return HttpResponse("This page is safe to load in a frame on any site.")
```

(2) 为每个视图设置 X-Frame-Options。

在 Django 框架中，提供了以下装饰器来为每个基础视图设置 X-Frame-Options 协议头。

```
from django.http import HttpResponse
from django.views.decorators.clickjacking import xframe_options_deny
from django.views.decorators.clickjacking import xframe_options_sameorigin

@xframe_options_deny
def view_one(request):
```

```
        return HttpResponse("I won't display in any frame!")

@xframe_options_sameorigin
def view_two(request):
    return HttpResponse("Display in a frame if it's from the same origin as me.")
```

注意，我们可以使用装饰器来覆盖中间件，这样可以更加方便地编写程序。另外，X-Frame-Options 协议头只在现代浏览器中保护点击劫持。老式的浏览器会忽视这个协议头，因而需要其他点击劫持防范技巧。下面是支持 X-Frame-Options 的浏览器。

- Internet Explorer 8+。
- Firefox 3.6.9+。
- Opera 10.5+。
- Safari 4+。
- Chrome 4.1 及更高版本。

8.4 跨站请求伪造保护

在 Django 项目中，可以使用内置的 CSRF 中间件和模板标签实现对跨站请求伪造的防护。防护 CSRF 攻击的第一道防线是保证 GET 请求不会产生副作用，例如对于 POST、PUT 和 DELETE 不安全的请求方法，可以通过以下步骤进行防护。

扫码观看本节视频讲解

8.4.1 在 Django 中使用 CSRF 防护的方法

要想在 Django 视图中使用 CSRF 防护，可按照以下步骤操作。

（1）在 MIDDLEWARE_CLASSES 设置中默认启用 CSRF 中间件，如果要覆盖这个设置，请记住 'django.middleware.csrf.CsrfViewMiddleware' 应该位于其他任何假设 CSRF 已经处理过的视图中间件之前。如果关闭了 MIDDLEWARE_CLASSES，也可以在想要保护的视图上使用 csrf_protect()。

（2）在使用 POST 表单的模板中，对于内部的 URL 访问，需要在<form>元素中使用 csrf_token 标签修饰：

```
<form action="." method="post">
```

然后手工导入并使用处理器来生成 CSRF token，并将它添加到模板上下文中：

```
from django.shortcuts import render_to_response
from django.template.context_processors import csrf

def my_view(request):
    c = {}
    c.update(csrf(request))
    return render_to_response("a_template.html", c)
```

不能使用 csrf_token 标签修饰目标是外部 URL 的 POST 表单,因为这将引起 CSRF 信息泄露而导致出现漏洞。

(3) 在对应的视图函数中,确保使用 'django.template.context_processors.csrf' Context 处理器。通常可以使用如下两种方法实现。

- 默认情况下,Django 采用参数 TEMPLATE_CONTEXT_PROCESSORS 指定默认处理器,意味着只要是调用的 RequestContext,那么默认处理器中返回的对象都就将存储在 context 中。
- 手工导入并使用处理器来生成 CSRF token(令牌),并将它添加到模板上下文中。

例如:

```
from django.shortcuts import render_to_response
from django.template.context_processors import csrf

def my_view(request):
    c = {}
    c.update(csrf(request))
    # ... view code here
    return render_to_response("a_template.html", c)
# 可以编写你自己的 render_to_response() 来处理这个步骤
```

8.4.2 装饰器方法

我们可以对需要保护的特定视图使用 csrf_protect 装饰器,而不是添加 CsrfViewMiddleware 对整个 Web 项目实现整体保护。对于在输出中插入 CSRF 令牌的视图以及接受 POST 表单数据的视图,都需要添加装饰器。(这些通常是相同的视图函数,但不总是相同)。

> **注 意**
> 在 Python 开发中,通常不推荐单独使用装饰器,这样很容易由于忘记使用而造成安全隐患。

在 Django 中使用 csrf_protect 装饰器的格式是:

```
csrf_protect(view)
```

csrf_protect 装饰器为视图提供了 CsrfViewMiddleware 保护,例如下面的用法:

```
from django.views.decorators.csrf import csrf_protect
from django.shortcuts import render

@csrf_protect
def my_view(request):
    c = {}
    # ...
    return render(request, "a_template.html", c)
```

> **注 意**
>
> 在默认情况下，如果传入请求未能通过 CsrfViewMiddleware 的验证，则会向用户发送 "403 禁止" 的响应。这通常只有在有一个真正的跨站点请求伪造，或由于编程错误，CSRF 令牌未包括在 POST 表单中的情况下才会看到。

8.4.3 实战演练：求和计时器

在下面的实例文件中，演示了使用 Django 开发一个求和计时器的过程。为了提高系统的安全性，使用内置的 CSRF 防护策略保护了站点安全。

 源码路径：**daima\8\Django_Csrf**

(1) 创建一个名为 Django_Csrf 的工程，然后在里面创建一个名字为 calc 的 app，然后在配置文件 settings.py 中引入 Django 的内置 Csrf 中间件。代码如下：

```
MIDDLEWARE = [
    'django.middleware.security.SecurityMiddleware',
    'django.contrib.sessions.middleware.SessionMiddleware',
    'django.middleware.common.CommonMiddleware',
    'django.middleware.csrf.CsrfViewMiddleware',
    'django.contrib.auth.middleware.AuthenticationMiddleware',
    'django.contrib.messages.middleware.MessageMiddleware',
    'django.middleware.clickjacking.XFrameOptionsMiddleware',
]
```

(2) app 中 URL 路径导航文件 urls.py 的代码如下：

```
from django.urls import path
from . import views

urlpatterns = [
    path('', views.index, name="index"),
    path('add', views.add, name="add")
]
```

(3) 编写视图文件 views.py，分别实现系统主页和求和运算功能对应的视图函数。代码如下：

```
def index(request):
    return render(request, 'home.html')

def add(request):
    num1 = request.POST["num1"]
    num2 = request.POST["num2"]
    res = int(num1) + int(num2)

    return render(request, 'results.html', {'results':res})
```

(4) 系统主页对应的模板文件是 home.html，在里面提供了两个输入数字的表单，在提交表单数据时运用了 Django 的内置 Csrf 保护机制。代码如下：

```html
<!DOCTYPE html>
<html lang="en">
<head>
    <meta charset="UTF-8">
    <meta name="viewport" content="width=device-width, initial-scale=1.0">
    <meta http-equiv="X-UA-Compatible" content="ie=edge">
    <title>Home page</title>
</head>
<body>
    <h1>Home PAge</h1>
    <form action="add" method="POST">
        {% csrf_token %}
        输入一个数字<input type="text" name="num1">
        输入一个数字<input type="text" name="num2">
        <button type="submit">求和</button>
    </form>
</body>
</html>
```

编写计算结果模板文件 results.html，能够计算在两个表单中输入数字的和，代码如下：

```html
<!DOCTYPE html>
<html lang="en">
<head>
    <meta charset="UTF-8">
    <meta name="viewport" content="width=device-width, initial-scale=1.0">
    <meta http-equiv="X-UA-Compatible" content="ie=edge">
    <title>Document</title>
</head>
<body>
    结果是:{{results}}
</body>
</html>
```

执行程序后可以计算两个表单中数字的和，如图 8-4 所示。如果在模板文件 home.html 删除 Csrf 标签{% csrf_token %}，则会出现验证错误，如图 8-5 所示。

图 8-4 计算两个数字的和

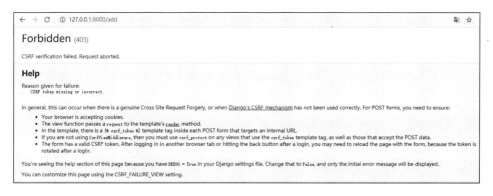

图 8-5 验证错误

8.4.4 实战演练：每日任务管理器

在下面的实例文件中，演示了使用 Django 开发一个每日任务管理器的过程。为了提高系统的安全性，使用内置的 CSRF 防护策略保护了站点安全。

> 源码路径：**daima\8\Django-Todo**

（1）创建一个名为 Todo 的工程，然后在里面创建一个名字为"tasks"的 app，然后在配置文件 settings.py 中引入 Django 的内置 Csrf 中间件。代码如下：

```python
MIDDLEWARE = [
    'django.middleware.security.SecurityMiddleware',
    'django.contrib.sessions.middleware.SessionMiddleware',
    'django.middleware.common.CommonMiddleware',
    'django.middleware.csrf.CsrfViewMiddleware',
    'django.contrib.auth.middleware.AuthenticationMiddleware',
    'django.contrib.messages.middleware.MessageMiddleware',
    'django.middleware.clickjacking.XFrameOptionsMiddleware',
]
```

（2）在 Django 工程的根目录中，URL 路径导航文件 urls.py 的代码如下：

```python
from django.db import models

class Task(models.Model):
    title = models.CharField(max_length=200)
    complete = models.BooleanField(default=False)
    created = models.DateTimeField(auto_now_add=True)

    def __str__(self):
        return self.title
```

在 app 目录下，URL 路径导航文件 urls.py 的代码如下：

```python
from django.urls import path
from . import views

urlpatterns = [
    path('', views.index, name='list'),
    path('update_task/<str:pk>/', views.updateTask, name='update_task'),
    path('delete/<str:pk>/', views.deleteTask, name='delete'),
]
```

（3）编写数据库模型文件 models.py，通过模型类 Task 表示系统中的所有任务信息。代码如下：

```python
from django.db import models

class Task(models.Model):
    title = models.CharField(max_length=200)
    complete = models.BooleanField(default=False)
    created = models.DateTimeField(auto_now_add=True)
```

```
def __str__(self):
    return self.title
```

(4) 编写视图文件 views.py，分别为三个页面创建对应的视图函数。
- index(request)：系统主页视图函数，功能是列表显示数据库中的任务信息。
- updateTask()：修改任务页面视图函数，功能是修改系统中已经存在的某条任务信息。
- deleteTask()：删除任务页面视图函数，功能是删除系统中已经存在的某条任务信息。

视图文件 views.py 的具体实现代码如下所示。

```
# 任务列表视图
def index(request):
    tasks = Task.objects.all()

    form = TaskForm()

    if request.method == 'POST':
        form = TaskForm(request.POST)
        if form.is_valid():
            form.save()
        return redirect('/')

    context = {'tasks':tasks, 'form':form}
    return render(request, 'tasks/list.html', context)

#修改任务视图
def updateTask(request, pk):
    task = Task.objects.get(id=pk)

    form = TaskForm(instance=task)

    if request.method == 'POST':
        form = TaskForm(request.POST, instance=task)
        if form.is_valid():
            form.save()
            return redirect('/')

    context = {'form':form}

    return render(request, 'tasks/update_task.html', context)

#删除任务视图
def deleteTask(request, pk):
    item = Task.objects.get(id=pk)

    if request.method == 'POST':
        item.delete()
        return redirect('/')
```

```
        context = {'item':item}
        return render(request, 'tasks/delete.html', context)
```

(5) 系统主页对应的模板文件是 list.html，功能是列表显示系统中已经存在的任务，为每个人物的签名都提供了"修改"和"删除"按钮。在用户单击"修改"和"删除"按钮时，使用 Django 内置的 CSRF 保护机制验证用户身份的合法性。代码如下：

```
<div class="main">
   <form method="POST" action="/">
      {% csrf_token %}
      {{form.title}}
      <input class="btn btn-info" type="submit" name="Create Task">
   </form>

   <div>
      {% for task in tasks %}
         <div class="item-row">
            <a class="btn btn-sm btn-info" href="{% url 'update_task' task.id %}">修改</a>
            <a class="btn btn-sm btn-danger" href="{% url 'delete' task.id %}">删除</a>

            {% if task.complete == True %}
            <strike>{{task}}</strike>
            {% else %}
            <span>{{task}}</span>
            {% endif %}
         </div>
      {% endfor %}
   </div>
</div>
```

任务修改页面对应的模板文件是 update_task.html，在修改时使用 Django 内置的 CSRF 保护机制验证用户身份的合法性。代码如下：

```
<h3>修改任务</h3>

<form method="POST" action="">
   {% csrf_token %}

   {{form}}
   <input type="submit" name="Update Task">
</form>
```

删除任务页面对应的模板文件是 delete.html，在删除时使用 Django 内置的 CSRF 保护机制验证用户身份的合法性。代码如下：

```
<p>你确定删除这个任务吗？"{{item}}"</p>
<a href="{% url 'list' %}">取消</a>

<form method="POST" action="">
   {% csrf_token %}
   <input type="submit" name="confirm">
</form>
```

执行后系统首页的运行效果如图 8-6 所示。

图 8-6　执行效果

8.5　加密签名

Web 应用安全的黄金法则是，永远不要相信来自不可信来源的数据。为了提高传递数据的安全性，用密码签名后的数据可以通过不受信任的途径进行传递，这样的传递是安全的，因为任何对传递数据的篡改都会被检测到。

在 Django 系统中提供了用于签名的底层 API，以及用于设置和读取被签名 Cookie 的上层 API，它们是 Web 应用中最常使用的签名工具之一。通过使用加密签名，可以完成很多有意义的事情：

扫码观看本节视频讲解

- 生成用于"重置我的账户"的 URL，并发送给丢失密码的用户。
- 确保存储在隐藏表单字段的数据不被篡改。
- 生成一次性的秘密 URL，用于暂时性允许访问受保护的资源，例如用户付费的下载文件。

当使用 startproject 创建新的 Django 项目时，自动生成的 settings.py 文件会得到一个随机的 SECRET_KEY 值。这个值是保护签名数据的密钥，至关重要，我们必须妥善保管，否则攻击者会使用它来生成自己的签名值。

1. 使用底层 API

Django 内置的签名方法存放于 django.core.signing 模块中，首先创建一个 Signer 实例用于对一个值进行签名。代码如下：

```
>>> from django.core.signing import Signer
>>> signer = Signer()
```

```
>>> value = signer.sign('My string')
>>> value
'My string:GdMGD6HNQ_qdgxYP8yBZAdAIV1w'
```

这个签名会附加到字符串末尾，紧跟在冒号后面。我们可以使用方法 unsign() 来获取这个字符串的原始值。代码如下：

```
>>> original = signer.unsign(value)
>>> original
'My string'
```

如果签名或者值以任何方式发生改变，则会抛出 django.core.signing.BadSignature 异常：

```
>>> from django.core import signing
>>> value += 'm'
>>> try:
...     original = signer.unsign(value)
... except signing.BadSignature:
...     print("Tampering detected!")
```

通常，签名类 Signer 使用 SECRET_KEY 来生成签名，我们可以通过向 Signer 构造器传递一个不同的密钥的方法来使用它：

```
>>> signer = Signer('my-other-secret')
>>> value = signer.sign('My string')
>>> value
'My string:EkfQJafvGyiofrdGnuthdxImIJw'
class Signer(key=None, sep=':', salt=None) [source]
```

这样会返回一个 signer，它使用 key 来生成签名，并且使用 sep 来分割值。sep 不能是 UR 安全的 base64 字母表中的字符。在字母表含有数字、字母、连字符和下划线。

2. 使用 salt 参数

如果不希望对每个特定的字符串都生成一个相同的签名散列值，那么可以在类 Signer 中使用可选的参数 salt。在使用 salt 参数时，会同时用它和 SECRET_KEY 初始化签名散列函数：

```
>>> signer = Signer()
>>> signer.sign('My string')
'My string:GdMGD6HNQ_qdgxYP8yBZAdAIV1w'
>>> signer = Signer(salt='extra')
>>> signer.sign('My string')
'My string:Ee7vGi-ING6n02gkcJ-QLHg6vFw'
>>> signer.unsign('My string:Ee7vGi-ING6n02gkcJ-QLHg6vFw')
'My string'
```

当以这种方法使用 salt 时，会把不同的签名放在不同的命名空间中。来自于单一命名空间(一个特定的 salt 值)的签名不能用于在不同的命名空间中验证相同的纯文本字符串。不同的命名空间使用不同的 salt 设置。这是为了防止攻击者使用在一个地方的代码中生成的签名后的字符串，作为使用不同 salt 来生成(和验证)签名的另一处代码的输入。

> **注 意**
>
> 参数 salt 不像 SECRET_KEY，可以不用保密。

3. 验证带有时间戳的值

类 TimestampSigner 是 Signer 的子类，能够向值中附加一个签名后的时间戳。这可以让我们确认一个签名后的值是否在特定时间段之内被创建：

```
>>> from datetime import timedelta
>>> from django.core.signing import TimestampSigner
>>> signer = TimestampSigner()
>>> value = signer.sign('hello')
>>> value
'hello:1NMg5H:oPVuCqlJWmChm1rA2lyTUtelC-c'
>>> signer.unsign(value)
'hello'
>>> signer.unsign(value, max_age=10)
...
SignatureExpired: Signature age 15.5289158821 > 10 seconds
>>> signer.unsign(value, max_age=20)
'hello'
>>> signer.unsign(value, max_age=timedelta(seconds=20))
'hello'
```

- sign(value)[source]：签名 value，并且附加当前的时间戳。
- unsign(value, max_age=None)[source]：检查 value 是否在少于 max_age 秒之前被签名，如果不是则抛出 SignatureExpired 异常。max_age 参数接收一个整数或者 datetime.timedelta 对象。

4. 保护复杂的数据结构

如果你希望保护一个列表、元组或字典，则可以使用签名模块的函数 dumps()和 loads()来实现。它们模仿了 Python 的 pickle 模块，但是在背后使用了 JSON 序列化。JSON 可以确保即使你的 SECRET_KEY 被盗取，攻击者也不能利用 pickle 的格式来执行任意的命令。代码如下：

```
>>> from django.core import signing
>>> value = signing.dumps({"foo": "bar"})
>>> value
'eyJmb28iOiJiYXIifQ:1NMg1b:zGcDE4-TCkaeGzLeW9UQwZesciI'
>>> signing.loads(value)
{'foo': 'bar'}
```

由于 JSON 的本质(列表和元组之间没有原生的区别)，如果你传进来一个元组，则会从 signing.loads(object)得到一个列表：

```
>>> from django.core import signing
>>> value = signing.dumps(('a','b','c'))
>>> signing.loads(value)
['a', 'b', 'c']
```

- dumps(obj, key=None, salt='django.core.signing', compress=False)[source]：返回 URL 安全、带有 SHA-1(是一种密码散列函数)签名的 base64 压缩的 JSON 字符串，序列化的对象使用 TimestampSigner 来签名。
- loads(string, key=None, salt='django.core.signing', max_age=None)[source]：是函数 dumps() 的反转，如果签名失败则会抛出 BadSignature 异常。如果提供了 max_age 则会检查它(以秒为单位)。

8.6 中间件

在 Django 框架中自带了很多实现安全验证功能的中间件组件，通过使用这些中间件可以提高 Web 网站的安全性。在 Django 中常用的中间件如下所示。

扫码观看本节视频讲解

(1) 缓存中间件。

```
class UpdateCacheMiddleware[source]
class FetchFromCacheMiddleware[source]
```

用于开启全站范围的缓存，如果开启了这些缓存，任何一个由 Django 提供的页面将会被缓存，缓存时长在 CACHE_MIDDLEWARE_SECONDS 配置中定义。

(2) "通用"的中间件。

```
class CommonMiddleware[source]
```

- 禁止访问 DISALLOWED_USER_AGENTS 中设置的用户代理，这项配置应该是一个已编译的正则表达式对象的列表。
- 基于 APPEND_SLASH 和 PREPEND_WWW 的设置来重写 URL。
- 如果 APPEND_SLASH 设为 True,并且初始 URL 没有以斜杠结尾以及在 URLconf 中没找到对应定义，那么形成一个以斜杠结尾的新 URL。如果这个新的 URL 存在于 URLconf，那么 Django 重定向请求到这个新 URL 上。否则，按正常情况处理初始的 URL。比如，如果没有为 foo.com/bar 定义有效的正则，但是为 foo.com/bar/ 定义了有效的正则，foo.com/bar 将会被重定向到 foo.com/bar/。如果将 PREPEND_WWW 设置为 True，前面缺少 "www." 的 url 将会被重定向到相同但是以一个 "www." 开头的 url。两种选项都是为了规范化 url。其中的道理就是，任何一个 url 应该在一个地方仅存在一个。从技术上来讲，URL foo.com/bar 区别于 foo.com/bar/ —— 搜索引擎索引会把这两种写法分开处理，因此最佳实践就是规范化 URL。
- 基于 USE_ETAGS 设置来处理 ETag。如果设置 USE_ETAGS 为 True，Django 会通过 MD5-hashing 处理页面的内容来为每一个页面请求计算 ETag，并且如果合适的话，它将会发送携带 Not Modified 的响应。

(3) GZip 中间件。

```
class GZipMiddleware[source]
```

安全研究员最近发现,当压缩技术(包括 GZipMiddleware)用于一个网站的时候,网站会受到一些可能的攻击。此外,这些方法可以用于破坏 Django 的 CSRF 保护。在你的站点使用 GZipMiddleware 之前,你应该先仔细考虑一下你的站点是否容易受到这些攻击。 如果你不确定是否会受到这些影响,应该避免使用 GZipMiddleware。

为了支持 GZip 压缩的浏览器(一些现代的浏览器)压缩内容,建议把这个中间件放到中间件配置列表的第一个,这样压缩响应内容的处理会到最后才发生。如果满足下列条件,内容不会被压缩。

- 消息体的长度小于 200 个字节。
- 响应已经设置了 Content-Encoding 协议头。
- 请求(浏览器)没有发送包含 GZip 的 Accept-Encoding 协议头。
- 你可以通过这个 gzip_page()装饰器使用独立的 GZip 压缩。

(4) 带条件判断的 GET 中间件。

```
class ConditionalGetMiddleware[source]
```

处理带有条件判断状态的 GET 操作,如果一个请求包含 ETag 或者 Last-Modified 协议头,并且请求包含 If-None-Match 或 If-Modified-Since,这时响应会被替换为 HttpResponseNotModified。

(5) 地域性中间件。

```
class LocaleMiddleware[source]
```

基于请求中的数据开启语言选择。它可以为每个用户进行定制。

(6) LocaleMiddleware.response_redirect_class。

默认为 HttpResponseRedirect。继承自 LocaleMiddleware 并覆写了属性来自定义中间件发出的重定向。

(7) 消息中间件。

```
class MessageMiddleware[source]
```

开启基于 Cookie 和会话的消息支持。

(8) 安全中间件。

如果我们的部署环境允许,让我们的前端 Web 服务器展示 SecurityMiddleware 提供的功能是一个好主意。这样一来,如果有任何请求没有被 Django 处理(比如静态媒体或用户上传的文件),它们会拥有和向 Django 应用的请求相同的保护。

(9) 会话中间件。

用于开启会话支持。

```
class SessionMiddleware[source]
```

(10) 站点中间件。

```
class CurrentSiteMiddleware[source]
```

用于向每个接收到的 HttpRequest 对象添加一个 site 属性，表示当前的站点。

(11) 认证中间件。

- class AuthenticationMiddleware[source]：向每个接收到的 HttpRequest 对象添加 user 属性，表示当前登录的用户。
- class RemoteUserMiddleware[source]：使用 Web 服务器提供认证的中间件。
- class SessionAuthenticationMiddleware[source]：当用户修改密码的时候使用户的会话失效。详见密码更改时的会话失效。在 MIDDLEWARE_CLASSES 中，这个中间件必须出现在 django.contrib.auth.middleware.AuthenticationMiddleware 之后。

(12) CSRF 保护中间件。

`class CsrfViewMiddleware[source]`

添加跨站点请求伪造的保护，向 POST 表单添加一个隐藏的表单字段，并检查请求中是否有正确的值。

(13) X-Frame-Options 中间件。

`class XFrameOptionsMiddleware[source]`

通过 X-Frame-Options 协议头进行简单的点击劫持保护。

8.7 实战演练：安全版的仿 CSDN 登录验证系统

在本节的内容中，将创建一个完整的仿 CSDN 的登录验证系统。我们没有使用 Django 内置的登录验证模块，而是使用自定义代码编写登录验证系统，并且使用了密码签名的方式将用户密码和 Cookie 进行加密，提高了系统的安全性。

扫码观看本节视频讲解

 源码路径：daima\8\bbs\

8.7.1 系统设置

在配置文件 settings.py 中首先设置 SECRET_KEY，然后在 MIDDLEWARE 中添加和安全相关的中间件，例如 CsrfViewMiddleware 和 XFrameOptionsMiddleware。代码如下：

```
SECRET_KEY = '3&2xq%nb^+4k%m2rgg-ry4ybh(-o6'
MIDDLEWARE = [
    'django.middleware.security.SecurityMiddleware',
    'django.contrib.sessions.middleware.SessionMiddleware',
    'django.middleware.common.CommonMiddleware',
    'django.middleware.csrf.CsrfViewMiddleware',
    'django.contrib.auth.middleware.AuthenticationMiddleware',
    'django.contrib.messages.middleware.MessageMiddleware',
    'django.middleware.clickjacking.XFrameOptionsMiddleware',
]
```

8.7.2 会员注册和登录验证模块

在本项目的"user"目录中保存了会员注册和登录验证模块的实现代码，具体实现流程如下。

(1) 在模型文件 models.py 中设置和会员用户有关的数据库表。代码如下：

```python
class UserManager(models.Manager):
    def all(self):
        return super().all().filter(is_delete=False)

    def create(self, username, password):
        user = self.model()
        user.username = username
        user.password = make_password(password)
        user.save()
        return user

    # 在这里添加模型管理方法


class User(models.Model):
    username = models.CharField(max_length=16, unique=True)
    password = models.CharField(max_length=256)
    post_count = models.IntegerField(default=0)
    comm_count = models.IntegerField(default=0)
    is_delete = models.BooleanField(default=False)

    objects = UserManager()

    class Meta:
        db_table = 'users'

    # 通过加密算法验证密码
    def valid_password(self, password):
        return check_password(password, self.password)
```

(2) 在文件 urls.py 中设置了相关页面的路径导航。代码如下：

```python
from django.urls import path
from user import views

urlpatterns = [
    path('register/', views.register, name='register'),
    path('register_handler/', views.register_handler,
name='register_handler'),
    path('login/', views.login, name='login'),
    path('login_handler/', views.login_handler, name='login_handler'),
    path('logout/', views.logout, name='logout'),
]
```

(3) 在表单文件 forms.py 中定义了两个类，分别实现新用户注册表单功能和会员登录表单功能。代码如下：

```python
class LoginForm(forms.Form):
    username = forms.CharField(max_length=16,
                        widget=forms.TextInput(attrs={'class': 'form-control', 'placeholder': '请输入用户名'}))
    password = forms.CharField(min_length=6, max_length=32,
                        widget=forms.PasswordInput(attrs={'class': 'form-control', 'placeholder': '请输入密码'}))
    remember_me = forms.BooleanField(required=False)

    def clean(self):
        cleaned_data = self.cleaned_data
        username = self.cleaned_data['username']
        pwd = self.cleaned_data['password']
        self.valid_username(username)
        self.valid_password(pwd)
        return cleaned_data

    def valid_username(self, username):
        try:
            self.user = User.objects.filter(username=username)[0]
        except IndexError:
            raise forms.ValidationError(_('该用户不存在'))

    def valid_password(self, pwd):
        if not self.user.valid_password(pwd):
            raise forms.ValidationError(_('密码错误'))


class RegisterForm(forms.Form):
    username = forms.CharField(max_length=16,
                        widget=forms.TextInput(attrs={'class': 'form-control'}))
    password = forms.CharField(min_length=6, max_length=32,
                        widget=forms.PasswordInput(attrs={'class': 'form-control'}))
    confirm_password = forms.CharField(min_length=6, max_length=32,
                        widget=forms.PasswordInput(attrs={'class': 'form-control'}))

    def clean(self):
        cleaned_data = self.cleaned_data
        username = self.cleaned_data['username']
        pwd1 = self.cleaned_data['password']
        pwd2 = self.cleaned_data['confirm_password']
        self.valid_username(username)
        self.valid_password(pwd1, pwd2)
        return cleaned_data

    def valid_username(self, username):
        if re.findall('[^0-9a-zA-Z_]', username):
            raise forms.ValidationError(_('用户名只允许使用数字字母或下划线'))
        if User.objects.filter(username=username):
```

```python
            raise forms.ValidationError(_('用户名%(username)s 已经被注册了'),
params={'username': username})

    def valid_password(self, pwd1, pwd2):
        if re.findall('[^0-9a-zA-Z_]', pwd1):
            raise forms.ValidationError(_('密码只允许使用数字字母或下划线'))
        if pwd1 != pwd2:
            raise forms.ValidationError(_('两次密码输入不一致'))
```

(4) 在视图文件 views.py 中分别编写视图处理函数，根据获取的注册表单数据实现新用户注册逻辑功能，根据从登录表单获取的表单数据实现登录验证功能。并且为了提高系统的安全性，特意定义了函数 cookie_handler(username)，能够将登录用户的 Cookie 数据进行加密。代码如下：

```python
def register(request):
    form = RegisterForm()
    return render(request, 'register.html', {'form': form})

def register_handler(request):
    if request.method == 'POST':
        form = RegisterForm(request.POST)
        if form.is_valid():
            User.objects.create(form.cleaned_data.get('username'),
form.cleaned_data.get('password'))
            return redirect('login')
        else:
            return render(request, 'register.html', {'form': form})
    raise Http404


def login(request):
    form = LoginForm()
    from_redirect = request.GET.get("from_redirect")
    if from_redirect is not None and from_redirect == "True":
        return render(request, 'login.html', {'form': form, "from_redirect": True})
    else:
        return render(request, 'login.html', {'form': form})


def login_handler(request):
    next_url = request.session.get("next_url")
    if request.method == 'POST':
        form = LoginForm(request.POST)
        if form.is_valid():
            username = form.cleaned_data['username']
            session_id = cookie_handler(username)
            # request.session[session_id] = username
            request.session['username'] = username
            if next_url != None:
                # 如果是被 login_required 拦截，登录后跳转回原来的操作
                # print(next_url)
```

```python
            response = HttpResponseRedirect(next_url)
            response.set_cookie('session_id', session_id)
            return response
        else:
            response = HttpResponseRedirect(reverse('index'))
            response.set_cookie('session_id', session_id)
            return response
    else:
        return render(request, 'login.html', {'form': form})
raise Http404


def logout(request):
    response = HttpResponseRedirect(reverse('index'))
    for key in request.COOKIES:
        response.delete_cookie(key)
    return response


def cookie_handler(username):
    timestamp_signing = signing.TimestampSigner()
    value1 = signing.dumps({"username": username})
    value2 = timestamp_signing.sign(value1)
    return value2
```

（5）编写系统后台文件 admin.py，在后台中注册添加博客信息管理功能。代码如下：

```python
from django.contrib import admin

from .models import User
admin.site.register(User)
```

执行效果如图 8-7 所示。

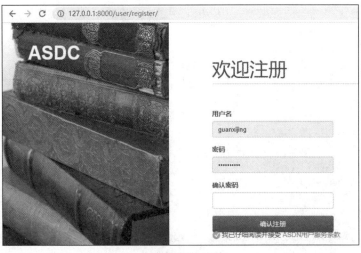

(a) 会员注册页面

图 8-7　执行效果

(b) 登录验证页面

(c) 会员登录成功时的首页效果

(d) 后台中的用户密码是加密的

图 8-7 执行效果(续)

```
Set-Cookie: session_id=eyJ1c2VybmFtZSI6ImdlYW54aWppbmcifQ:1kCPPu:T8ZRJBdaiQx3riKw1b6pXc-9e2qz
Fnxspj3T-qgJMjpQvn3N45-oWpo_3o; Path=/
Set-Cookie: sessionid=yiraj3pqw1d28m7ob59nkouz28dakabc; expires=Sun, 13 Sep 2020 15:38:50 GMT
SameSite=Lax
Vary: Cookie
X-Content-Type-Options: nosniff
X-Frame-Options: DENY
```

(e) 在浏览器中保存的 Cookie(名字是 sessionid)也是加密的

图 8-7　执行效果(续)

8.7.3　博客发布模块

在本项目的"post"目录中保存了博客发布模块的实现代码，具体实现流程如下。

(1) 在模型文件 models.py 中设置和会员用户有关的数据库表。代码如下：

```python
# 帖子管理
class PostManager(models.Manager):
    def all(self):
        return super().all().filter(is_delete=False)

    def create(self, user, title, author, cont_html, cont_str):
        post = self.model()
        post.user = user
        post.author = author
        post.title = title
        post.cont_html = cont_html
        post.cont_str = cont_str
        post.save()
        return post

    # 在这里添加模型管理方法


# 帖子
class Post(models.Model):
    title = models.CharField(max_length=50)
    author = models.CharField(max_length=16)
    cont_html = models.TextField()
    cont_str = models.TextField()
    timestamp = models.DateTimeField(auto_now_add=True)
    view_count = models.IntegerField(default=0)
    like_count = models.IntegerField(default=0)
    coll_count = models.IntegerField(default=0)
    comm_count = models.IntegerField(default=0)
    is_delete = models.BooleanField(default=False)

    user = models.ForeignKey(User,on_delete=models.CASCADE)

    likes = models.ManyToManyField(User, through='Like', related_name='likes')
    colls = models.ManyToManyField(User, through='Collection', related_name='colls')
```

```
    comms = models.ManyToManyField(User, through='Comment',
related_name='comms')

    objects = PostManager()

    class Meta:
        db_table = 'posts'


# 点赞
class Like(models.Model):
    is_list = models.BooleanField(default=True)
    timestamp = models.DateTimeField(auto_now_add=True)
    uid = models.ForeignKey(User,
related_name='like',on_delete=models.CASCADE)
    pid = models.ForeignKey(Post,
related_name='like',on_delete=models.CASCADE)


# 收藏
class Collection(models.Model):
    is_coll = models.BooleanField(default=True)
    timestamp = models.DateTimeField(auto_now_add=True)
    uid = models.ForeignKey(User,
related_name='coll',on_delete=models.CASCADE)
    pid = models.ForeignKey(Post,
related_name='coll',on_delete=models.CASCADE)


# 评论
class Comment(models.Model):
    cont_str = models.CharField(max_length=256)
    is_delete = models.BooleanField(default=False)
    timestamp = models.DateTimeField(auto_now_add=True)
    uid = models.ForeignKey(User,
related_name='comm',on_delete=models.CASCADE)
    pid = models.ForeignKey(Post,
related_name='comm',on_delete=models.CASCADE)
    replys = models.ManyToManyField(User, through='Reply',
related_name='replys')
```

(2) 在文件 urls.py 中设置相关页面的路径导航。代码如下：

```
urlpatterns = [
    url(r'^$', views.index, name='index'),
    url(r'^post_edit$', views.post_edit, name='post_edit'),
    url(r'^post_detail/(\d+)', views.post_detail, name='post_detail'),
    url(r'^search_results/$', views.search_results, name='search_results'),
    url(r'^get_posts/', views.get_posts, name='get_posts'),
]
```

(3) 在表单文件 forms.py 中创建类 PostEditForm，用于获取博客表单中的信息，包括博客标题和内容。代码如下：

```
class PostEditForm(forms.Form):
    post_title = forms.CharField(max_length=50,
```

```python
                            widget=forms.TextInput(attrs={'class':
'form-control', 'placeholder': '标题'}))
    post_content = forms.CharField(widget=forms.Textarea(attrs={'class':
'form-control', 'placeholder': '正文'}))

    def clean(self):
        cleaned_data = self.cleaned_data
        title = self.cleaned_data['post_title']
        cont_str = self.cleaned_data['post_content']
        cont_html = cont_str
        return cleaned_data
```

(4) 在视图文件 views.py 中编写视图处理函数，首先判断用户是否登录系统，如果没有登录则不能发布博客；如果已经登录，则根据获取的表单数据实现博客发布功能。并且为了提高系统的安全性，特意使用底层 API 中的 signing 将用户的登录信息进行加密。代码如下：

```python
# 登录认证装饰器
def login_required(view_fun):
    def valid_cookie_and_session(request):
        try:
            next_url = request.path_info
            request.session['next_url'] = next_url  # 直接写入 next 中

            session_id = request.COOKIES.get('session_id')
            username_session = request.session.get('username')
            timestamp_signing = signing.TimestampSigner()
            result = timestamp_signing.unsign(session_id, max_age=60 * 60 * 24)
            username = signing.loads(result)['username']  # 解析 cookie 中的用户名
            print(next_url)
            if username and username_session == username:
                return view_fun(request)
            else:
                # return redirect('login')
                return HttpResponseRedirect('/user/login/?from_redirect=True')
        except Exception as e:
            print(e)
            return HttpResponseRedirect('/user/login/?from_redirect=True')

    return valid_cookie_and_session

# 将经过加密的 session_id 解密为 username
def get_username(request):
    session_id = request.COOKIES.get('session_id', 0)
    if not session_id:
        return session_id
    timestamp_signing = signing.TimestampSigner()
    result = timestamp_signing.unsign(session_id, max_age=60 * 60 * 24)
    username = signing.loads(result)['username']
    return username


# @login_required
def index(request, **kwargs):
```

```python
    username = get_username(request)
    posts = Post.objects.all().order_by('-timestamp')[:20]
    if kwargs != None and "from_redirect" in kwargs.keys():
        return render(request, 'index.html', {'username': username, 'posts': posts, 'from_redirect': True})  # 跳转过来的
    else:
        return render(request, 'index.html', {'username': username, 'posts': posts, 'from_redirect': False})  # 跳转过来的


# 登录以后才可以发表帖子
@login_required
def post_edit(request):
    username = get_username(request)
    form = PostEditForm()

    if request.method == 'POST':
        form = PostEditForm(request.POST)
        if form.is_valid():
            user = User.objects.filter(username=username)[0]
            title = form.cleaned_data['post_title']
            cont_str = form.cleaned_data['post_content']
            cont_html = cont_str
            post = Post.objects.create(user=user, author=username, title=title, cont_str=cont_str, cont_html=cont_html)
            return redirect(reverse('post_detail', args=(post.id,)))
    return render(request, 'post_edit.html', {'form': form, 'username': username})


def post_detail(request, pid):
    post = get_object_or_404(Post, pk=pid)
    username = get_username(request)
    return render(request, "post_detail.html", {'post': post, 'username': username})


def search_results(request):
    username = get_username(request)
    return render(request, 'search_results.html', {'username': username})


# js异步获取帖子列表
def get_posts(request):
    if request.method == 'GET':
        offset = int(request.GET.get('offset'))
        start = int(request.GET.get('start'))
        queryset = Post.objects.order_by('-timestamp').filter(pk__in=range(start, start + offset))
        posts = serialize('json', queryset)  # 将查询集序列化成json
        response = JsonResponse({'data': posts})
        return response
    raise Http404
```

(5) 编写系统后台文件 admin.py，在后台中注册添加博客信息管理功能。代码如下：

```python
from django.contrib import admin
from .models import Post

class PostAdmin(admin.ModelAdmin):
    fields = ['title', 'author']

admin.site.register(Post, PostAdmin)
```

执行效果如图 8-8 所示。

图 8-8　只有登录后才能写博客

第 9 章

站 点 管 理

在开发 Web 网站的过程中，开发者不但需要为网站的功能而工作，而且还需要为整个网站的架构和运营而工作，例如实现 Web 国际化和站点地图等功能。在本章的内容中，将详细讲解在 Django Web 项目中实现站点管理的知识，为读者步入本书后面知识的学习打下基础。

9.1 Django Web 国际化

国际化和本地化的目标是让同一个 Web 网站为不同国家和地区的用户提供定制化的语言和格式服务。国际化是一个设计过程，能够确保我们的产品(通常是软件应用程序)可以适应各种语言和地区，而不需要对源代码进行工程更改。Django 内置了 Web 国际化和本地化功能，完美支持文本翻译、日期格式化、时间格式化、数字格式化以及时区设置功能。在本节的内容中，将讲解 Django 国际化的知识。

扫码观看本节视频讲解

9.1.1 Django 中 Python 程序的国际化

为了使 Django 项目支持翻译，需要在 Python 代码和模板中添加少量的钩子。这些钩子被称为 Translation Strings，能够告知 Django：如果这个文本有对应的翻译，那么应该根据用户的国别使用对应的翻译。标记字符串是程序员的职责，系统只会翻译它知道的字符串。

在 Django 中提供了一些工具，可以提取翻译字符串到消息文件中，这个文件方便翻译人员提供翻译字符串的目标语言。翻译人员填充完消息文件后，必须编译它。这个过程依赖 GNU gettext 工具集。当完成这些事情之后，Django 将负责根据用户的语言偏好将网页翻译成对应的语言。

Django 的国际化钩子默认是打开的，这表示在框架中的某些地方已经有相关的 I18N 钩子。如果不需要使用国际化，可以在设置文件中设 USE_I18N = False。这样 Django 将做一些优化，而不加载国际化的机制。

1. 标准翻译

在 Python 语言中，使用内置函数 ugettext()指定标准的翻译，这在习惯上通常会将它导入成一个别名"_"以节省打字。

我们需要注意的是，在 Python 的标准库模块 gettext 中，将_() 安装进全局命名空间中，并作为 gettext() 的别名。在 Django 中，我们不使用这种习惯做法，主要原因有两点：

(1) 有时对于特定文件来说，应该使用函数 gettext_lazy()作为默认翻译方法。如果在全局命名空间里没有_()，开发者必须考虑哪个是最合适的翻译函数。

(2) 下划线("_")用于表示在 Python 的交互式终端和 doctest 测试中的"上一个结果"。安装全局_()函数会引发冲突。此时如果导入 gettext()替换_()则可以避免这个问题。

例如在下面的示例中，将文本"Welcome to my site." 标记为一个翻译字符串：

```
from django.utils.translation import ugettext as _
from django.http import HttpResponse

def my_view(request):
    output = _("Welcome to my site.")
    return HttpResponse(output)
```

很明显，我们可以不用别名的方式来编写这段代码。例如下面代码的功能与上面的完全一样：

```
from django.utils.translation import ugettext
from django.http import HttpResponse

def my_view(request):
    output = ugettext("Welcome to my site.")
    return HttpResponse(output)
```

具体的翻译工作也可以在计算生成的值上进行，例如下面代码的功能与前面两个完全一样：

```
def my_view(request):
    words = ['Welcome', 'to', 'my', 'site.']
    output = _(' '.join(words))
    return HttpResponse(output)
```

具体的翻译工作也可以在变量上进行，例如下面代码的功能与前面的完全一样：

```
def my_view(request):
    sentence = 'Welcome to my site.'
    output = _(sentence)
    return HttpResponse(output)
```

在上面的例子中，当使用变量或计算值时会产生警告：Django 的翻译字符检测实用程序 django-admin makemessages 不能找到这些字符串。传递给_() 或 ugettext()的字符串，可以通过 Python 标准的命名字符串插值语法接收占位符。例如：

```
def my_view(request, m, d):
    output = _('Today is %(month)s %(day)s.') % {'month': m, 'day': d}
    return HttpResponse(output)
```

这种技术可以让语言相关的翻译重新排序，例如，英语翻译可能是"Today is November 26."，而西班牙语可能是"Hoy es 26 de Noviembre." —— 月份和天数的占位符交换位置了。由于这个原因，每当有多个参数的时候，都应该使用命名的字符串插值(例如，%(day)s)而不是位置插值(例如，%s 或%d)。如果使用位置插值，翻译将不能重新排序占位符。

2. 标记不用翻译的字符

通过使用 django.utils.translation.gettext_noop()，可以将字符串标记为不用翻译的翻译字符串，这个字符会稍后使用变量来翻译。使用这个方法的场景是：如果有一个常量字符串，该字符串以源语言存储，它们通过系统或用户进行交换(比如数据库里的字符串)，但最后应该在某个可能的时间点进行翻译，比如当字符串展示给某个用户时进行翻译。

9.1.2　Django 中模板的国际化

在 Django 模板中提供了两个模板标签实现翻译功能，其语法与在 Python 代码中使用的语法有稍许不同。为了让模板能够访问这些标签,需要将{% load i18n %}放置在模板的顶部。

和所有模板标签一样,需要在所有使用翻译的模板中加载这个标签,甚至从其他模板继承过来的模板也要继承 i18n 标签。

1. 模板标签 trans

在 Django 模板中,标签{% trans %}能够翻译一个常量字符串(位于单引号或双引号中)或变量:

```
<title>{% trans "This is the title." %}</title>
<title>{% trans myvar %}</title>
```

如果使用了 noop 选项,则变量的查找工作仍然继续,但是会忽略翻译功能。这对于需要在未来进行翻译的内容非常有用,例如:

```
<title>{% trans "myvar" noop %}</title>
```

在内部,使用函数 ugettext()调用内联的翻译。

当将一个模板变量(上面的 myvar)传递给该标签时,该标签首先会在运行时将变量解析成一个字符串,然后在消息目录中查找这个字符串。

标签{% trans %} 不可以将模板标签嵌入字符串中,如果我们的翻译字符串需要带有变量(占位符),可以使用{% blocktrans %}来实现。

如果想检索翻译的字符串而不显示它,可以使用下面的语法实现:

```
{% trans "This is the title" as the_title %}
<title>{{ the_title }}</title>
<meta name="description" content="{{ the_title }}">
```

在实际应用中,可以使用这个办法来获取一个在模板中多处使用的字符串,或将输出作为其他模板标签或过滤器的参数:

```
{% trans "starting point" as start %}
{% trans "end point" as end %}
{% trans "La Grande Boucle" as race %}

<h1>
  <a href="/" title="{% blocktrans %}Back to '{{ race }}' homepage{% endblocktrans %}">{{ race }}</a>
</h1>
<p>
{% for stage in tour_stages %}
    {% cycle start end %}: {{ stage }}{% if forloop.counter|divisibleby:2 %}<br />{% else %}, {% endif %}
{% endfor %}
</p>
```

通过使用关键字 context,{% trans %} 还支持 contextual markers(上下文标记)功能:

```
{% trans "May" context "month name" %}
```

2. 模板标签 blocktrans

标签 blocktrans 允许使用占位符标记由文字和可变内容组成的复杂句子,这些复杂的句子可以被国际化识别并翻译处理：

```
{% blocktrans %}This string will have {{ value }} inside.{% endblocktrans %}
```

要翻译模板表达式，例如访问对象属性或使用模板过滤器，需要将表达式绑定到本地变量，这样可以在翻译块中使用。例如：

```
{% blocktrans with amount=article.price %}
That will cost $ {{ amount }}.
{% endblocktrans %}

{% blocktrans with myvar=value|filter %}
This will have {{ myvar }} inside.
{% endblocktrans %}
```

可以在单个 blocktrans 标记中使用多个表达式：

```
{% blocktrans with book_t=book|title author_t=author|title %}
This is {{ book_t }} by {{ author_t }}
{% endblocktrans %}
```

3. 模板标签 language

通过使用标签 language 可以在模板中切换语言，例如：

```
{% load i18n %}

{% get_current_language as LANGUAGE_CODE %}
<!-- Current language: {{ LANGUAGE_CODE }} -->
<p>{% trans "Welcome to our page" %}</p>

{% language 'en' %}
   {% get_current_language as LANGUAGE_CODE %}
   <!-- Current language: {{ LANGUAGE_CODE }} -->
   <p>{% trans "Welcome to our page" %}</p>
{% endlanguage %}
```

当第一次执行这个模板时，使用当前语言显示文本，在第二次执行时将始终用英语显示文本"Welcome to our page"。

4. 其他标签

接下来介绍的其他标签需要用到{% load i18n %}。

(1) {get_available_languages。

例如{% get_available_languages as LANGUAGES %}用于返回一个元组列表，其中第一个元素是：term:language code，第二个元素是语言名称(翻译成当前活动的语言环境)。

(2) get_current_language。

例如{% get_current_language as LANGUAGE_CODE %}用于返回字符串类型的当前用户首选语言，类似于 en-us。

(3) get_current_language_bidi。

例如{% get_current_language_bidi as LANGUAGE_BIDI %}用于返回当前语言文字的阅

读方向。如果是 True，则是从右向左阅读的语言，比如希伯来语、阿拉伯语；如果是 False，则是从左向右阅读的语言，比如英语、法语、德语等。

(4) i18n 上下文处理器。

如果启用了 django.template.context_processors.i18n 上下文处理器，则每个 RequestContext 都可以访问上面所定义的 LANGUAGES、LANGUAGE_CODE 和 LANGUAGE_BIDI。

(5) get_language_info。

我们可以查找关于任何使用支持模板标签和过滤器的可用语言的信息，为了得到单一语言的信息，可以使用{% get_language_info %}标签实现：

```
{% get_language_info for LANGUAGE_CODE as lang %}
{% get_language_info for "pl" as lang %}
```

然后，可以访问以下信息：

```
Language code: {{ lang.code }}<br>
Name of language: {{ lang.name_local }}<br>
Name in English: {{ lang.name }}<br>
Bi-directional: {{ lang.bidi }}
Name in the active language: {{ lang.name_translated }}
get_language_info_list
```

我们也可以使用模板标签{% get_language_info_list %}来检索语言列表的信息(例如，LANGUAGES 中指定的有效语言)。

另外，除了元组的 LANGUAGES 样式列表之外，{% get_language_info_list %} 也支持语言代码列表。如果在视图中这么做：

```
context = {'available_languages': ['en', 'es', 'fr']}
return render(request, 'mytemplate.html', context)
```

那么可以在模板中遍历这些语言：

```
{% get_language_info_list for available_languages as langs %}
{% for lang in langs %} ... {% endfor %}
```

9.1.3 Django 中 URL 模式的国际化

在 Django Web 程序中，可通过如下两种方式来国际化 URL 模式。

(1) 将语言前缀添加到 URL 模式的根，使 LocaleMiddleware 从请求的 URL 中检测要激活的语言。

(2) 通过函数 django.utils.translation.gettext_lazy()使得 URL 模式的本身可翻译。

> **注意**
> 要想使用上述特性，需要在 MIDDLEWARE 设置中包含 django.middleware.locale.LocaleMiddleware。

1. URL 模式中的语言前缀

在 Django Web 项目中，可以在根 URLconf 中使用函数 i18n_patterns(*urls,

prefix_default_language=True），Django 会自动将当前激活的语言代码添加到 i18n_patterns()
中定义的所有 URL 的模式中。如果将参数 prefix_default_language 设置为 False，则会从默
认语言(LANGUAGE_CODE)中删除前缀。这种用法在向现有网站添加翻译时非常有用，这
样当前的网址就不会发生改变。

例如下面是在 URL 中使用函数 i18n_patterns()的例子：

```
from django.conf.urls.i18n import i18n_patterns
from django.urls import include, path

from about import views as about_views
from news import views as news_views
from sitemap.views import sitemap

urlpatterns = [
   path('sitemap.xml', sitemap, name='sitemap-xml'),
]

news_patterns = ([
   path('', news_views.index, name='index'),
   path('category/<slug:slug>/', news_views.category, name='category'),
   path('<slug:slug>/', news_views.details, name='detail'),
], 'news')

urlpatterns += i18n_patterns(
   path('about/', about_views.main, name='about'),
   path('news/', include(news_patterns, namespace='news')),
)
```

在定义上述 URL 模式后，Django 会自动将语言前缀添加到由函数 i18n_patterns()添加
的 URL 模式中。例如：

```
>>> from django.urls import reverse
>>> from django.utils.translation import activate
>>> activate('en')
>>> reverse('sitemap-xml')
'/sitemap.xml'
>>> reverse('news:index')
'/en/news/'
>>> activate('nl')
>>> reverse('news:detail', kwargs={'slug': 'news-slug'})
'/nl/news/news-slug/'
```

如果 prefix_default_language=False，并且 LANGUAGE_CODE='en'，URLs 将会是：

```
>>> activate('en')
>>> reverse('news:index')
'/news/'

>>> activate('nl')
>>> reverse('news:index')
'/nl/news/'
```

> **注意**
> 只允许在根 URLconf 中运行函数 i18n_patterns()。如果在包含的 URLconf 中使用函数 i18n_patterns(),则会弹出 ImproperlyConfigured 异常。

2. 翻译 URL 模式

也可以使用函数 gettext_lazy()标记将 URL 标记为翻译模式,例如:

```
from django.conf.urls.i18n import i18n_patterns
from django.urls import include, path
from django.utils.translation import gettext_lazy as _

from about import views as about_views
from news import views as news_views
from sitemaps.views import sitemap

urlpatterns = [
    path('sitemap.xml', sitemap, name='sitemap-xml'),
]

news_patterns = ([
    path('', news_views.index, name='index'),
    path(_('category/<slug:slug>/'), news_views.category, name='category'),
    path('<slug:slug>/', news_views.details, name='detail'),
], 'news')

urlpatterns += i18n_patterns(
    path(_('about/'), about_views.main, name='about'),
    path(_('news/'), include(news_patterns, namespace='news')),
)
```

然后可以创建翻译,函数 reverse()将返回 activate()设置的语言的 URL,例如:

```
>>> from django.urls import reverse
>>> from django.utils.translation import activate

>>> activate('en')
>>> reverse('news:category', kwargs={'slug': 'recent'})
'/en/news/category/recent/'

>>> activate('nl')
>>> reverse('news:category', kwargs={'slug': 'recent'})
'/nl/nieuws/categorie/recent/'
```

注意,在大多数情况下,最好只在带有语言代码前缀的模式块中使用翻译后的网址(使用 i18n_patterns()),以避免无意中翻译的网址与未翻译的网址模式冲突。

3. 在模板中反向解析 URL

如果在模板中解析本地化的 URLs,那么它们会始终使用当前的语言。要链接其他语言中的 URL,需要使用模板标签 language,它可以在对应的模板中启用给定的语言。例如:

```
{% load i18n %}
{% get_available_languages as languages %}
```

```
{% translate "View this category in:" %}
{% for lang_code, lang_name in languages %}
   {% language lang_code %}
   <a href="{% url 'category' slug=category.slug %}">{{ lang_name }}</a>
   {% endlanguage %}
{% endfor %}
```

标签 language 将语言代码作为唯一的参数。

9.2　Django Web 本地化

Django 框架对于国际化和本地化功能都提供了很好的支持，本地化主要是指将标记的待翻译字符串翻译成本地语言的过程。在本节的内容中，将详细讲解 Django Web 本地化的知识和具体用法。

扫码观看本节视频讲解

9.2.1　Message File(消息文件)

一旦应用程序中的字符串文字被标记为以后翻译,那么就可以写入到 Message File(消息文件)中进行记录。首先需要为新语言创建消息文件。消息文件是一个纯文本文件，代表一种语言，它包含所有可用的翻译字段以及如何以给定语言表示。消息文件的扩展名是".po"。

要想创建或更新消息文件，首先需执行下面的命令：

```
django-admin makemessages -l de
```

其中 de 表示要创建的消息文件的名字。例如，pt_BR 是葡萄牙语，de_AT 是奥地利德语，id 是印度尼西亚语。

然后从以下两个位置之一来运行脚本：

- Django 项目的根目录(就是包含 manage.py 的那个目录)。
- Django app 的根目录。

脚本会遍历项目源代码树或者应用程序源代码库，并抽出所有需要被翻译的字符串。在 locale/LANG/LC_MESSAGES 目录中创建(或更新)消息文件。以德语为例，这个文件会是 locale/de/LC_MESSAGES/django.po。

当在项目的根目录中执行 makemessages 命令时，提取的字符串将自动分发到合适的消息文件。也就是说，从包含 locale 目录的 app 文件中提取的字符串将进入该目录下的消息文件中。从不包含任何 locale 目录的 app 文件中提取的字符串，将进入 LOCALE_PATHS 中列出的第一个目录下的消息文件，如果 LOCALE_PATHS 为空则会报错。

在默认情况下，django-admin makemessages 会检查每一个以".html"".txt"或".py"为后缀的文件。如果想覆盖默认值，需要使用--extension 或-e 选项来指定要检查的文件扩展名：

```
django-admin makemessages -l de -e txt
```

可使用逗号和(或)多次使用-e 或--extension 来分隔多个扩展名：

```
django-admin makemessages -l de -e html,txt -e xml
```

> **注 意**
> 当从 JavaScript 源码中创建消息文件时，需要使用特别的 djangojs 域，而不是-e js。

每一个".po"文件包含少量的元数据(例如翻译维护者的联系方式等)以及大量的翻译文件：要翻译的字符串以及实际翻译的字段之间的映射。例如，如果在我们的 Djanog 程序中包含一段 "Welcome to my site." 的翻译字符串：

```
_("Welcome to my site.")
```

然后使用 django-admin makemessages 创建一个包含以下代码片段的".po"文件：

```
#: path/to/python/module.py:23
msgid "Welcome to my site."
msgstr ""
```

对上述代码的具体说明如下。

- ▶ msgid：是显示在源代码中需要翻译的字符串，不要改动它。
- ▶ msgstr：是翻译后的字符串。一开始它是空的，因此需要填充它。确保在翻译中保留引号。

长消息是一种特殊情况，紧跟 msgstr(或 msgid)之后的第一个字符串是空字符串。然后内容本身将被写在下几行作为每行一个字符串。这些字符串是直接连接的。不要漏写字符串中的尾随空格；否则，它们将被连在一起没有空格！

9.2.2 编译消息文件

在创建消息文件后，当每次对其进行更改时，都需要将其编译为更有效的形式，以供 gettext 使用。可使用 django-admin compilemessages 工具来编译消息文件，此工具可以运行所有可用的".po"文件，并创建".mo"文件，这些文件是为 gettext 使用而优化的二进制文件。可在运行 django-admin makemessages 的同一目录中运行 django-admin compilemessages：

```
django-admin compilemessages
```

9.2.3 本地格式化

通过使用 Django 的格式化系统，可以在模板中使用指定的格式化工具为当前的 locale(Django 的本地化系统)设置显示日期、时间和数字。在启用本地格式化后，访问相同内容的两个用户可能会看见不同格式的日期、时间和数字，这取决于他们本地的语言环境。

在 Django 项目中，格式化系统默认是关闭的。如果要启用它，需要在配置文件中将 USE_L10N 的值设置为 True。

1. 表单中感知本地语言环境输入

在开启格式化功能后,当在表单里解析日期、时间和数字时,Django 可以使用本地化格式进行解析。这意味着当猜测用户在表单输入内容所使用的格式化时,会对不同语言环境尝试不同的格式化操作。

> **注意**
>
> Django 使用与解析数据不同的格式来显示数据。最值得注意的是,解析日期的格式化不能使用 %a (缩写的日期)、%A (完整的日期)、%b(缩写的月份)、%B (完整的月份),或者 %p (AM/PM)。

例如在下面使用参数 localize 可以使表单字段能本地化输入和输出数据:

```
class CashRegisterForm(forms.Form):
    product = forms.CharField()
    revenue = forms.DecimalField(max_digits=4, decimal_places=2, localize=True)
```

2. 在模板中控制本地化

在使用 USE_L10N 启用本地格式化功能后,每当 Django 在模板中输入数值时,它会试着使用本地语言环境指定的格式化。但是,使用本地化值可能并不总是合适的。例如,如果你正在输出机器可读的 JavaScript 或 XML,则始终需要非本地化的值。我们也可以在已选的模板里使用本地化,而不是在所有地方使用。

为了能更好地控制使用本地化功能,Django 提供了 l10n 模板库,在里面包含了如下 tags(标签)和 filters(过滤器)。

(1) 模板标签 localize。

在模板中启用和禁用模板变量本地化功能,和 USE_L10N 相比,模板标签 localize 可以对本地化进行更精细的控制。要想为模板激活或禁用本地化功能,可以用下面的代码实现。

```
{% load l10n %}

{% localize on %}
    {{ value }}
{% endlocalize %}

{% localize off %}
    {{ value }}
{% endlocalize %}
```

(2) 模板过滤器。

▶ localize:用于实现对单一值的强制本地化功能。例如:

```
{% load l10n %}
{{ value|localize }}
```

要想对单一值取消本地化,需要使用 unlocalize 实现。要想控制大部分模板的本地化,可使用模板标签 localize 实现。

- unlocalize：强制地让单一值不被本地化。例如：

```
{% load l10n %}
{{ value|unlocalize }}
```

(3) 创建自定义的格式化文件。

虽然 Django 为本地环境提供了很多内置的格式化定义功能，但是有时因为项目的需求，需要自己创建一个格式化文件，因为格式化文件不包括我们的本地环境，又或者开发者想重写一些值。

要想使用自定义格式化功能，首先需要指定格式化文件的路径，将 FORMAT_MODULE_PATH 设置为格式化文件所在的包路径。例如：

```
FORMAT_MODULE_PATH = [
    'mysite.formats',
    'some_app.formats',
]
```

不能直接将自定义的格式化文件放在这个目录里，而是要放在本地语言环境的目录中，而且必须命名为 formats.py。并且还需要注意，不要在这些文件里放置敏感信息，因为当把字符串传递给函数 django.utils.formats.get_format()(使用 date 模板过滤器)时，其内部值会被暴露出来。

例如想要自定义英语的格式化文件，需要下面这样的工程结构：

```
mysite/
    formats/
        __init__.py
        en/
            __init__.py
            formats.py
```

在文件 formats.py 中包含了自定义的格式化信息，例如：

```
THOUSAND_SEPARATOR = '\xa0'
```

我们需要使用不换行空格(硬空格 Unicode 00A0)作为千位分隔符，而不是英语中的默认逗号。

9.3 国际化和本地化的应用

经过本书前面内容的学习，已经了解了 Django 实现国际化和本地化的基本知识。在本节的内容中，将通过具体实例来讲解在 Django 中实现国际化和本地化的方法。

9.3.1 实战演练：展示法语环境

在下面的实例代码中，演示了使用国际化和本地化功能展示法语问候语过程。

源码路径：daima\9\Django-i18n\

(1) 新建一个名为"PND"的工程，然后定位到工程根目录，新建一个名为"France"的 app。

(2) 在配置文件 settings.py 的 INSTALLED_APPS 中加入 France，并开启 USE_L10N、USE_I18N 和 USE_TZ 等国际化功能，设置默认语言环境为法语。代码如下：

```
INSTALLED_APPS = [
    'django.contrib.admin',
    'django.contrib.auth',
    'django.contrib.contenttypes',
    'django.contrib.sessions',
    'django.contrib.messages',
    'django.contrib.staticfiles',
    'France'
]
LANGUAGE_CODE = 'fr-FR'
TIME_ZONE = 'UTC'

USE_I18N = True

USE_L10N = True

USE_TZ = True
```

(3) 编写路径导航文件 urls.py。代码如下：

```
urlpatterns = [
    path('admin/', admin.site.urls),
    path('', views.index)
]
```

(4) 编写视图文件 views.py，设置主页对应的视图函数 index()，设置调用的本地化文件，并设置转向为模板文件 Welcome.html。代码如下：

```
from django.shortcuts import render
from django.utils.translation import ugettext as _

def index(request):
    data = {
        'title': _('Petnet Direct')
    }
    return render(request, 'Welcome.html', data)
```

(5) 在"locale"目录下创建 Message File(消息文件)：

```
django-admin makemessages -l fr_FR
```

这样会生成文件 django.po，在此我们设置法语环境显示的内容。代码如下：

```
msgid ""
msgstr ""
"Project-Id-Version: PACKAGE VERSION\n"
"Report-Msgid-Bugs-To: \n"
"POT-Creation-Date: 2020-06-23 16:01-0400\n"
```

```
"PO-Revision-Date: YEAR-MO-DA HO:MI+ZONE\n"
"Last-Translator: FULL NAME <EMAIL@ADDRESS>\n"
"Language-Team: LANGUAGE <LL@li.org>\n"
"Language: \n"
"MIME-Version: 1.0\n"
"Content-Type: text/plain; charset=UTF-8\n"
"Content-Transfer-Encoding: 8bit\n"

#: France/templates/Welcome.html:11
msgid "Welcome to the home site"
msgstr "Bienvenue sur le site d'accueil"

#: France/views.py:7
msgid "Petnet Direct"
msgstr "Petnet Direct (FR)"
```

（6）在模板文件 Welcome.html 中调用国际化和本地化功能。代码如下：

```
{% load i18n %}
<!DOCTYPE html>
<html lang="en">
<head>
    <meta charset="UTF-8">
    <meta name="viewport" content="width=device-width, initial-scale=1.0">
    <title>{{ title }}</title>
</head>
<body>
    <h1>{{ title }}</h1>
    <h2> {% trans "Welcome to the home site" %} </h2>
</body>
</html>
```

因为在配置文件 settings.py 中设置的是法语环境，所以执行后的效果如图 9-1 所示。

图 9-1　法语环境的执行效果

如果在配置文件 settings.py 中设置为美国环境：

```
LANGUAGE_CODE = 'en-us'
TIME_ZONE = 'America/Chicago'
```

此时执行后会展示英语信息，执行效果如图 9-2 所示。

图 9-2　英语环境的执行效果

9.3.2 实战演练：创建多语言环境

在下面的实例代码中，演示了使用国际化和本地化功能创建多种语言环境的过程。

> 源码路径：**daima\9\django_internationalization_example**

(1) 通过如下命令新建一个名为"django_internationalization_example"的工程，然后定位到工程根目录，新建一个名为"internationalization_languages"的 app。

(2) 在配置文件 settings.py 的 INSTALLED_APPS 中加入 internationalization_languages，设置使用 SQLite3 数据库，并开启 USE_L10N、USE_I18N 和 USE_TZ 等国际化功能，设置默认语言环境为美国。代码如下：

```
INSTALLED_APPS = [
    'django.contrib.admin',
    'django.contrib.auth',
    'django.contrib.contenttypes',
    'django.contrib.sessions',
    'django.contrib.messages',
    'django.contrib.staticfiles',
    'internationalization_languages',
]

DATABASES = {
    'default': {
        'ENGINE': 'django.db.backends.sqlite3',
        'NAME': os.path.join(BASE_DIR, 'db.sqlite3'),
    }
}

LANGUAGE_CODE = 'en-us'

TIME_ZONE = 'UTC'

USE_I18N = True

USE_L10N = True

USE_TZ = True
```

(3) 编写路径导航文件 urls.py。代码如下：

```
from django.contrib import admin
from django.urls import path, include
from internationalization_languages import views

urlpatterns = [
    path('admin/', admin.site.urls),
    path('', views.home, name='index'),
    path('i18n/', include('django.conf.urls.i18n')),
]
```

(4) 编写视图文件 views.py，设置主页对应的视图函数 index()，并设置转向为模板文件 index.html。代码如下：

```python
from django.shortcuts import import render

def home(request):
    return render(request, "index.html")
```

(5) 在locale目录下，使用如下命令创建Message File(消息文件)：

```
django-admin makemessages -l 名字
```

这样会生成文件django.po，在此我们设置多语环境显示的内容。在本实例中我们创建了多个国家和地区的消息文件，具体说明如下。

- 目录"fr"：法语环境。代码如下：

```
msgid ""
msgstr ""
"Project-Id-Version: PACKAGE VERSION\n"
"Report-Msgid-Bugs-To: \n"
"POT-Creation-Date: 2020-04-08 13:50+0530\n"
"PO-Revision-Date: YEAR-MO-DA HO:MI+ZONE\n"
"Last-Translator: FULL NAME <EMAIL@ADDRESS>\n"
"Language-Team: LANGUAGE <LL@li.org>\n"
"Language: \n"
"MIME-Version: 1.0\n"
"Content-Type: text/plain; charset=UTF-8\n"
"Content-Transfer-Encoding: 8bit\n"
"Plural-Forms: nplurals=2; plural=(n > 1);\n"
#: .\internationalization_languages\templates\index.html:10
msgid "msg"
msgstr "Comment ça va"
```

- 目录"en"：英语环境。代码如下：

```
msgid ""
msgstr ""
"Project-Id-Version: PACKAGE VERSION\n"
"Report-Msgid-Bugs-To: \n"
"POT-Creation-Date: 2020-04-08 13:50+0530\n"
"PO-Revision-Date: YEAR-MO-DA HO:MI+ZONE\n"
"Last-Translator: FULL NAME <EMAIL@ADDRESS>\n"
"Language-Team: LANGUAGE <LL@li.org>\n"
"Language: \n"
"MIME-Version: 1.0\n"
"Content-Type: text/plain; charset=UTF-8\n"
"Content-Transfer-Encoding: 8bit\n"
"Plural-Forms: nplurals=2; plural=(n != 1);\n"
#: .\internationalization_languages\templates\index.html:10
msgid "msg"
msgstr "how are you"
```

- 目录"hi"：印地语环境。代码如下：

```
msgid ""
msgstr ""
"Project-Id-Version: PACKAGE VERSION\n"
"Report-Msgid-Bugs-To: \n"
```

```
"POT-Creation-Date: 2020-04-08 13:50+0530\n"
"PO-Revision-Date: YEAR-MO-DA HO:MI+ZONE\n"
"Last-Translator: FULL NAME <EMAIL@ADDRESS>\n"
"Language-Team: LANGUAGE <LL@li.org>\n"
"Language: \n"
"MIME-Version: 1.0\n"
"Content-Type: text/plain; charset=UTF-8\n"
"Content-Transfer-Encoding: 8bit\n"
"Plural-Forms: nplurals=2; plural=(n != 1);\n"
#: .\internationalization_languages\templates\index.html:10
msgid "msg"
msgstr "क्या हाल है"
```

（6）在模板文件 index.html 中调用国际化和本地化功能。首先创建另一个表单，然后设置一个下拉框，选择不同的下拉框可以显示不同语言环境的问候语。代码如下：

```
<!DOCTYPE html>
{% load i18n %}
<html lang="en">
<head>
    <meta charset="UTF-8">
    <title>Internationalization Languages and Time zones</title>
</head>
<body>

<h1>{% trans 'msg' %}</h1>

<form action="{% url 'set_language' %}" method="post">
    {% csrf_token %}
    <input name="next" type="hidden" value="{{ redirect_to }}" />
    <select name="language">
        {% get_current_language as LANGUAGE_CODE %}
        {% get_available_languages as LANGUAGES %}
        {% for lang in LANGUAGES %}
            <option value="{{ lang.0 }}" {% if lang.0 == LANGUAGE_CODE %}selected="selected"{% endif %}>
                {{ lang.1 }} ({{ lang.0 }})
            </option>
        {% endfor %}
    </select>
    <input type="submit" value="Go" />
</form>
</body>
</html>
```

执行后会根据用户选择的下拉框选项显示对应的问候语，执行效果如图 9-3 所示。

英语环境　　　　　　　　法语环境　　　　　　　　印地语环境

图 9-3　执行效果

9.4 网站地图 sitemap

网站地图是根据网站的结构、框架和内容生成的导航网页，是一个网站所有链接的容器。很多网站的链接层次比较深，蜘蛛很难抓取到，网站地图可以方便搜索引擎或者网络蜘蛛抓取网站页面、了解网站的架构，为网络蜘蛛指路，增加网站内容页面的收录概率。在Django Web 项目中，通常将网站地图存放在域名根目录下，并命名为"sitemap"。

扫码观看本节视频讲解

9.4.1 安装 sitemap

Django 框架自带了一个高级的生成网站地图的框架，开发者可以很容易地创建出 XML 格式的网站地图。在 Django Web 中创建网站地图时需要编写一个 Sitemap 类，并在 URLconf 中编写对应的访问路由。在 Django 工程中安装 sitemap 的步骤如下。

（1）在 INSTALLED_APPS 设置中添加"django.contrib.sitemaps"。

（2）确认在配置文件 settings.py 的 TEMPLATES 设置中包含 DjangoTemplates 后端，并将 APP_DIRS 选项设置为 True。其实，默认配置就是这样的，只有当曾经修改过这些设置时才需要调整过来。

（3）确认已经安装 sites 框架，我们需要注意，网站地图 APP 并不需要在数据库中建立任何数据库表。修改 INSTALLED_APPS 的唯一原因是，方便在 oader()加载模板时找到默认模板。

9.4.2 sitemap 的初始化

为了在网站上激活 sitemap 功能，需要将以下代码添加到 URLconf 中：

```
from django.contrib.sitemaps.views import sitemap
path('sitemap.xml', sitemap, {'sitemaps': sitemaps},
    name='django.contrib.sitemaps.views.sitemap')
```

当用户访问/sitemap.xml 时，Django 将生成并返回一个网站地图。

网站地图的文件名并不重要，重要的是文件的位置。搜索引擎只会索引网站的当前 URL 层级及下属层级。例如，如果地图文件 sitemap.xml 位于根目录中，它会引用网站中的任何 URL。但是如果站点地图位于/content/sitemap.xml，则它只能引用以/content/开头的网址。

sitemap 视图需要使用一个额外的必需参数：{'sitemaps': sitemaps}。sitemaps 是一个字典，将部门的标签(例如 news 或 blog)映射到其 Sitemap 类(例如，NewsSitemap 或 BlogSitemap)，也可以映射到 Sitemap 类的实例(例如，BlogSitemap(some_var))。

假设现在有一个博客系统，拥有 Entry 模型，并且希望在站点地图中包含指向每篇博客文章的所有链接。下面是类 Sitemap 的代码：

```
from django.contrib.sitemaps import Sitemap
from blog.models import Entry

class BlogSitemap(Sitemap):
    changefreq = "never"
    priority = 0.5

    def items(self):
        return Entry.objects.filter(is_draft=False)

    def lastmod(self, obj):
        return obj.pub_date
```

对上述代码的具体说明如下。
- changefreq 和 priority：分别对应于 HTML 页面中的<changefreq>和<priority>标签。
- items()：只是一个返回对象列表的方法。
- lastmod()：返回一个 datetime 时间对象。

在上述代码中没有编写方法 location()，但是可以自己增加此方法来指定对象的 URL。在默认情况下，location()在每个对象上调用 get_absolute_url()并将返回结果作为对象的 url。也就是说，使用站点地图的模型，比如 Entry，需要在模型内部实现方法 get_absolute_url()。

9.4.3 类 Sitemap 的成员

在 Django 框架中，类 Sitemap 的原型如下：

```
class Sitemap[source]
```

在类 Sitemap 中可以定义如下所示的方法/属性。

(1) items[source]：必需定义，用于返回对象列表的方法。框架不关心它们是什么类型，这些对象会被传递到 location()、lastmod()、changefreq()和 priority()方法。

(2) location[source]：可选项，其值可以是一个方法或属性。如果它是一个方法，它的值应该为 items()返回的对象的绝对路径。如果它是一个属性，它的值应该是一个字符串，表示 items()返回的每个对象的绝对路径。

注意，上面所说的"绝对路径"表示不包含协议和域名的 URL。例如：
- 正确：'/foo/bar/'
- 错误：'example.com/foo/bar/'
- 错误：'https://example.com/foo/bar/'

如果没有提供 location，Django 将调用 items()返回的每个对象上的 get_absolute_url()方法。

该属性最终反映到 HTML 页面上的<loc></loc>标签。

(3) lastmod：可选项，其值可以是一个方法或属性，表示当前条目最后的修改时间。如果它是一个方法，它应该接收一个参数(由 items()返回的对象)，并返回对象的最后修改日期/时间，例如 Python datetime.datetime。

(4) changefreq：可选项，其值可以是一个方法或属性，表示当前条目最后的修改时间。如果它是一个方法，它应该接收一个参数(由 items()返回的对象)，并将对象的更改频率作为 Python 字符串返回。如果它是一个属性，则其值应为表示 items()返回的每个对象的更改频率的字符串。

不管使用方法还是属性，changefreq 的可能值为：

- 'always'
- 'hourly'
- 'daily'
- 'weekly'
- 'monthly'
- 'yearly'
- 'never'

(5) priority：可选项，其值可以是一个方法或属性，表示当前条目在网站中的权重系数、优先级。如果它是一个方法，它应该接收一个参数(由 items()返回的对象)，并返回对象的优先级，如字符串或浮点数。如果它是一个属性，它的值应该是一个字符串，表示 items()返回的每个对象的优先级。

priority 的示例值有 0.4 或 1.0。页面的默认优先级为 0.5。

(6) protocol：可选属性，用于定义网站地图中的网址的协议(http 或 https)。如果未设置此属性，则使用请求站点地图的协议。如果 Sitemap 是在请求的上下文之外构建的，则默认为"http"。

(7) limit：可选属性，用于定义网站地图的每个网页上包含的最大超级链接数。其值不能超过默认值 50000，这是 Sitemaps 协议中允许的上限。

(8) i18n：可选的 boolean 属性，用于定义是否应使用所有语言生成此网站地图，默认值为 False。

9.4.4 快捷类 GenericSitemap

在 sitemap 框架中提供了一个快捷类，能够帮助开发者迅速生成网站地图：

```
class GenericSitemap[source]
```

通过使用快捷类，我们无须为 sitemap 编写单独的视图模块，可以直接在 URLconf 中获取对象和参数，传递参数，设置 url。例如下面的演示代码实现了这些功能：

```python
from django.contrib.sitemaps.views import sitemap
from django.urls import path
from blog.models import Entry

info_dict = {
    'queryset': Entry.objects.all(),
    'date_field': 'pub_date',
}
```

```python
urlpatterns = [
    # 一些使用信息字典的通用视图
    # ...

    # the sitemap
    path('sitemap.xml', sitemap,
        {'sitemaps': {'blog': GenericSitemap(info_dict, priority=0.6)}},
        name='django.contrib.sitemaps.views.sitemap'),
]
```

9.4.5 静态视图的 Sitemap

有时不希望在站点地图中出现一些静态页面，比如商品的详细信息页面，这应该如何实现呢？解决方案是在 items 中显式列出这些页面的网址名称，并在网站地图的 location 方法中调用 reverse()。例如下面的代码：

```python
from django.contrib import sitemaps
from django.urls import reverse

class StaticViewSitemap(sitemaps.Sitemap):
    priority = 0.5
    changefreq = 'daily'

    def items(self):
        return ['main', 'about', 'license']

    def location(self, item):
        return reverse(item)

# urls.py
from django.contrib.sitemaps.views import sitemap
from django.urls import path

from .sitemaps import StaticViewSitemap
from . import views

sitemaps = {
    'static': StaticViewSitemap,
}

urlpatterns = [
    path('', views.main, name='main'),
    path('about/', views.about, name='about'),
    path('license/', views.license, name='license'),
    # ...
    path('sitemap.xml', sitemap, {'sitemaps': sitemaps},
        name='django.contrib.sitemaps.views.sitemap')
]
```

上面做法的本质是先找出不想展示的页面，然后反向选择，获取想生成站点条目的对象，最后展示到站点地图中。

9.4.6 创建网站地图索引

可以通过如下视图函数创建网站地图索引：

```
views.index(request, sitemaps,
template_name='sitemap_index.xml',
content_type='application/xml',
sitemap_url_name='django.contrib.sitemaps.views.sitemap'
)
```

另外，还可以使用站点地图框架创建引用单个站点地图文件的站点地图索引，在 sitemaps 字典中定义每个站点地图文件的一个部分。两者唯一的区别是：

- 在 URLconf 中使用两个视图：django.contrib.sitemaps.views.index() 和 django.contrib.sitemaps.views.sitemap()。
- django.contrib.sitemaps.views.sitemap()视图应该采用 section 关键字参数。

例如下面是上面的例子的相关 URLconf 行：

```
from django.contrib.sitemaps import views

urlpatterns = [
    path('sitemap.xml', views.index, {'sitemaps': sitemaps}),
    path('sitemap-<section>.xml', views.sitemap, {'sitemaps': sitemaps},
        name='django.contrib.sitemaps.views.sitemap'),
]
```

这样将自动生成引用地图文件 sitemap-flatpages.xml 和 sitemap-blog.xml 中的文件 sitemap.xml。类 Sitemap 和类 sitemaps 根本不会更改。

假设在一个 Sitemap 中包含了 50000 多个网址，此时应该建立索引。在这种情况下，Django 会自动对网站地图分页，索引会反映出来。

如果不使用 vanilla 网站地图视图(例如，如果使用缓存装饰器修饰)，则必须为网站地图视图命名，并将 sitemap_url_name 传递到索引视图：

```
from django.contrib.sitemaps import views as sitemaps_views
from django.views.decorators.cache import cache_page

urlpatterns = [
    path('sitemap.xml',
        cache_page(86400)(sitemaps_views.index),
        {'sitemaps': sitemaps, 'sitemap_url_name': 'sitemaps'}),
    path('sitemap-<section>.xml',
        cache_page(86400)(sitemaps_views.sitemap),
        {'sitemaps': sitemaps}, name='sitemaps'),
]
```

9.4.7 模板定制

如果希望为网站上可用的每个站点地图或站点地图索引使用不同的模板，可以通过将参数 template_name 传递到 sitemap 和 index 视图的方式实现，例如：

```
from django.contrib.sitemaps import views

urlpatterns = [
    path('custom-sitemap.xml', views.index, {
        'sitemaps': sitemaps,
        'template_name': 'custom_sitemap.html'
    }),
    path('custom-sitemap-<section>.xml', views.sitemap, {
        'sitemaps': sitemaps,
        'template_name': 'custom_sitemap.html'
    }, name='django.contrib.sitemaps.views.sitemap'),
]
```

这些视图会返回 TemplateResponse 实例，允许我们在渲染之前轻松自定义响应数据。

9.4.8 实战演练：在 Django 博客系统中创建网站地图

在下面的实例代码中，演示了使用 Sitemap 功能创建地图文件的过程。

源码路径：**daima\9\Django_Sitemap_Tutorial**

(1) 新建一个名为"config"的工程，然后定位到工程根目录，新建一个名为"recipes"的 app。

(2) 在配置文件 settings.py 的 INSTALLED_APPS 中加入 recipes，设置使用 SQLite3 数据库，添加地图文件功能 django.contrib.sitemaps，并开启 USE_L10N、USE_I18N 和 USE_TZ 等国际化功能。代码如下：

```
INSTALLED_APPS = [
    "django.contrib.admin",
    "django.contrib.auth",
    "django.contrib.contenttypes",
    "django.contrib.sessions",
    "django.contrib.messages",
    "django.contrib.staticfiles",
    "django.contrib.sites",  # new
    "django.contrib.sitemaps",  # new
    "recipes.apps.RecipesConfig",  # new
]

DATABASES = {
    "default": {
        "ENGINE": "django.db.backends.sqlite3",
        "NAME": os.path.join(BASE_DIR, "db.sqlite3"),
    }
}

LANGUAGE_CODE = "zh_Hans"

TIME_ZONE = "sia/Shanghai"
```

```
USE_I18N = True

USE_L10N = True

USE_TZ = True
```

(3) 编写路径导航文件 urls.py，调用快捷类 GenericSitemap 快速生成网站地图。代码如下：

```
from django.contrib import admin
from django.contrib.sitemaps import GenericSitemap  # new
from django.contrib.sitemaps.views import sitemap   # new
from django.urls import path, include

from recipes.models import Recipe  # new

info_dict = {
    "queryset": Recipe.objects.all(),
}

urlpatterns = [
    path("admin/", admin.site.urls),
    path("", include("recipes.urls")),
    path(
        "sitemap.xml",
        sitemap,  # new
        {"sitemaps": {"blog": GenericSitemap(info_dict, priority=0.6)}},
        name="django.contrib.sitemaps.views.sitemap",
    ),
]
```

(4) 编写模型文件 models.py，功能是创建保存博客信息的数据模型。代码如下：

```
from django.db import models
from django.urls import reverse

class Recipe(models.Model):
    title = models.CharField(max_length=50)
    description = models.TextField()

    def __str__(self):
        return self.title

    def get_absolute_url(self):
        return reverse("recipe_detail", args=[str(self.id)])
```

(5) 编写视图文件 views.py，分别设置系统主页(博客列表)和某条博客详情页对应的模板文件。代码如下：

```
from django.db import models
from django.urls import reverse

class Recipe(models.Model):
    title = models.CharField(max_length=50)
    description = models.TextField()
```

```python
    def __str__(self):
        return self.title

    def get_absolute_url(self):
        return reverse("recipe_detail", args=[str(self.id)])
```

(6) 在模板文件 recipe_list.html 中调用数据库中保存的博客信息，列表显示每一条博客的标题。代码如下：

```
<h1>Recipes</h1>
{% for recipe in object_list %}
<ul>
    <li><a href="{{ recipe.get_absolute_url }}">{{ recipe.title }}</a></li>
</ul>
{% endfor %}
```

(7) 在模板文件 recipe_detail.html 中调用数据库中保存的博客信息，展示指定编号博客的详细信息。代码如下：

```
<div>
    <h2>{{ object.title }}</h2>
    <p>{{ object.description }}</p>
</div>
```

开始运行程序，首先通过如下命令同步数据库：

```
python manage.py makemigrations
python manage.py migrate
```

然后通过如下命令创建一个管理员账号信息：

```
python manage.py createsuperuser
```

登录系统后台页面 http://127.0.0.1:8000/admin/，在里面添加两条博客信息，如图 9-4 所示。

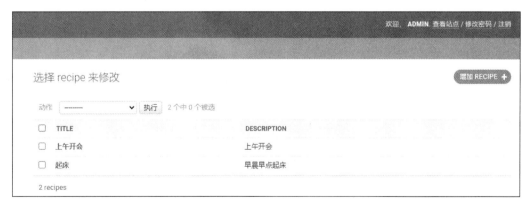

图 9-4　添加两条博客信息

此时登录系统主页 http://127.0.0.1:8000/，会列表显示上面刚添加的两条博客信息的标题，如图 9-5 所示。

图 9-5 系统主页

登录 http://127.0.0.1:8000/sitemap.xml 可以查看使用类 GenericSitemap 快速创建的站点地图文件，如图 9-6 所示。

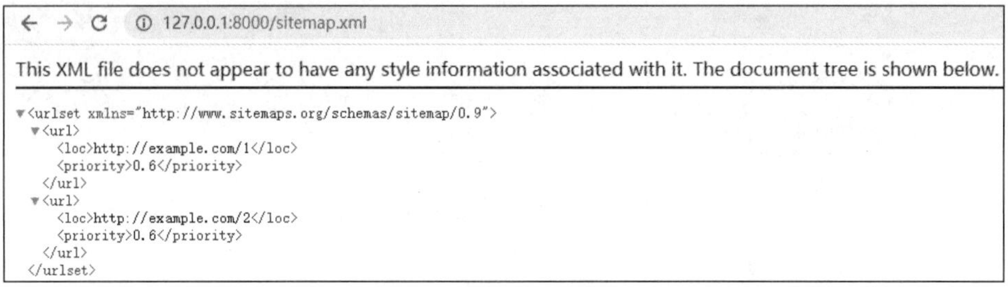

图 9-6 使用类 GenericSitemap 快速创建的站点地图文件

第 10 章

系统优化、调试和部署

在开发软件项目的过程中,系统的优化、调试和部署是整个软件开发流程中的重要一环。在本章的内容中,将详细讲解在 Django Web 程序中实现系统优化、调试和部署的知识,为读者步入本书后面知识的学习打下基础。

10.1　Django 性能与优化

在软件开发过程中，通常人们首先关心的是编写的代码可以正常运行，代码可以按照要求工作，并产生预期的执行效果。但是，这些还不足以使代码像我们希望的那样有效地工作。一名优秀的程序员可以很方便地改善代码的性能，并且不会影响原有代码的运行结果，或者说是最小程度地影响其行为。在上面提到的"改善代码的性能"，就是我们接下来将要学习的性能与优化。

扫码观看本节视频讲解

10.1.1　什么是优化？

我们需要优化的是性能，"性能"不只有一个指标，例如提高运行速度可能是程序设计的最明显的目标，但有时可能需要寻求其他性能改进，例如更低的内存消耗或更少的对数据库或网络的需求。

在大多数情况下，一个领域的改进往往会带来另一个领域的改进。但是有的时候，可能需要牺牲另一个领域，例如在提高某个程序的运行速度时可能会导致这个程序使用更多的内存。更糟糕的是，虽然运行速度提升了，但是需要占据大量的内存，可能会导致整个系统的内存不足，这样反而更加不好。

在优化性能的时候需要做到权衡考虑，例如我们的时间是宝贵的资源，比 CPU 的时间更加珍贵。一些性能方面的改进可能难以实现，或者可能会影响代码的可移植性或可维护性，并非所有的性能改进都值得付出努力。所以，作为一名优秀的程序开发者，需要知道你的目标是改进什么方面的性能，然后瞄准这个方向进行优化。

10.1.2　Django 中的性能优化技术

1. 性能基准测试

如果只是猜测或者假设你的某段代码性能不好，这是不对的，我们可以进行性能基准测试。

(1) Django tools。

django-debug-toolbar 是一个非常方便的测试工具，可以深入了解代码的工作过程以及运行花费的时间。特别是它可以显示在你的页面中生成的所有 SQL 查询，以及每个人花了多长时间。

(2) Third-party services。

通过使用一些免费服务，可以从远程 HTTP 客户端的角度分析和报告你的网站的页面性能，可以在模拟实际用户的体验，观察网站的整体性能。常用的免费服务有：

- 雅虎 Yslow；

- Google PageSpeed。

另外,还有一些付费服务执行类似的分析,甚至可以与我们的 Django 代码库集成,这样可以更加广泛地、深入地分析性能。

2. 从一开始就把事情做好

很多优化工作涉及解决性能缺陷问题,但是这需要耗费一定的精力和时间。对于一名优秀的开发者来说,在编写代码之前就需要考虑性能问题。

在性能方面,Python 是一门很好的编程语言。而 Django 提供了许多不同的方法来处理事情,但是用不同的方式做某件事的效率是不同的,我们必须选择一种合适的方式。例如可能会发现可以用多种方式做同样的事情:计算集合中的数据数量,我们可以在 QuerySet 中计算,也可以是在 Python 程序中计算。在 Django 架构模式下,在较低级别的模块下执行此计算工作总是最快的。因为在更高的层次中,系统必须通过多级抽象和多层机械来处理对象,所经历的环节更多,所以速度就会慢下来。也就是说,数据库通常可以比 Python 程序能够更快地完成某件事情,例如下面这样做可以比模板语言更快:

```
#在数据库上快速执行 QuerySet 操作,因为这是数据库擅长的
my_bicycles.count()

#计算 Python 对象的速度较慢,因为它无论如何都需要数据库查询和 Python 对象的处理
len(my_bicycles)
# Django 模板过滤器的速度仍然较慢,因为无论如何,
#它都必须在 Python 中对它们进行计数,并且还因为模板语言的开销大
{{ my_bicycles|length }}
```

注意,上述演示代码仅仅是说明性的。首先,在现实情况下,需要考虑在计数之前和之后发生的情况,以确定在特定上下文中执行的最佳方式。其次,还需要考虑其他选择,在现实生活中,直接从模板调用 QuerySet count()方法可能是最合适的选择。

3. Caching(缓存)

通常,计算值是最耗费电脑资源(即资源匮乏和缓慢)的,因此将值保存到可以快速访问的缓存中会有巨大的好处,这样可以为下一次需要做好准备。Caching 是一个重要的强大的技术,在 Django 中内置了一个功能强大的 The caching framework(缓存框架),以及其他较小的缓存功能。

> **注 意**
> 有关 Django 内置缓存框架的知识,将在本章后面的内容中进行讲解。

Django 的缓存框架通过保存动态内容以便不需要计算每个请求,从而为性能提升提供了非常重要的机会。为了方便起见,Django 提供了不同级别的缓存粒度:可以输出缓存的特定视图,或仅缓存难以生成的片段,甚至整个网站。

> **注 意**
> 不应该将缓存视为改进代码性能的核心手段,这通常是实在找不到代码优化后使用的优化手段。缓存是设计性能良好的代码的最后一步,而不是一个捷径。

4. HTTP 性能优化

(1) Middleware 中间件。

Django 内置了一些有用的中间件，可以帮助我们优化网站的性能，这些中间件包括：

- ConditionalGetMiddleware：添加了对现代浏览器的支持，可以根据 Entity Tag(实体标记)和 Last-Modified 标头有条件地获取响应。
- GZipMiddleware：压缩所有浏览器的响应，节省带宽和传输时间。读者需要注意，GZipMiddleware 方式目前被认为存在安全隐患，并且容易受到由 TLS / SSL 提供的保护无效的攻击。

(2) Sessions。

Django 建议使用缓存会话功能，将从像数据库这种较慢的存储源加载会话数据的方式修改为使用内存存储会话数据的方式，这样可以大大提高网站的性能。

(3) CachedStaticFilesStorage(缓存静态文件存储)。

通过使用 Web 浏览器的缓存功能，可以在初始下载完成后消除给定文件的网络匹配。CachedStaticFilesStorage 会将一个内容相关的标记附加到 Static Files(静态文件)文件名中，以确保浏览器能够安全地对其进行长期缓存，而不会丢失未来的更改，当文件更改时会做一个标记。

(4) Minification。

一些第三方 Django 工具和包提供了"缩小"HTML、CSS 和 JavaScript 的功能，能够删除这些文件里不必要的空格、换行符和注释，缩短变量名，从而减少网站发布的文档的大小。

5. Template(模板)性能优化

在 Template(模板)中，使用 {% block %} 比使用 {% include %} 的速度更快；启用 Cached Template Loader(缓存模板加载器)会大幅提高性能，因为它能够避免每次编译每个模板时渲染的步骤。

在绝大多数情况下，Django 的内置模板系统是完全够用的，但是如果我们的 Django 项目的瓶颈是模板系统，并且我们已经用尽了所有的方案来优化，这个时候可以考虑使用第三方产品来替代。例如 Jinja2 可以提高性能，特别是在速度方面。

> **注意**
> 如果在模板中遇到性能问题，首先要做的是弄清楚这是为什么。使用替代模板系统可能会更快，但是要确保不会遇到额外的麻烦。

6. 版本更新

市面中的 Python 库和 Django 多种多样，在使用这些库之前建议检查不同版本的适应性和性能是否合适。一些版本的升级可能针对的是更高级的用户，他们希望提高已经优化的 Django 站点的性能。

在大多数情况下，新版本的效率会更高。但是在有些时候，新版本并不一定是最好的选择，所以虽然知道较新的版本可能性能更好，但是不要简单地假设新版本一定会对你的项目的性能提升有好处。

例如 Django 的版本更新速度很快，在连续发布的新版本中提供了许多改进，但我们仍应检查应用程序的真实性能。因为在某些情况下，你可能会发现这些新版本的改进会造成性能更差的问题，而不是更好。

> **注 意**
> 在绝大多数情况下，较新的 Python 版本以及 Python 软件包往往会表现出更好的性能。除非我们在新版本中遇到了异常的性能问题，否则建议使用新的版本，因为在新版本中通常会找到更好的功能、可靠性和安全性。

10.1.3 实战演练：在 Django 博客系统中添加 django-debug-toolbar 面板

在下面的实例代码中，演示了使用 Django 国际化和本地化功能创建一个博客系统的过程，并在里面添加了 django-debug-toolbar 调试面板。

源码路径：daima\10\django-sitemaps\

（1）新建一个 Django 工程，然后定位到工程根目录，新建一个名为"sitemap"的 app。

（2）在配置文件 settings.py 的 INSTALLED_APPS 中加入 sitemap，设置使用 SQLite3 数据库，添加地图文件功能 django.contrib.sitemaps，并开启 USE_L10N、USE_I18N 和 USE_TZ 等国际化功能。代码如下：

```
INSTALLED_APPS = [
    'django.contrib.admin',
    'django.contrib.auth',
    'django.contrib.sites',
    'django.contrib.sitemaps',
    'django.contrib.contenttypes',
    'django.contrib.sessions',
    'django.contrib.messages',
    'django.contrib.staticfiles',
    'sitemap',
]

DATABASES = {
    "default": {
        "ENGINE": "django.db.backends.sqlite3",
        "NAME": os.path.join(BASE_DIR, "db.sqlite3"),
    }
}

LANGUAGE_CODE = 'en-us'
```

```
TIME_ZONE = 'UTC'

USE_I18N = True

USE_L10N = True

USE_TZ = True
```

(3) 编写路径导航文件 urls.py，调用快捷类 GenericSitemap 快速生成网站地图。代码如下：

```
from django.contrib import admin
from django.urls import path, include
from django.contrib.sitemaps.views import sitemap
from sitemap.sitemaps import BlogSitemap

sitemaps = {
   'blogs': BlogSitemap,
}

urlpatterns = [
   path('admin/', admin.site.urls),
   path('sitemap.xml', sitemap, {'sitemaps': sitemaps}),
   path('', include('sitemap.urls')),
]
```

(4) 编写模型文件 models.py，功能是创建保存博客信息的数据模型。代码如下：

```
from django.db import models
from django.urls import reverse

class Blog(models.Model):
   title = models.CharField(max_length=200, null=True)
   identifier = models.CharField(max_length=200, null=True)
   body = models.CharField(max_length=200, null=True)

   def get_absolute_url(self):
      return reverse("blog", args=[(self.identifier)])
```

(5) 编写视图文件 views.py，分别设置系统主页显示的内容和某条博客详情页对应的模板文件。代码如下：

```
from django.shortcuts import get_object_or_404, render
from django.http import HttpResponse
from sitemap.models import Blog


def index(request):
   return HttpResponse("Hello, world. You're at the polls index.")

def blog(request, identifier):
   item = get_object_or_404(Blog, identifier=identifier)

   context = {'blog': item}
   return render(request, 'sitemap/blog.html', context)
```

(6) 在模板文件 sitemap/blog.html 中调用数据库中保存的博客信息,展示指定编号博客的详细信息。代码如下:

```
<hr>
{{blog.title}}
<hr>
{{blog.identifier}}
<hr>
{{blog.body}}
<hr>
```

开始运行程序,首先通过如下命令同步数据库:

```
python manage.py makemigrations
python manage.py migrate
```

然后通过如下命令创建一个管理员账号信息:

```
python manage.py createsuperuser
```

登录系统后台页面 http://127.0.0.1:8000/admin/,在里面添加多条博客信息,如图 10-1 所示。

图 10-1 添加一些博客信息

此时登录系统主页 http://127.0.0.1:8000/blog/111,会显示上面某条博客的详细信息,其中 URL 参数 "111" 是这条博客的 identifier,如图 10-2 所示。

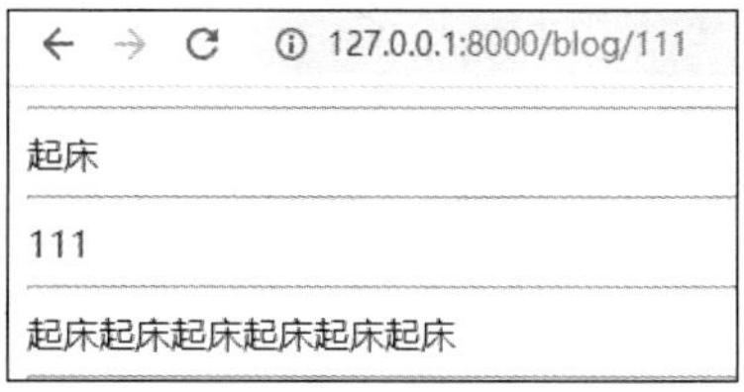

图 10-2 系统主页

登录 http://127.0.0.1:8000/sitemap.xml 可以查看使用类 GenericSitemap 快速创建的站点地图文件,如图 10-3 所示。

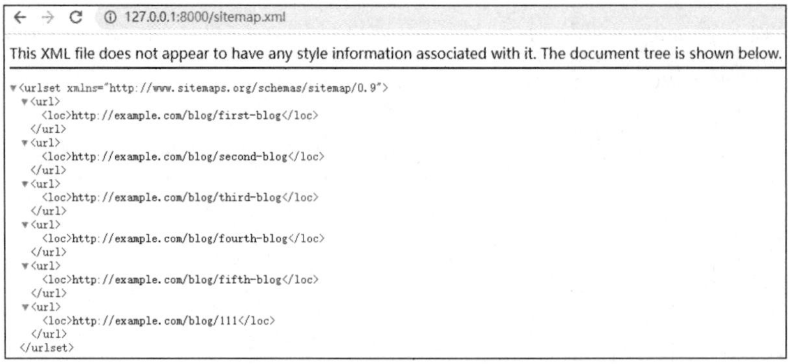

图 10-3 使用类 GenericSitemap 快速创建的站点地图文件

（7）开始添加 django-debug-toolbar 功能，首先通过如下命令安装 django-debug-toolbar：

```
pip install django-debug-toolbar
```

（8）在配置文件 settings.py 的 INSTALLED_APPS 中加入'debug_toolbar'，代码如下：

```
INSTALLED_APPS = [
    'django.contrib.admin',
    'django.contrib.auth',
    'django.contrib.sites',
    'django.contrib.sitemaps',
    'django.contrib.contenttypes',
    'django.contrib.sessions',
    'django.contrib.messages',
    'django.contrib.staticfiles',
    'sitemap',
    'debug_toolbar',
]
```

（9）在配置文件 settings.py 的 MIDDLEWARE 中加入'debug_toolbar.middleware.DebugToolbarMiddleware'，代码如下：

```
MIDDLEWARE = [
    'debug_toolbar.middleware.DebugToolbarMiddleware',
    'django.middleware.security.SecurityMiddleware',
    'django.contrib.sessions.middleware.SessionMiddleware',
    'django.middleware.common.CommonMiddleware',
    'django.middleware.csrf.CsrfViewMiddleware',
    'django.contrib.auth.middleware.AuthenticationMiddleware',
    'django.contrib.messages.middleware.MessageMiddleware',
    'django.middleware.clickjacking.XFrameOptionsMiddleware',
]
```

（10）在配置文件 settings.py 中设置访问 IP，我们设置的本地测试 IP：

```
INTERNAL_IPS=['127.0.0.1',]
```

接下来使用如下命令重新运行这个 Django 工程：

```
python manage.py runserver
```

此时会发现在我们的这个 Django 博客系统中已经成功添加了 django-debug-toolbar 功能，例如访问系统后台的页面效果如图 10-4 所示。

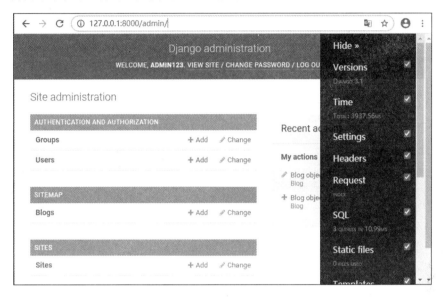

图 10-4　后台主页

在图 10-4 中，右侧黑色的带有复选框的面板便是 django-debug-toolbar 面板。通过这个面板可以查看和系统调试有关的信息，勾选每个复选框可以查看对应的信息。例如勾选 Versions 复选框后可以查看当前 Django、Python 和 debug_toolbar 的版本信息，如图 10-5 所示。

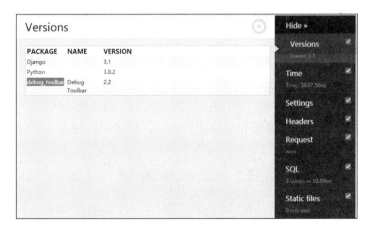

图 10-5　Versions 面板信息

在 django-debug-toolbar 面板中，常用复选框选项的说明如下。
- Versions：查看 Django、Python 和 debug_toolbar 的版本信息。
- Time：用来计时的，计算加载当前页面总共花的时间。
- Settings：读取 Django 中的配置信息。
- Headers：当前请求头和响应头的信息。

- Request：当前请求的信息(视图函数、Cookie 信息、Session 信息等)。
- SQL：查看当前界面执行的 SQL 语句。
- Static files：当前界面加载的静态文件。
- Templates：当前界面用的模板。
- Cache：缓存信息。
- Signals：信号。
- Logging：当前界面的日志信息。
- Redirects：当前界面的重定向信息。

10.2 Django 缓存处理

扫码观看本节视频讲解

Django Web 是动态网站，它展示的内容是动态的。每当用户请求页面，服务器会重新计算。在正常情况下，客户端的每次请求都会经历如下过程：

接收请求 -> url 路由 -> 视图处理 -> 数据库读写 -> 视图处理 -> 模板渲染 -> 返回请求

假设有这么一个场景，有很多用户都对某个页面非常感兴趣，于是发出了大量内容相同或相似的请求，如果每次请求都采取上面的流程进行，将会出现大量的重复工作，尤其是大量无谓的数据库读写工作。有很多解决这个问题的办法，其中一个就是使用缓存。缓存的基本思路是，既然某个请求已经被处理过一次，并且得到了处理结果，就把当前结果缓存下来。这样当下次再请求时，直接返回缓存的处理结果。这样可以极大地减少重复工作，降低数据库的负载。

10.2.1 缓存的思路

在 Web 程序中，缓存内容是为了保存那些需要很多计算资源的结果，这样就不必在下次消耗重复计算的资源。下面是缓存思路的伪代码，用来解释缓存是如何在动态生成的网页中的工作过程。

```
给定一个 URL，  试图在缓存中查询是否有这个 URL 对应的页面

如果缓存中有该页面：
    返回这个缓存的页面
否则：
    生成页面
    将生成的页面保存到缓存中 (用作以后)
    返回这个生成的页面
```

Django 作为一款功能强大的 Web 开发框架，内置了功能强大的缓存功能。为了方便起见，Django 提供了不同层级的缓存：

- 可以缓存指定的页面。

- 可以仅仅缓存那些很难生产出来的部分。
- 可以缓存整个网站。

除此之外，Django 也可以很好地配合那些"下游"缓存，比如 Squid 和基于浏览器的缓存。

10.2.2 设置缓存

Django 支持基于数据库的、文件的和内存的缓存，在使用之前首先要对其进行设置。在 Django Web 中，和缓存相关的设置都位于配置文件 settings.py 中的 CACHES 配置项中。具体来说，Django 支持如下所示的几种缓存系统。

1. Memcached

Memcached 是 Django 原生支持的缓存系统，优点是速度快、效率高。Memcached 是一种基于内存的缓存服务，起初是为了解决网站 LiveJournal.com 的负载问题而开发的，后来由 Danga 开源。目前 Memcached 被 Facebook 和维基百科等大型网站使用，用来减少数据库访问次数，极大地提高了网站的性能。

Memcached 的原理是：启动一个守护进程并分配单独的内存块，为缓存提供一个快速的添加、检索、删除的接口。所有的数据直接存储在内存中，所以它不能取代数据库或者文件系统的功能。

> **注 意**
>
> Memcached 不是 Django 自带的软件，而是一个独立的软件，需要用户自己安装、配置和启动服务。在安装 Memcached 后还要安装 Python 操作 Memcached 的依赖库，其中最常用的依赖库有 python-memcached 和 pylibmc。

在安装 Memcached 和相关的依赖库后，接下来需要在 Django 中进行配置。

(1) 根据安装 Python 依赖库的不同，将 CACHES 的 BACKEND 设置为如下两者之一。
- django.core.cache.backends.memcached.MemcachedCache。
- django.core.cache.backends.memcached.PyLibMCCache。

(2) 设置 LOCATION 为我们的 Memecached 守护进程所在的主机 IP 和进程端口，格式为：

```
ip:port
```

在上述格式中，ip 是 Memcached 守护进程的 IP 地址，port 是 Memcached 运行的端口。在 UNIX 操作系统中，设置格式为：

```
unix:path
```

在上述格式中，path 是 Memcached Unix 套接字文件的路径。

例如在下面的例子中，Memcached 运行在本地(127.0.0.1)的 11211 端口，使用 python-memcached(也就是需要这么一个 Python 插件)绑定：

```
CACHES = {
    'default': {
        'BACKEND': 'django.core.cache.backends.memcached.MemcachedCache',
        'LOCATION': '127.0.0.1:11211',
    }
}
```

而在下面的例子中，Memcached 通过一个本地的 UNIX 端口文件 file/tmp/memcached.sock 来交互，也需要使用 python-memcached 进行绑定。

```
CACHES = {
    'default': {
        'BACKEND': 'django.core.cache.backends.memcached.MemcachedCache',
        'LOCATION': 'unix:/tmp/memcached.sock',
    }
}
```

Memcached 有一个非常好的特点就是可以让几个服务的缓存共享。这就意味着我们可以在几个机器上运行 Memcached 服务，这些程序将这几个机器当作同一个缓存，从而不需要复制每个缓存的值在每个机器上。为了使用这个特性，把所有的服务地址放在 LOCATION 里面，用分号隔开或者当作一个 list 列表。

例如在下面的例子中，在两个 Memcached 实例中实现缓存共享，两个 IP 地址为 172.19.26.240 和 172.19.26.242，端口同为 11211。

```
CACHES = {
    'default': {
        'BACKEND': 'django.core.cache.backends.memcached.MemcachedCache',
        'LOCATION': [
            '172.19.26.240:11211',
            '172.19.26.242:11211',
        ]
    }
}
```

例如在下面的例子中，通过几个 Memcached 实例共享缓存，IP 地址分别为 172.19.26.240 (端口 11211)、172.19.26.242 (端口 11212) 和 172.19.26.244 (端口 11213)。

```
CACHES = {
    'default': {
        'BACKEND': 'django.core.cache.backends.memcached.MemcachedCache',
        'LOCATION': [
            '172.19.26.240:11211',
            '172.19.26.242:11212',
            '172.19.26.244:11213',
        ]
    }
}
```

注意

基于内存的缓存有一个缺点：因为缓存数据是存储在内存中的，所以如果你的服务器宕机，数据就会丢失。读者需要明确的是，内存不能替代常驻的数据存储，所以不要把基于内存的缓存当成唯一的数据存储方式。其实对于当下，建议读者选择 Redis 作为缓存。

2. 数据库缓存

Django 可以把缓存保存在数据库里,如果有一个快速的、专业的数据库服务器,那么可以方便地把缓存保存在数据库中。为了把数据表当作缓存后台,可以执行下面的操作。

(1) 把 BACKEND 设置为 django.core.cache.backends.db.DatabaseCache。

(2) 把 LOCATION 设置为 tablename,表示数据表的名称。这个名字可以是任何想要的名字,只要它是一个合法的表名并且在你的数据库中没有被使用过。

例如在下面的演示中,缓存表的名字是 my_cache_table。

```
CACHES = {
    'default': {
        'BACKEND': 'django.core.cache.backends.db.DatabaseCache',
        'LOCATION': 'my_cache_table',
    }
}
```

> **注 意**
>
> 我们使用缓存的主要原因是要减少数据库的操作,如果将缓存又存到数据库岂不是自相矛盾?所以,建议读者尽量不要使用基于数据库的缓存。

1) 创建缓存表

使用数据库缓存之前,必须用下面的命令来创建缓存表。

```
python manage.py createcachetable
```

这将在数据库中创建一个基于数据库缓存系统预期的特定格式的数据表。表名会从 LOCATION 中获得。

2) 多个数据库的情况

如果在多数据库的情况下使用数据库缓存,还必须为你的数据库缓存表设置路由说明。出于路由的目的,数据库缓存表会在一个名为 django_cache 的应用下的名为 CacheEntry 的模型中出现。这个模型不会出现在模型缓存中,但这个模型的细节可以在路由中使用。例如,下面的路由会分配所有缓存读操作到 cache_replica,以及所有写操作到 cache_primary,缓存表只会同步到 cache_primary。

```
class CacheRouter(object):
    """A router to control all database cache operations"""

    def db_for_read(self, model, **hints):
        "All cache read operations go to the replica"
        if model._meta.app_label == 'django_cache':
            return 'cache_replica'
        return None

    def db_for_write(self, model, **hints):
        "All cache write operations go to primary"
        if model._meta.app_label == 'django_cache':
            return 'cache_primary'
        return None
```

```python
def allow_migrate(self, db, app_label, model_name=None, **hints):
    "Only install the cache model on primary"
    if app_label == 'django_cache':
        return db == 'cache_primary'
    return None
```

如果没有指定路由路径给数据库缓存 model，那么缓存就会使用默认的数据库。如果你不使用数据库做缓存，就无须担心要提供路由结构给数据库缓存模型。

3. 文件系统缓存

为了使用文件缓存，需要将 BACKEND 设置为：

```
django.core.cache.backends.filebased.FileBasedCache
```

并且将 LOCATION 设置为一个合适的目录，例如把缓存存储在 /var/tmp/django_cache，则用下面的代码进行设置。

```
CACHES = {
    'default': {
        'BACKEND': 'django.core.cache.backends.filebased.FileBasedCache',
        'LOCATION': '/var/tmp/django_cache',
    }
}
```

如果是在 Windows 系统中，需要将硬盘驱动器名放在路径的开头，例如：

```
CACHES = {
    'default': {
        'BACKEND': 'django.core.cache.backends.filebased.FileBasedCache',
        'LOCATION': 'c:/foo/bar',
    }
}
```

> **注　意**
>
> 这里的路径应该是绝对路径，并且在这个路径下，你要具有系统用户的足够的读写权限。继续上面的例子，如果你是一个名叫 apache 的用户，需要确保在"/var/tmp/django_cache"这个路径上具有存取、读和写的权限。

4. 本地内存缓存

这是 Django 默认的缓存方式，在使用本地内存缓存时，请将 BACKEND 设置为：

```
django.core.cache.backends.locmem.LocMemCache
```

例如下面的配置代码：

```
CACHES = {
    'default': {
        'BACKEND': 'django.core.cache.backends.locmem.LocMemCache',
        'LOCATION': 'unique-snowflake',
    }
}
```

5. 虚拟缓存(开发用的缓存)

Django 很贴心地为我们设计了一个开发用的缓存。当你的生产环境是个大型的缓存系统，在开发软件的时候没有相应的缓存系统支持，或者不想用大型的缓存系统进行开发，而实际软件开发过程中，你又不得不接入缓存系统，使用缓存的 API，这种情况下开发用的缓存就很有用了。要想使用虚拟缓存，需设置 BACKEND 为：

```
CACHES = {
  'default': {
    'BACKEND': 'django.core.cache.backends.dummy.DummyCache',
  }
}
```

6. 使用自定义缓存后端

Django 包含了一定数量的对外部缓存后端的支持，例如有时可能想要使用一个自定义的缓存后端。想要使用外部的缓存，就像 Python 语音中的 import 语句那样，例如：

```
CACHES = {
  'default': {
    'BACKEND': 'path.to.backend',
  }
}
```

如果建立自己的缓存后端，那么可以使用标准的缓存后端作为参考实现，具体可以在 Django 源码的 django/core/cache/backends/ 目录中找到代码。

7. 缓存参数

上面介绍的每一个缓存后端都可以设置一些额外的参数来控制缓存的行为，我们可以在 CACHES 设置中以额外键值对的形式设置这些参数。开发者可以设置如下参数。

(1) TIMEOUT：缓存的默认过期时间，以秒为单位，默认是 300 秒，None 表示永远不会过期。如果设置成 0 将造成缓存立即失效(缓存就没有意义了)。可以设置 TIMEOUT 为 None，这样缓存将永远不会过期。

(2) OPTIONS：可选参数，根据缓存后端的不同而不同。可以跟项目情况来选择缓存的后端，可以是下面的这些选项。

- MAX_ENTRIES：高速缓存允许的最大条目数，超出这个数则旧值将被删除，这个参数默认是 300。
- CULL_FREQUENCY：当达到 MAX_ENTRIES 的时候，被删除的条目比率。实际比率是 1 / CULL_FREQUENCY，所以将 CULL_FREQUENCY 设置为 2 会在达到 MAX_ENTRIES 所设置值时删去一半的缓存。这个参数应该是整数，默认值为 3。如果把 CULL_FREQUENCY 的值设置为 0，意味着当达到 MAX_ENTRIES 时缓存将被清空。某些缓存后端(例如：database)将以丢失很多缓存为代价，大大提高了接受访问的速度。

(3) KEY_PREFIX：Django 服务器使用的所有缓存键的字符串。

(4) VERSION：由 Django 服务器生成的缓存键的默认版本号。

(5) KEY_FUNCTION：一个字符串，其中包含一个函数的点路径，该函数定义了如何将前缀、版本和密钥组合成最终缓存密钥。

例如在下面的例子中配置了一个基于文件系统的缓存后端，将缓存过期时间设置为 60 秒，最大条目数为 1000。

```
CACHES = {
    'default': {
        'BACKEND': 'django.core.cache.backends.filebased.FileBasedCache',
        'LOCATION': '/var/tmp/django_cache',
        'TIMEOUT': 60,
        'OPTIONS': {
            'MAX_ENTRIES': 1000
        }
    }
}
```

在下面配置了一个基于 python-memcached 库的后端，将对象大小限制为 2MB。

```
CACHES = {
    'default': {
        'BACKEND': 'django.core.cache.backends.memcached.MemcachedCache',
        'LOCATION': '127.0.0.1:11211',
        'OPTIONS': {
            'server_max_value_length': 1024 * 1024 * 2,
        }
    }
}
```

下面是基于 pylibmc 库的后端配置，该后端启用二进制协议、SASL 认证和 ketama 行为模式。

```
CACHES = {
    'default': {
        'BACKEND': 'django.core.cache.backends.memcached.PyLibMCCache',
        'LOCATION': '127.0.0.1:11211',
        'OPTIONS': {
            'binary': True,
            'username': 'user',
            'password': 'pass',
            'behaviors': {
                'ketama': True,
            }
        }
    }
}
```

10.2.3 站点级缓存

在使用 Web 缓存时，最简单的方法是存缓存整个网站，此时需要在 MIDDLEWARE_CLASSES 中添加如下两个选项。

- 'django.middleware.cache.UpdateCacheMiddleware'
- 'django.middleware.cache.FetchFromCacheMiddleware'

例如下面的代码：

```
MIDDLEWARE_CLASSES = (
   'django.middleware.cache.UpdateCacheMiddleware',
   'django.middleware.common.CommonMiddleware',
   'django.middleware.cache.FetchFromCacheMiddleware',
)
```

在上述设置中，必须将中间件 update 放在列表的开始位置，将中间件 fectch 放在最后，然后在 settings 文件中添加下面这些需要的参数。

- CACHE_MIDDLEWARE_ALIAS：用于存储的缓存的别名。
- CACHE_MIDDLEWARE_SECONDS：每个 page 需要被缓存多少秒。
- CACHE_MIDDLEWARE_KEY_PREFIX：如果缓存被多个网站所共享，那么可以把这个值设成当前网站名，或其他能代表这个 Django 实例的唯一字符串，以避免 key 发生冲突。也可以设成空字符串。

FetchFromCacheMiddleware 缓存 GET 状态和 HEAD 状态为 200 的回应，用不同的参数请求相同的 url 也会被视为独立的页面，每个独立页面的缓存是分开的。这个中间件期望 HEAD 请求用与相应的 GET 请求相同的响应头来应答；在这种情况下，它可以返回 HEAD 请求的缓存的 GET 响应。

10.2.4 缓存单个 view 视图

在 Django Web 中，可以使用方法 django.views.decorators.cache.cache_page()对单个有效视图的输出进行缓存。django.views.decorators.cache 定义了一个自动缓存视图响应的 cache_page 装饰器，使用方法非常简单，例如：

```
from django.views.decorators.cache import cache_page

@cache_page(60 * 15)
def my_view(request):
    ...
```

函数 cache_page()的参数是 timeout，单位为秒。在上述代码中，"my_view()"视图的结果将被缓存 15 分钟（注意为了提高可读性，我们写了 60 * 15。60 * 15 等于 900。也就是说，15 分钟等于 60 秒乘以 15。）

和站点缓存一样，视图缓存与 URL 无关。如果多个 URL 指向同一视图，每个 URL 将会分别缓存。继续以上面的视图函数 my_view 为例，如果 URLconf 如下所示：

```
urlpatterns = [
   path('foo/<int:code>/', my_view),
]
```

那么正如我们所期待的那样，发送到/foo/1/和/foo/23/会被分别缓存。但是一旦一个明确的 URL（例如/foo/23/）已经被请求过了，之后再度发出的指向该 URL 的请求将使用缓存。

另外，函数 cache_page()也可以使用一些额外的参数，设置修饰符缓存某部分页面内容。默认会被使用参数 default cache，但是我们可以特别指定你要用的缓存。例如：

```
@cache_page(60 * 15, cache="special_cache")
def my_view(request):
    ...
```

还可以在每个视图的基础上覆盖缓存前缀，函数 cache_page()采用可选的关键字参数 key_prefix，其工作方式与中间件的 CACHE_MIDDLEWARE_KEY_PREFIX 设置相同。它可以这样使用：

```
@cache_page(60 * 15, key_prefix="site1")
def my_view(request):
    ...
```

参数 key_prefix 和 cache 可以被一起指定，将连接 key_prefix 参数和 caches 下指定的 KEY_PREFIX。

10.2.5 在 URLconf 中指定视图缓存

在前面演示视图被缓存的过程，这种方法将视图和缓存系统耦合起来，但是这样并不理想。例如，你可能想在其他没有缓存的站点上重用这个视图函数，或者可能想分发这个视图给那些想使用视图但不想缓存它们的人员。解决这些问题的办法是在 URLconf 中指定视图缓存，而不是使用视图函数指定。

当在 URLconf 中使用 cache_page 时，可以这样包装视图函数，这就是之前提到的 URLconf：

```
urlpatterns = [
    path('foo/<int:code>/', my_view),
]
```

将 my_view 包含在 cache_page 中：

```
from django.views.decorators.cache import cache_page

urlpatterns = [
    path('foo/<int:code>/', cache_page(60 * 15)(my_view)),
]
```

10.2.6 模板片段缓存

在 Django Web 中，可以使用标签 "cache" 来缓存模板片段。要使你的模板能够访问这个标签，需将 {% load cache %} 放在模板顶部。要想使用模板标签{% cache %}在给定的时间里缓存片段内容，需要设置至少两个参数：

- 缓存时效时间(以秒为单位)。
- 缓存片段的名称。

如果缓存失效时间被设置为 None，那么这个片段将被永久缓存。其名称不能使用变量名。例如：

```
{% load cache %}
{% cache 500 sidebar %}
   .. sidebar ..
{% endcache %}
```

有时想缓存某个片段的多个副本，例如可能想为站点内每个用户分别独立缓存上面例子中的使用的 sidebar 副本，此时可以通过传递一个或多个附加参数的方式实现这一功能，这些参数可以带有或不带有过滤器的变量，模板标签{% cache %}必须在缓存片段中被唯一识别：

```
{% load cache %}
{% cache 500 sidebar request.user.username %}
   .. sidebar for logged in user ..
{% endcache %}
```

如果 USE_I18N 被设为 True，那么站点中间件缓存将支持多语言(respect the active language)。对于 cache 模板标签来说，你可以使用模板中可用的特定翻译变量之一来达到同样的结果：

```
{% load i18n %}
{% load cache %}

{% get_current_language as LANGUAGE_CODE %}

{% cache 600 welcome LANGUAGE_CODE %}
   {% translate "Welcome to example.com" %}
{% endcache %}
```

缓存失效时间可以是模板变量，只要模板变量能解析为一个整数值即可。例如，如果模板变量 my_timeout 被设置成 600，那么下面两行代码的功能是一样的：

```
{% cache 600 sidebar %} ... {% endcache %}
{% cache my_timeout sidebar %} ... {% endcache %}
```

用模板变量设置缓存失效时间的这一用法可以避免在模板中重复，可以在某处设置缓存失效时间，然后复用这个值。

在默认情况下，缓存标签会先尝试使用名为"template_fragments"的缓存。如果这个缓存不存在，它将回退使用默认缓存。我们可以选择一个备用缓存后端与关键字参数 using 一起使用，这个参数必须是标签的最后一个参数。例如：

```
{% cache 300 local-thing ...  using="localcache" %}
```

10.2.7 实战演练：在上传系统中使用 Redis 缓存

在下面的实例代码中，演示了在 Django 文件上传系统中使用缓存的过程。

源码路径：daima\10\django_file

（1）新建一个 Django 工程，然后定位到工程根目录，新建两个分别名为"cache1"和

"Upload"的应用app。

(2) 在配置文件 settings.py 的 INSTALLED_APPS 中加入上面创建的应用 cache1 和 Upload，设置使用 SQLite3 数据库，在 CACHES 中配置两个缓存，其中在 default 中使用 Memcached 设置默认缓存，将 session 设置为登录信息缓存。代码如下：

```python
INSTALLED_APPS = [
    'cache1',
    'Upload',
    'django.contrib.admin',
    'django.contrib.auth',
    'django.contrib.contenttypes',
    'django.contrib.sessions',
    'django.contrib.messages',
    'django.contrib.staticfiles',
]

DATABASES = {
    "default": {
        "ENGINE": "django.db.backends.sqlite3",
        "NAME": os.path.join(BASE_DIR, "db.sqlite3"),
    }
}

CACHES = {
    "default": {
        # 使用redis做缓存
        'BACKEND': 'django_redis.cache.RedisCache',
        # 将缓存的数据保存在该目录下
        # 缓存的地址
        'LOCATION': 'redis://127.0.0.1:6379/1',
        # rediss: //[:password]@localhost:6379 / 0
        'TIMEOUT': 300,
        'OPTIONS': {
            # "PASSWORD": ""
            # 是否压缩缓存数据
            # "COMPRESSOR": "django_redis.compressors.zlib.ZlibCompressor",
            # 配置连接池
            "CONNECTION_POOL_KWARGS": {"max_connections": 100, "retry_on_timeout": True}
        }
    },
    'session': {
        # 指定缓存的类型是文件缓存
        'BACKEND': 'django_redis.cache.RedisCache',
        # 将缓存的数据保存在该目录下
        'LOCATION': 'redis://127.0.0.1:6379/15',
        'TIMEOUT': 300,
        'OPTIONS': {
            # "PASSWORD": ""
            # 是否压缩缓存数据(非必要)
            "COMPRESSOR": "django_redis.compressors.lzma.LzmaCompressor",
            # 配置连接池
            "CONNECTION_POOL_KWARGS": {"max_connections": 100, "retry_on_timeout": True}
```

```
        }
    },
}
# session 使用 redis 作为缓存
SESSION_ENGINE = "django.contrib.sessions.backends.cache"
SESSION_CACHE_ALIAS = "session"
```

(3) 开始实现"cache1"模块，首先编写路径导航文件 urls.py，在 "index/"中调用 cache_page 显示缓存信息，设置缓存时间为 1 个小时。代码如下：

```python
from django.conf.urls import url
from django.contrib import admin
from django.conf.urls.static import static
from django.views.decorators.cache import cache_page

from Upload import views
from cache1 import views as v
from django_file import settings

urlpatterns = [
            url('admin/', admin.site.urls),
            url('upload/', views.UploadView.as_view()),
            url('index/', cache_page(10 * 60)(v.index)),
            url('index1/', v.index1),
        ] + static(settings.MEDIA_URL, document_root=settings.MEDIA_ROOT)
```

然后编写视图文件 views.py，分别为 index/和 index1/编写对应的视图函数。代码如下：

```python
def index(request):
    now = datetime.datetime.now().strftime('%Y-%m-%d %H:%M:%S')
    return render(request, 'cache.html', locals())

def index1(request):
    now = datetime.datetime.now().strftime('%Y-%m-%d %H:%M:%S')
    return render(request, 'cache.html', locals())
```

最后编写模板文件 cache.html，功能是使用缓存显示当前的时间。代码如下：

```html
<!DOCTYPE html>
{% load cache %}
<html lang="en">
<head>
    <meta charset="UTF-8">
    <title>Title</title>
</head>
<body>
{% cache 5000 'now' %}
    <p>缓存的时间:{{ now }}</p>
{% endcache %}

<p>不使用缓存:{{ now }}</p>
<p>咋就不变化呢</p>
</body>
</html>
```

（4）开始实现"Upload"模块，首先编写模型文件 models.py，功能是创建保存上传文件信息的数据模型。代码如下：

```python
import datetime
import os
import random

from django.core.files.storage import FileSystemStorage
from django.db import models


# 图片+图片的格式
class ImageFilesStorage(FileSystemStorage):
    # /upload_to/img/图片的名字
    def _save(self, name, content):
        old_name = name.split('/')[-1]
        # 获取图片后缀名
        suffix_name = old_name.split('.')[-1]
        prefix_name = f"IMG_{datetime.datetime.now().strftime('%Y%m%d%H%M%S')}{str(random.randint(100000,999999))}"
        image_path = os.path.dirname(name)
        name = os.path.join(image_path, f'{prefix_name}.{suffix_name}')
        return super()._save(name, content, )


class Shop(models.Model):
    name = models.CharField(max_length=100)
    desc = models.CharField(max_length=100)
    # 框架自动配置了media
    # img = models.ImageField(upload_to='shop/img/', storage=ImageFilesStorage(), max_length=255)

    class Meta:
        db_table = 'shop'


class Img(models.Model):
    img = models.ImageField(upload_to='shop/img', storage=ImageFilesStorage(), max_length=255)
    type = models.SmallIntegerField()
    name = models.ForeignKey(Shop, on_delete=models.CASCADE)

    class Meta:
        db_table = 'images'
```

然后编写视图文件 views.py，功能是获取上传表单中的信息，将表单中的数据添加到数据库，并将文件上传到本地服务器。代码如下：

```python
class UploadView(View):
    def get(self, request):
        return render(request, 'upload.html')

    def post(self, request):
```

```python
# POST 获取字符的部分信息
desc = request.POST.get('desc')
name = request.POST.get('name')
# 获取文件部分信息
# img = request.FILES.get('img')
shop = Shop(desc=desc, name=name)
shop.save()

# 获取多个文件
img_list = request.FILES.getlist('img')
images = []
for image in img_list:
    images.append(Img(img=image, name=shop, type=1))
# 批量保存
Img.objects.bulk_create(images)
return HttpResponse('上传成功')
```

最后编写模板文件 upload.html，功能是实现一个基本的文件上传表单界面。代码如下：

```html
<!DOCTYPE html>
<html lang="en">
<head>
    <meta charset="UTF-8">
    <title>Title</title>
</head>
<body>
<form action="#" method="post" enctype="multipart/form-data">
    {% csrf_token %}
    <p><input type="text" name="name"></p>
    <p><input type="file" name="img" multiple accept="image/*"></p>
    <p><textarea name="desc" id="" cols="10" rows="10"></textarea></p>
    <p><input type="submit" name="上传图片"></p>
</form>
</body>
</html>
```

开始运行程序，首先通过如下命令同步数据库：

```
python manage.py makemigrations
python manage.py migrate
```

然后通过如下命令创建一个管理员账号信息：

```
python manage.py createsuperuser
```

通过如下命令启动 Redis：

```
redis-server.exe redis.windows.conf
```

通过如下命令运行所创建的 Django 工程：

```
python manage.py runserver
```

浏览 http://127.0.0.1:8000/index/的效果如图 10-6 所示，由此可以看出，使用缓存后保存了缓存时间，在上方显示的是缓存中保存的时间。在本页面中，设置的缓存时间是 1 小时。浏览 http://127.0.0.1:8000/upload/显示文件上传表单页面，如图 10-7 所示。

图 10-6　缓存显示主页　　　　　　图 10-7　文件上传表单页面

10.3　日志系统

Django 使用 Python 的内置模块 logging 来处理系统日志。模块 logging 由四个部分组成，分别是 Loggers(记录器)、Handlers(处理器)、Filters(过滤器)和 Formatters(格式化器)。接下来将详细讲解在 Django Web 中使用日志模块 logging 的知识。

扫码观看本节视频讲解

10.3.1　在 Django 视图中使用 logging

在 Django 中使用 logging 的方法非常简单，例如：

```
# 导入logging库
import logging

# 获取一个logger对象
logger = logging.getLogger(__name__)

def my_view(request, arg1, arg):
    ...
    if bad_mojo:
        # 记录一个错误日志
        logger.error('Something went wrong!')
```

这样每当满足 bad_mojo 条件一次，就写入一条错误日志。

在上述代码中，方法 logging.getLogger()调用获取(如有必要则创建)一个 logger 的实例。Logger 实例通过名字进行标识，使用名称的目的是标识其配置。在习惯上，Logger 的名称通常使用__name__格式，即包含该 logger 的 Python 模块的名字，以便我们基于模块 filter 和 handle 日志进行调用。如果想使用其他方式组织日志消息，可以用点号分隔的名称来标识 logger，例如：

```
logger = logging.getLogger('project.interesting.stuff')
```

点号分隔的 logger 名称用于定义一个层级，在上述代码中，project.interesting logger 被认为是 project.interesting.stuff logger 的上一级。project logger 是 project.interesting logger 的上一级。层级为何如此重要呢？因为可以设置 logger 传播它们的 logging 调用给它们的上一级。利用这种方式，可以在根 logger 上定义一系列的 handler，并捕获子 logger 中的所有 logging 调用。在 project 命名空间中定义的 handler 将捕获 project.interesting 和 project.interesting.stuff logger 上的所有日志消息。

这种传播行为可以基于每个 logger 进行控制。如果不想让某个 logger 传播消息给它的上一级，我们可以关闭这个行为。

Logger 实例为每个默认的日志级别提供一个入口方法：

```
logger.debug()
logger.info()
logger.warning()
logger.error()
logger.critical()
```

另外还有两个调用方法：

- logger.log()：打印消息时手工指定日志级别。
- logger.exception()：创建一个 ERROR 级别的日志消息，用于封装当前异常栈的帧。

10.3.2 在 Django 中配置 logging

通常，只是像上面的例子那样简单地使用 logging 模块是远远不够的，我们一般都要对 logging 的四个部分(Loggers、Handlers、Filters 和 Formatters)进行一一配置。在 Python 的 logging 模块中提供了几种配置方式。在默认情况下，Django 使用 dictConfig format 进行配置，也就是字典方式，例如下面的配置能够将来自 django.request logger 的所有日志请求写入到一个本地文件中。

```
LOGGING = {
    'version': 1,
    'disable_existing_loggers': False,
    'handlers': {
        'file': {
            'level': 'DEBUG',
            'class': 'logging.FileHandler',
            'filename': '/path/to/django/debug.log',
        },
    },
    'loggers': {
        'django': {
            'handlers': ['file'],
            'level': 'DEBUG',
            'propagate': True,
        },
    },
}
```

如果使用上面的配置，请确保 Django 用户对'filename'对应目录和文件的写入权限。

再看下面的配置代码，让 Django 将日志打印到控制台，通常用作开发期间的信息展示。

```python
import os

LOGGING = {
    'version': 1,
    'disable_existing_loggers': False,
    'handlers': {
        'console': {
            'class': 'logging.StreamHandler',
        },
    },
    'loggers': {
        'django': {
            'handlers': ['console'],
            'level': os.getenv('DJANGO_LOG_LEVEL', 'INFO'),
        },
    },
}
```

在上述代码中，django.request 和 django.security 不会传播日志给上一级，它在本地开发期间可能有用。在默认情况下，这个配置只会将 INFO 和更高级别的日志发送到控制台。Django 中这样的日志信息不多。可以设置环境变量 DJANGO_LOG_LEVEL=DEBUG 来查看 Django 的 debug 日志，它包含所有的数据库查询，所以非常详尽。

再看下面的配置代码，这是一个相当复杂的 logging 配置。

```
LOGGING = {
    'version': 1,
    'disable_existing_loggers': False,
    'formatters': {
        'verbose': {
            'format':
'%(levelname)s %(asctime)s %(module)s %(process)d %(thread)d %(message)s'
        },
        'simple': {
            'format': '%(levelname)s %(message)s'
        },
    },
    'filters': {
        'special': {
            '()': 'project.logging.SpecialFilter',
            'foo': 'bar',
        },
        'require_debug_true': {
            '()': 'django.utils.log.RequireDebugTrue',
        },
    },
    'handlers': {
        'console': {
            'level': 'INFO',
            'filters': ['require_debug_true'],
            'class': 'logging.StreamHandler',
            'formatter': 'simple'
        },
        'mail_admins': {
            'level': 'ERROR',
```

```
            'class': 'django.utils.log.AdminEmailHandler',
            'filters': ['special']
        }
    },
    'loggers': {
        'django': {
            'handlers': ['console'],
            'propagate': True,
        },
        'django.request': {
            'handlers': ['mail_admins'],
            'level': 'ERROR',
            'propagate': False,
        },
        'myproject.custom': {
            'handlers': ['console', 'mail_admins'],
            'level': 'INFO',
            'filters': ['special']
        }
    }
}
```

上面 logging 配置的主要功能如下。

(1) 定义了配置文件的版本，当前版本号为 1.0。

(2) 定义了两个 formatter：simple 和 format，分别表示两种文本格式。

- simple：只输出日志的级别(例如 DEBUG)和日志消息。format 字符串是一个普通的 Python 格式化字符串，用于描述每行日志的细节。输出的完整细节可以在 formatter 文档中找到。
- verbose：输出日志级别、日志消息，以及时间、进程、线程和生成日志消息的模块。

(3) 定义了两个过滤器：SpecialFilter 和 RequireDebugTrue。

(4) 定义了两个处理器：

- console：一个 StreamHandler，它将打印 DEBUG(和更高级)的消息传到 stderr。这个处理器使用 simple 输出格式。
- mail_admins：一个 AdminEmailHandler，它将用邮件发送 ERROR(和更高级)的消息传到站点管理员的邮箱。这个 handler 使用特定的过滤器。

(5) 配置了三个 logger：

- django：传递所有 INFO 和更高级的消息给 null handler。
- django.request：传递所有 ERROR 消息给 mail_admins handler。另外，标记这个 logger 不向上传播消息。这表示写入 django.request 的日志信息将不会被 django logger 处理。
- myproject.custom：传递所有 INFO 和更高级的消息并通过 special filter 的消息给两个 handler：console 和 mail_admins。这表示所有 INFO(和更高级)的消息将打印到控制台上，ERROR 和 CRITICAL 消息还会通过邮件发送出来。

10.3.3 自定义 logging 配置和禁用 logging 配置

(1) 自定义 logging 配置。

如果不想使用 Python 的 dictConfig 格式配置 logger，则可以指定自己的配置模式。可通过 LOGGING_CONFIG 设置定义一个可调用对象，将它用来配置 Django 的 logger。在默认情况下，它指向 Python 的 logging.config.dictConfig()函数。但是，如果想使用不同的配置过程，你可以使用其他只接收一个参数的可调用对象。在配置 logging 时，将使用 LOGGING 的内容作为参数的值。

(2) 禁用 logging 配置。

如果你完全不想配置 logging(或者你想使用自己的方法手工配置 logging)，则可以设置 LOGGING_CONFIG 为 None。这样会禁用 Django 默认 logging 的配置过程。例如在下面 settings.py 文件的代码中，使用 Django 配置禁用 logging。

```
LOGGING_CONFIG = None

import logging.config
logging.config.dictConfig(...)
```

在上述代码中，将 LOGGING_CONFIG 设置为 None，表示只禁用自动配置过程，而不是禁用 logging 本身。如果禁用配置过程，Django 仍然执行 logging 调用，只是调用的是默认定义的 logging 行为。

10.3.4 Django 对 logging 模块的扩展

Django 对 logging 模块进行了一定的扩展，用来满足 Web 服务器专门的日志记录需求。

(1) 记录器 Loggers。

Django 额外提供了如下几个内建的 logger。

- django：不建议使用这个记录器，建议使用下面的记录器。
- django.request：记录与处理请求相关的消息。5XX 错误被记录为 ERROR 消息；4XX 错误记录为 WARNING 消息。接收额外参数：status_code 和 request。
- django.server：记录在 Django 服务器下处理请求相关的消息，只用于开发阶段。
- django.template：记录与渲染模板相关的日志。
- django.db.backends：与数据库交互的代码相关的消息。
- django.security：记录任何与安全相关的错误。
- django.security.csrf：记录 CSRF 验证失败日志。
- django.db.backends.schema：记录查询导致数据库修改的日志。

(2) 处理器 Handlers。

Django 额外提供了一个 handler——AdminEmailHandler，这个处理器将它收到的每个日志信息用邮件发送给站点管理员。

(3) 过滤器 Filters。

Django 还额外提供如下所示的两个过滤器。

- CallbackFilter(callback)[source]：这个过滤器接受一个回调函数，并对每个传递给过滤器的记录调用这个回调函数。如果回调函数返回 False，将不会进行记录的处理。
- RequireDebugFalse[source]：这个过滤器只会在 settings.DEBUG==False 时传递。

10.3.5 实战演练：在日志中记录用户的访问操作

在下面的实例代码中，演示了在日志文件中记录用户访问 Django Web 的过程。

源码路径：**daima\10\Django-Logging**

(1) 新建一个 Django 工程，然后定位到工程根目录，新建一个名为 api 的应用 app。

(2) 在配置文件 settings.py 的 INSTALLED_APPS 中加入上面创建的应用"api"，然后添加第三方应用"django_extensions"和"rest_framework"，设置使用 SQLite3 数据库，在 LOGGING 中使用字典方式配置两日志信息，设置了保存日志信息的文件是./logs/fileengine.log，设置日志的格式是{levelname} {asctime} {module} {message}。代码如下：

```python
INSTALLED_APPS = [
    "django.contrib.admin",
    "django.contrib.auth",
    "django.contrib.contenttypes",
    "django.contrib.sessions",
    "django.contrib.messages",
    "django.contrib.staticfiles",
    "django_extensions",
    "rest_framework",
    "api",
]

DATABASES = {
    "default": {
        "ENGINE": "django.db.backends.sqlite3",
        "NAME": BASE_DIR / "db.sqlite3",
    }
}

LOGGING = {
    "version": 1,
    "loggers": {
        "django": {"handlers": ["django_console"], "level": "DEBUG"},
        "api.views":{"handlers":["console","file_engine"],"level":"DEBUG"},
    },
    "handlers": {
        "console": {
            "level": "DEBUG",
            "class": "logging.StreamHandler",
```

```
            "formatter": "simpleRe",
        },
        "django_console": {
            "level": "INFO",
            "class": "logging.StreamHandler",
            "formatter": "simpleRe",
        },
        "file_engine": {
            "level": "INFO",
            "class": "logging.FileHandler",
            "filename": "./logs/fileengine.log",
            "formatter": "simpleRe",
        },
    },
    "formatters": {
        "simpleRe": {
            "format": "{levelname} {asctime} {module} {message}",
            "style": "{",
        }
    },
}
```

(3) 编写路径导航文件 urls.py。代码如下:

```
from django.contrib import admin
from django.urls import path, include

urlpatterns = [
    path("admin/", admin.site.urls),
    path("api/", include("api.urls", namespace="api")),
]
```

(4) 开始实现"api"模块,编写视图文件 views.py,为链接"API/"编写对应的视图函数,使用 rest_framework 框架将链接"API/"指向对应的 API 接口界面。代码如下:

```
import logging
from rest_framework import response, views, permissions, status
logger = logging.getLogger(__name__)

class HomeView(views.APIView):
    permission_classes = [permissions.AllowAny]

    def get(self, request):
        print(__name__)
        logger.info("TESTE")
        logger.debug("TESTADO")
        return response.Response(status=status.HTTP_200_OK)
```

开始运行程序,首先通过如下命令同步数据库:

```
python manage.py makemigrations
python manage.py migrate
```

然后通过如下命令创建一个管理员账号信息:

```
python manage.py createsuperuser
```

通过如下命令运行所创建的 Django 工程：

```
python manage.py runserver
```

在网页中输入 http://127.0.0.1:8000/api/ 后显示 rest_framework 的 API 页面，如图 10-8 所示。

图 10-8　页面效果

每当我们单击 OPTIONS 和 GET 按钮都会向日志文件 ./logs/fileengine.log 中写入操作信息，如图 10-9 所示。

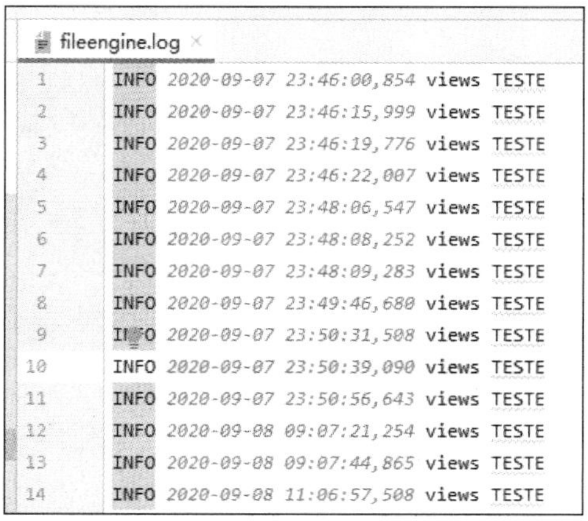

图 10-9　写入日志信息

第 11 章

邮件发送模块

在开发动态 Web 网站的过程中，发送和接收邮件功能是使用非常频繁的一个模块。在本章的内容中，将详细讲解在 Django Web 程序中开发发送邮件和接收邮件的知识，为读者步入本书后面知识的学习打下基础。

11.1 实战演练：使用 smtplib 发送邮件

SMTP(Simple Mail Transfer Protocol)即简单邮件传输协议，它是一组用于由源地址到目的地址传送邮件的规则，由它来控制信件的中转方式。Python 语言内置了模块 smtplib，对 SMTP 协议进行了简单的封装，可以直接调用内置模块 smtplib 发送邮件。例如在下面的实例中，演示了使用 Django 中的 auth 模块开发一个简易新闻系统的过程。本实例具有如下所示的功能。

扫码观看本节视频讲解

源码路径：daima\11\youjian\

（1）通过如下命令创建一个名为"youjian"的工程，然后在工程目录下新建一个名为"lizi"的 app。

```
django-admin.py startproject youjian
cd youjian
python manage.py startapp lizi
```

（2）编写视图文件 views.py，获取 forms 表单中的发送信息，然后通过 smtplib 模块发送邮件，主要实现代码如下所示。

```
def index(request):
    response = HttpResponse()
    response.write("<a href=\"contact\"><font color=red>联系我吧</font></a>")
    return response

def contact(request):
    form_class = ContactForm(request.POST or None)
    if form_class.is_valid():
        from_email = request.POST.get('frommail')
        password = request.POST.get('mima')
        to_list = request.POST.get('tomail')
        aa = request.POST.get('content')
        msg = MIMEText(aa)
        zhuzhu=request.POST.get('zhuti')
        msg['Subject'] = zhuzhu
        smtp_server = 'smtp.qq.com'
        server = smtplib.SMTP(smtp_server, 25)
        server.login(from_email, password)
        server.sendmail(from_email, to_list, msg.as_string())
        server.quit()
        return HttpResponseRedirect('thankyou')
    return render(request, 'form.html', {'form': form_class})

def thankyou(request):
    response = HttpResponse()
    response.write('<body bgcolor=silver><center><h1><font color =red>谢谢你,</font><br><font color=blue>刚刚发了一封邮件给你！<blue></h1></center></body>')
    return response
```

(3) 编写表单处理文件 forms.py，设置发送邮件表单，主要实现代码如下所示。

```
class ContactForm(forms.Form):
    frommail = forms.CharField(label='你的邮箱',required=True)
    mima = forms.CharField(label='邮箱密码', required=True)
    zhuti = forms.CharField(label='邮箱主题',required=True)
    content = forms.CharField(label='邮件内容',required=True,
widget=forms.Textarea)
    tomail = forms.CharField(label='发送给谁', required=True)
```

(4) URL 路径导航文件 urls.py 的主要实现代码如下所示。

```
from django.contrib import admin
from django.urls import path
from lizi.views import index, contact, thankyou

urlpatterns = [
    path('admin/', admin.site.urls),
    path('', index),
    path('contact', contact),
    path('thankyou/' , thankyou ),
]
```

(5) 在模板文件 form.html 中调用视图表单，显示完整的发送邮件表单，具体实现代码如下所示。

```
{% block content %}
    <h1><font color="#d2691e">邮件发送系统</font></h1>
    <body bgcolor="#ffe4b5">
<form name="form" action="" method="post">
    {% csrf_token %}
    {{form.as_p}}
    <button type="submit">发送</button>
</form>
    </body>
{% endblock %}
```

在浏览器中输入 http://127.0.0.1:8000/contact 后会显示邮件发送表单，如图 11-1 所示。填写表单信息，单击"发送"按钮后会发送邮件。

图 11-1 执行效果

11.2 使用django.core.mail发送邮件

除了使用smtplib模块发送电子邮件外，在Django框架中还提供了轻量级包django.core.mail，利用它可以更加快捷地发送电子邮件。在本节的内容中，将详细讲解在Django Web中发送电子邮件的过程。

11.2.1 django.core.mail基础

扫码观看本节视频讲解

在Django框架中，在django.core.mail模块中集成了发送邮件功能。创建一个Django工程和app项目后，在配置文件settings.py中可以设置邮件服务器的信息。例如：

```
EMAIL_USE_SSL = True
EMAIL_HOST = 'smtp.qq.com'  # smtp.163.com smtp.qq.com
EMAIL_PORT = 25
EMAIL_HOST_USER = '150649826@qq.com'  # 账号
EMAIL_HOST_PASSWORD = ''  # 密码
DEFAULT_FROM_EMAIL = EMAIL_HOST_USER
```

可以通过EMAIL_HOST和EMAIL_PORT设置发送邮箱的SMTP主机和端口，通过EMAIL_HOST_USER和EMAIL_HOST_PASSWORD设置发送邮箱的邮箱名和密码，通过EMAIL_USE_TLS和EMAIL_USE_SSL设置是否使用安全连接功能。

在模块django.core.mail中包含如下所示的成员。

(1) send_mail()：这是发送电子邮件最简单的方法。

```
send_mail(subject, message, from_email, recipient_list, fail_silently = False, auth_user = None, auth_password = None, connection = None, html_message = None)
```

其中参数subject、message、from_email和recipient_list是必需的。

- subject：邮件主题，是一个字符串。
- message：邮件正文内容，是一个字符串。
- from_email：发送者邮箱，是一个字符串。
- recipient_list：邮件接收者的邮件地址，这是一个字符串列表，每个字符串都是电子邮件地址。每个成员都将在电子邮件的"收件人："字段中看到其他收件人。
- fail_silently：发送异常参数，是一个布尔值。如果值是False，send_mail会显示一个smtplib.SMTPException异常。
- auth_user：向SMTP服务器进行身份验证的可选用户名。如果没有提供，Django将使用EMAIL_HOST_USER设置的值。
- auth_password：向SMTP服务器进行身份验证的可选密码。如果没有提供，Django将使用EMAIL_HOST_PASSWORD设置的值。
- html_message：如果提供了此参数，发送的将是一个嵌入式HTML类型电子邮件。
- connection：用于设置发送邮件的可选电子邮件后端。如果未指定，将使用默认的后端实例。

(2) send_mass_mail()：群发电子邮件。

```
send_mass_mail(datatuple, fail_silently = False, auth_user = None, auth_password
= None, connection = None)
```

参数 datatuple 是一个元组，其中每个元素都采用以下格式：

```
(subject, message, from_email, recipient_list)
```

参数 fail_silently、auth_user 和 auth_password 的功能与前面的方法 send_mail()相同。例如，以下代码将向两组不同的收件人发送两条不同的邮件，但是只打开一个与邮件服务器的连接。

```
message1 = ('Subject here', 'Here is the message', 'from@example.com',
['first@example.com', 'other@example.com'])
message2 = ('Another Subject', 'Here is another message', 'from@example.com',
['second@test.com'])
send_mass_mail((message1, message2), fail_silently=False)
```

(3) django.core.mail.mail_admins()：按照 ADMINS 的设置向网站管理员发送电子邮件。

(4) mail_managers()：功能类似于 mail_admins()，向在 MANAGERS 中定义的每个管理员发送邮件。

例如下面的代码会向邮箱 john @ example 和 jane @ example.com 发送电子邮件，这两个邮箱都会收到一封单独的电子邮件。

```
send_mail(
    'Subject',
    'Message.',
    'from@example.com',
    ['john@example.com', 'jane@example.com'],
)
```

下面的代码会向邮箱 john@example.com 和 jane@example.com 发送电子邮件，这两个邮箱都会收到一封单独的电子邮件。

```
datatuple = (
    ('Subject', 'Message.', 'from@example.com', ['john@example.com']),
    ('Subject', 'Message.', 'from@example.com', ['jane@example.com']),
)
send_mass_mail(datatuple)
```

例如下面的代码演示了一次性发送多个邮件的过程。

```
from django.core.mail import send_mass_mail

message1 = ('Subject here', 'Here is the message', 'from@example.com',
['first@example.com', 'other@example.com'])
message2 = ('Another Subject', 'Here is another message', 'from@example.com',
['second@test.com'])

send_mass_mail((message1, message2), fail_silently=False)
```

> **注 意**
>
> 方法 send_mail 每次发送邮件都会建立一个连接,发送多封邮件时建立多个连接。而方法 send_mass_mail 是建立单个连接发送多封邮件,所以一次性发送多封邮件时 send_mass_mail 要优于 send_mail。

如果想在邮件中添加附件,发送 HTML 格式的内容,可以通过如下代码实现。

```python
from django.conf import settings
from django.core.mail import EmailMultiAlternatives

from_email = settings.DEFAULT_FROM_EMAIL
# subject 为主题, content 为内容, to_addr 是一个列表,表示发送给哪些人
msg = EmailMultiAlternatives(subject, content, from_email, [to_addr])
msg.content_subtype = "html"

# 添加附件(可选)
msg.attach_file('./twz.pdf')

# 发送
msg.send()
```

在下面的代码中演示了使用两种版本发送邮件的过程,一种为默认的纯文本(text/plain)方式,另一种是 HTML 版本的。

```python
from __future__ import unicode_literals
from django.conf import settings
from django.core.mail import EmailMultiAlternatives
subject = '来自自强学堂的问候'
text_content = '这是一封重要的邮件.'
html_content = '<p>这是一封<strong>重要的</strong>邮件.</p>'
msg = EmailMultiAlternatives(subject, text_content, from_email, [to@youemail.com])
msg.attach_alternative(html_content, "text/html")
msg.send()
```

11.2.2 实战演练:使用 django.core.mail 实现一个邮件发送程序

在下面的实例代码中,演示了使用 Django 框架的内置模块 django.core.mail 开发一个邮件发送程序的过程。

> 源码路径:**daima\11\email_sending**

(1) 通过如下命令创建一个名为 "sending_mail" 的工程,然后在工程目录下新建一个名为 "mytest" 的 app。

```
django-admin.py startproject sending_mail
cd sending_mail
python manage.py startapp mytest
```

(2) 在文件 settings.py 中设置发件箱的账号信息,主要实现代码如下所示。

```
EMAIL_HOST ='smtp.qq.com'
EMAIL_HOST_USER ='729017304@qq.com'
EMAIL_HOST_PASSWORD = 'xxx'
EMAIL_USE_TLS = True
EMAIL_PORT = 25
```

(3) 编写视图文件 views.py，获取 forms 表单中的发送信息，然后通过 django.core.mail 模块发送邮件，主要实现代码如下所示。

```
def index(request):
    response = HttpResponse()
    response.write("<a href=\"contact\"><font color=red>联系我吧</font></a>")
    return response

def contact(request):
    form_class = ContactForm(request.POST or None)
    if form_class.is_valid():
        subject = '我好想你'
        messege = '你的信息是：' + request.POST.get('content')
        from_email = settings.EMAIL_HOST_USER
        usermail = request.POST.get('contact_email')
        to_list = [usermail, settings.EMAIL_HOST_USER]
        send_mail(subject, messege, from_email, to_list, fail_silently=False)
        return HttpResponseRedirect('thankyou')
    return render(request, 'form.html', {'form': form_class})

def thankyou(request):
    response = HttpResponse()
    response.write('<body bgcolor=silver><center><h1><font color =red>谢谢你,</font><br><font color=blue>刚刚发了一封邮件给你！<blue></h1></center></body>')
    return response
```

(4) 编写表单处理文件 forms.py，设置发送邮件表单，主要实现代码如下所示。

```
from django import forms
class ContactForm(forms.Form):
    contact_name = forms.CharField(label='名字',required=True)
    contact_email = forms.EmailField(label='邮箱',required=True)
    content = forms.CharField(label='内容',required=True,
widget=forms.Textarea)
```

(5) URL 路径导航文件 urls.py 的主要实现代码如下所示。

```
urlpatterns = [
    url(r'^admin/', admin.site.urls),
    url(r'^$', index),
    url(r'^contact$', contact),
    url(r'^thankyou/$' , thankyou ),
]
```

(6) 在模板文件 form.html 中调用视图表单，显示完整的发送邮件表单，具体实现代码如下所示。

```
{% block content %}
    <h1><font color="#d2691e">邮件发送系统</font></h1>
```

```
    <body bgcolor="#ffe4b5">
<form name="form" action="" method="post">
    {% csrf_token %}
    {{form.as_p}}
    <button type="submit">发送</button>
</form>
    </body>
{% endblock %}
```

在浏览器中输入 http://127.0.0.1:8000/contact 后会显示邮件发送表单,如图 11-2 所示。填写表单信息,单击"发送"按钮后会发送邮件。

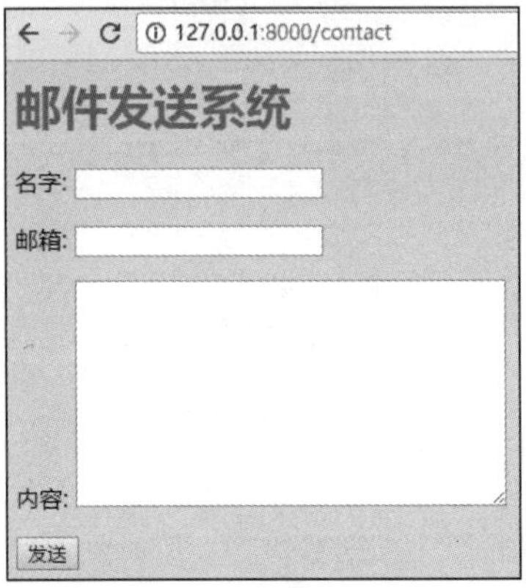

图 11-2　执行效果

11.3　实战演练:使用邮箱发送验证码的用户注册、登录验证系统

在下面的实例中,演示了使用 Django 开发一个会员用户注册和登录验证系统的过程,在注册过程中使用邮箱发送验证码。本实例具有如下所示的功能。

扫码观看本节视频讲解

- ▶ 前台会员注册:用户可以注册成为系统会员。
- ▶ 发送验证码:向注册用户提交的邮箱中发送验证码。
- ▶ 登录验证:验证会员登录信息是否合法。
- ▶ 后台管理:管理员可以查看、修改或删除系统内的会员信息。

源码路径:daima\11\email-verification-in-django\

(1) 新建一个名为 testemail 的工程,在 testemail 目录下新建一个名为 temail 的 app 项目。

(2) 在文件 settings.py 中,将上面创建的 app 项目名"temail"添加到 INSTALLED_APPS 中,然后设置使用 SQLite3 数据库,并且设置服务器邮箱的参数,在本项目中使用 QQ 邮箱向注册会员的邮箱中发送验证信息。代码如下:

```
DATABASES = {
    'default': {
        'ENGINE': 'django.db.backends.sqlite3',
        'NAME': os.path.join(BASE_DIR, 'db.sqlite3'),
    }
}
EMAIL_BACKEND = 'django.core.mail.backends.smtp.EmailBackend'
EMAIL_USE_SSL = True
EMAIL_HOST = 'smtp.qq.com'
EMAIL_HOST_USER = '371972484@qq.com'
EMAIL_HOST_PASSWORD = '密码'
EMAIL_PORT = 465
DEFAULT_FROM_EMAIL = '371972484@qq.com'
```

(3) 在路径导航文件 urls.py 中设置 URL 链接，主要实现代码如下所示。

```
from django.conf.urls import url, include
from django.contrib import admin
from django.urls import path
urlpatterns = [
    path('admin/', admin.site.urls),
    path('', include('temail.urls')),
]
```

(4) 开始实现应用程序 "tmail" 中的功能。首先在模型文件 models.py 中创模型类 UserData，用于保存注册成功用户的信息。代码如下：

```
from django.db import models

class UserData(models.Model):
    email = models.EmailField(max_length=200, help_text='Required')
    name = models.CharField(max_length=200)
    password= models.CharField(max_length=200)
```

(5) 编写视图文件 views.py 分别实现各个页面的视图，具体实现流程如下所示。
- 编写视图函数 signup() 实现新用户注册功能，获取并验证用户在注册表单中输入的信息，如果信息合法则向邮箱中发送验证信息。代码如下：

```
def signup(request):
    if request.method == 'POST':
        form = SignupForm(request.POST)
        if form.is_valid():
            user = form.save(commit=False)
            user.is_active = False
            user.save()
            current_site = get_current_site(request)
            message = render_to_string('acc_active_email.html', {
                'user':user,
                'domain':current_site.domain,
                'uid': urlsafe_base64_encode(force_bytes(user.pk)).decode(),
                'token': account_activation_token.make_token(user),
            })
            user.save()
            mail_subject = 'Activate your blog account.'
            to_email = form.cleaned_data.get('email')
```

```
            email = EmailMessage(mail_subject, message, to=[to_email])
            email.send()
            return HttpResponse('Please confirm your email address to complete the registration')
        else:
            try:
                u=User.objects.get(username=name)
                u.delete()
            except User.DoesNotExist:
                return HttpResponse('The email given is invalid please check it ')
            except Exception as e:
                return render(request, 'signup.html', {'form': form})
            return HttpResponse('The email given is invalid please check it ')


    else:
        form = SignupForm()

    return render(request, 'signup.html', {'form': form})
```

- 编写视图函数 activate()实现邮箱激活动能。代码如下：

```
def activate(request, uidb64, token):
    try:
        uid = force_text(urlsafe_base64_decode(uidb64))
        user = User.objects.get(pk=uid)
    except(TypeError, ValueError, OverflowError, User.DoesNotExist):
        user = None
    if user is not None and account_activation_token.check_token(user, token):
        user.is_active = True
        user.save()
        # login(request, user)
        # return redirect('home')
        # return render(request, 'login.html')
        return HttpResponse('Thank you for your email confirmation. Now you can login your account.')
        # return render(request, 'login.html', {'form': form})
    else:
        return HttpResponse('Activation link is invalid!')
```

- 编写视图函数 handlelogin()实现登录验证功能，确保只有在数据库中保存的合法会员用户才能登录系统。代码如下：

```
def handlelogin(request):
    if request.method =='POST':
        loginusername = request.POST['uname']
        loginpassword = request.POST['psw']

        user = authenticate(request,username=loginusername,password=loginpassword)
        if user is not None:
            login(request,user)
            return redirect('student')
        else:
            return render(request,'login.html')
```

```
    return render(request,'login.html')


    return render(request,'login.html')
```

- 编写视图函数 student(request)，当用户登录成功后指向模板文件 student.html。代码如下：

```
def student(request):
    return render(request, 'student.html')
```

(6) 编写文件 tokens.py 实现账户激活令牌生成器。代码如下：

```
from django.contrib.auth.tokens import PasswordResetTokenGenerator
from django.utils import six

class AccountActivationTokenGenerator(PasswordResetTokenGenerator):
    def _make_hash_value(self, user, timestamp):
        return (six.text_type(user.pk) + six.text_type(timestamp)) +
six.text_type(user.is_active)

account_activation_token = AccountActivationTokenGenerator()
```

(7) 编写表单文件 forms.py，功能是验证在表单中输入数据的合法性，如果合法则将获取的数据保存到数据库中。代码如下：

```
from django import forms
from django.contrib.auth.models import User
from django.contrib.auth.forms import UserCreationForm

class SignupForm(UserCreationForm):
    email = forms.EmailField(max_length=200, help_text='Required')

    class Meta:
        model = User
        fields = ('username', 'email', 'password1', 'password2')

    def save(self, commit=True):
        user = super(SignupForm, self).save(commit=False)
        user.email = self.cleaned_data["email"]
        if commit:
            user.save()
        return user
```

(8) 编写后台管理文件 admin.py，将会员用户模型注册到后台中。代码如下：

```
from django.contrib import admin

from .models import UserData
admin.site.register(UserData)
```

(9) 在模板文件 signup.html 中实现新用户注册表单效果，代码如下所示。

```
{% block content %}
 <h2>Sign up</h2>
 <form method="post">
```

```
    {% csrf_token %}
     {% for field in form %}
     <p>
      {{ field.label_tag }}<br>
      {{ field }}
      {% if field.help_text %}
       <small style="display: none">{{ field.help_text }}</small>
      {% endif %}
      {% for error in field.errors %}
       <p style="color: red">{{ error }}</p>
      {% endfor %}
     </p>
     {% endfor %}

    <button type="submit">Sign up</button> <!--<a
href="temail/login.html"><button type="submit">LogIn</a></button>-->
   </form>
{% endblock %}
```

模板文件 login.html 实现用户登录验证表单界面效果。代码如下:

```
<html>
<head>
 <title> Login Page </title>
</head>
<body>
    {% comment %} <h3>Thank you for your email confirmation. Now you can login
your account.<h3> {% endcomment %}
    <h2>Login </h2>
    <form action="{% url 'handlelogin' %}" method="post">
     {% csrf_token %}
    <label><b>Username</b></label>
    <input type="text" placeholder="Enter Username" name="uname"
required><br><br>

    <label><b>Password</b></label>
    <input type="password" placeholder="Enter Password" name="psw"
required><br><br>

    <button type="submit">Login</button>
   </form>
</body>
</html>
```

编写模板文件 acc_active_email.html，如果注册信息合法则生成一个激活链接，并将这个链接发送到用户的邮箱中。代码如下：

```
{% autoescape off %}
Hi {{ user.username }},
Please click on the link to confirm your registration,

http://{{ domain }}{% url 'activate' uidb64=uid token=token %}
{% endautoescape %}
```

开始运行程序，首先通过如下命令同步数据库：

```
python manage.py makemigrations
python manage.py migrate
```

然后通过如下命令创建一个管理员账号信息：

```
python manage.py createsuperuser
```

通过如下命令运行我们所创建的这个 Django 工程：

```
python manage.py runserver
```

在网页中输入 http://127.0.0.1:8000/signup/ 后显示新用户注册表单页面，如果输入的邮箱、用户名和密码的格式不合法，则会显示对应的提示信息。效果如图 11-3 所示。注册成功后会提示发送了一封邮件到用户的邮箱中，如图 11-4 所示。

图 11-3　用户注册表单页面效果

图 11-4　提示发送了一封邮件

在注册用户的邮箱中会收到本项目发送的邮件，邮件内容是一条激活链接，如图 11-5 所示。

图 11-5　邮件中的激活链接

第 12 章

用户登录验证模块

在开发动态 Web 网站的过程中，用户登录验证功能是使用非常频繁的一个模块。在本章的内容中，将详细讲解在 Django Web 程序中开发用户登录验证、会员注册等功能知识，为读者步入本书后面知识的学习打下基础。

12.1 使用 auth 实现登录验证系统

模块 auth 是 Django 框架提供的标准权限管理系统，利用它可以实现用户身份认证、用户组和权限管理。模块 auth 可以和后台管理模块 admin 配合使用，快速建立 Web 的管理系统。在创建一个 Django 工程后，会默认使用模块 auth，在 INSTALLED_APPS 中默认显示模块 auth 对应的选项：

扫码观看本节视频讲解

```
django.contrib.auth
```

在本节的内容中，将详细讲解使用 auth 模块的知识。

12.1.1 auth 模块基础

1. 对象 User

对象 User 表示 auth 模块中维护用户信息的关系模式(继承了 models.Model)，该表在数据库中被命名为 auth_user。表 User 的 SQL 描述如下所示。

```
CREATE TABLE "auth_user" (
  "id" integer NOT NULL PRIMARY KEY AUTOINCREMENT,
  "password" varchar(128) NOT NULL, "last_login" datetime NULL,
  "is_superuser" bool NOT NULL,
  "first_name" varchar(30) NOT NULL,
  "last_name" varchar(30) NOT NULL,
  "email" varchar(254) NOT NULL,
  "is_staff" bool NOT NULL,
  "is_active" bool NOT NULL,
  "date_joined" datetime NOT NULL,
  "username" varchar(30) NOT NULL UNIQUE
)
```

User 是一个表示用户的对象，在创建好 User 对象后，Django 会自动生成对应的表，表名为 auth_user，包含上述 SQL 中的字段。在 User 对象中包含如下所示的成员。

- 类 models.User：定义对象 User 的名字。
- username：用户名，必选成员。用户名可以包含字母、数字、_、@、+、. 和 - 字符，少于等于 30 个字符。
- first_name：可选成员，少于等于 30 个字符。
- last_name：可选成员，少于 30 个字符。
- email：可选成员，表示邮箱地址。
- password：必选成员，表示密码的散列数据及元数据。为了系统的安全性，Django 不保存明文格式的密码。原始密码可以无限长，而且可以包含任意字符。
- groups：与 Group 之间的多对多关系。
- user_permissions：与 Permission 之间的多对多关系。

- is_staff：是一个布尔值，设置用户是否可以访问 Admin 站点。
- is_active：是一个布尔值，设置是否激活用户的账号。
- is_superuser：是一个布尔值，只是这个用户拥有所有的权限而不需要被分配某个明确的权限。
- last_login：用户最后一次登录的时间。
- date_joined：创建账户的时间。当创建账号时，默认设置为当前的 date/time。

在模块 auth 中提供了很多 API 来管理用户信息，我们在必要时可以通过如下代码导入表 User 进行操作，例如当其他表需要与 User 建立关联时。

```
from django.contrib.auth.models import User
```

2. 新建用户

通过如下代码可以新建一个用户：

```
user = User.objects.create_user(username, email, password)
```

在建立好 User 对象后需要调用 save() 方法将新用户数据写入到数据库中。例如：

```
user.save()
```

> **注意**
> 为了提高 Web 的安全性，auth 模块不会存储用户密码的明文，而是存储一个散列值，比如迭代使用 Md5 算法。

例如在会员注册操作中会用到 save() 方法，在使用函数 User.objects.create_user() 新建用户时需要判断用户是否存在，通过 User.objects.get(username=xxx) 去获取一个用户 User 对象，如果用户不存在则抛出 User.DoesNotExist 异常，在这个异常中进行创建用户的操作。具体代码如下：

```
# 注册操作
from django.contrib.auth.models import User
try:
  User.objects.get(username=username)
  data = {'code': '-7', 'info': u'用户已存在'}
except User.DoesNotExist:
  user = User.objects.create_user(username, email, password)
  if user is not None:
    user.is_active = False
    user.save()
```

3. 认证用户

在使用模块 auth 认证用户时，需要先导入方法 authenticate：

```
from django.contrib.auth import authenticate
```

然后使用关键字参数传递用户名和密码：

```
user = authenticate(username=username, password=password)
```

需要确认用户的用户名和密码是否有效，如果有效则返回代表该用户的 user 对象，如果无效则返回 None。方法 authenticate 不会检查 is_active 标志位。

4. 修改密码

修改密码功能通过 User 对象的实例方法 set_password 实现，该方法不验证用户身份。

```
user.set_password(new_password)
```

在现实应用中，方法 set_password 通常需要和 authenticate 配合使用：

```
user = auth.authenticate(username=username, password=old_password)
if user is not None:
    user.set_password(new_password)
    user.save()
```

5. 登录

在使用模块 auth 实现登录功能时，需要先使用 import 导入方法 login：

```
from django.contrib.auth import login
```

方法 login 向 session 中添加 SESSION_KEY，便于对用户进行跟踪：

```
'login(request, user)'
```

方法 login 不进行认证，也不检查 is_active 标志位，一般和 authenticate 配合使用。

```
user = authenticate(username=username, password=password)
if user is not None:
    if user.is_active:
        login(request, user)
```

例如下面是实现验证的演示代码：

```
from django.contrib.auth import authenticate, login, logout
#认证操作
ca = Captcha(request)
if ca.validate(captcha_code):
  user = authenticate(username=username, password=password)
  if user is not None:
    if user.is_active:
     # 登录成功
     login(request, user)  # 登录用户
     data = {'code': '1', 'info': u'登录成功', 'url': 'index'}
    else:
     data = {'code': '-5', 'info': u'用户未激活'}
  else:
     data = {'code': '-4', 'info': u'用户名或密码错误'}
else:
  data = {'code': '-6', 'info': u'验证码错误'}
```

6. 退出登录

在模块 auth 中，方法 logout 会移除 request 中的 user 信息，并刷新 session。例如：

```
from django.contrib.auth import logout
def logout_view(request):
    logout(request)
from django.contrib.auth import authenticate, login, logout
def logout_system(request):
    """
    退出登录
    :param request:
    :return:
    """
    logout(request)
    return HttpResponseRedirect('/')
```

7. 只允许登录的用户访问

通过使用装饰器@login_required 装饰的 view 函数，会先通过 session key 检查用户是否登录，已登录用户可以正常地执行操作，未登录用户将被重定向到 login_url 指定的位置。如果未指定 login_url 参数，则重定向到 settings.LOGIN_URL。例如：

```
from django.contrib.auth.decorators import login_required
@login_required(login_url='/accounts/login/')
def my_view(request):
    ...
from django.contrib.auth.decorators import login_required
@login_required
def user_index(request):
    """
    用户管理首页
    :param request:
    :return:
    """
    if request.method == "GET":
        # 用户视图实现
Group
```

在模块 auth 中，django.contrib.auth.models.Group 定义了用户组的模型，每个用户组拥有 id 和 name 两个字段，该模型在数据库中被映射为数据表 auth_group。

在 User 对象中有一个名为 groups 的多对多字段，由数据表 auth_user_groups 维护多对多关系。Group 对象可以通过 user_set 反向查询用户组中的用户。可以通过创建、删除 Group 对象的方式来添加或删除用户组，例如下面的演示代码：

```
# 添加
group = Group.objects.create(name=group_name)
group.save()
# 删除
group.delete()
```

我们可以通过标准的多对多字段操作管理用户与用户组的关系：
- 用户加入用户组：user.groups.add(group)或 group.user_set.add(user)。
- 用户退出用户组：user.groups.remove(group)或 group.user_set.remove(user)。

- 用户退出所有用户组：user.groups.clear()。
- 用户组中所有用户退出组：group.user_set.clear()。

8. Permission

在模块 auth 中提供了模型级的权限控制，可以检查用户是否对某个数据表拥有增加 (add)、修改 (change) 和删除 (delete) 权限。模块 auth 中无法提供对象级的权限控制，即无法检查用户是否对数据表中某条记录拥有增改删的权限。假设在博客系统中通过数据库表 article 来存放博文信息，auth 可以检查某个用户是否拥有对所有博文的管理权限，但无法检查用户对某一篇博文是否拥有管理权限。

（1）检查用户权限。

使用方法 user.has_perm 可以检查用户是否拥有操作某个模型的权限。例如：

```
user.has_perm('blog.add_article')
user.has_perm('blog.change_article')
user.has_perm('blog.delete_article')
```

上述代码的功能是，检查用户是否拥有 blog 这个 app 中 article 模型的添加、修改和删除权限，如果拥有则返回 True。

方法 has_perm 仅仅是进行权限检查，即使用户没有权限也不会阻止程序员执行相关操作。

通过使用修饰器 permission_required 可以代替方法 has_perm，并在用户没有相应权限时重定向到登录页或者抛出异常。例如：

```
# permission_required(perm[, login_url=None, raise_exception=False])

@permission_required('blog.add_article')
def post_article(request):
    pass
```

每个模型默认拥有增加 (add)、修改 (change) 和删除 (delete) 权限。在模型 django.contrib.auth.models.Permission 中保存了项目中的所有权限，该模型在数据库中被保存为 auth_permission 数据表，每条权限拥有 id、name、content_type_id 和 codename 四个字段。

（2）管理用户权限。

User 和 Permission 通过多对多字段 user.user_permissions 进行关联，在数据库中由数据表 auth_user_user_permissions 维护。

- user.user_permissions.add(permission)：添加权限。
- user.user_permissions.delete(permission)：删除权限。
- user.user_permissions.clear()：清空权限。

（3）自定义权限。

在定义 Model 时可以使用 Meta 自定义权限：

```
class Discussion(models.Model):
    ...
    class Meta:
```

```
    permissions = (
        ("create_discussion", "Can create a discussion"),
        ("reply_discussion", "Can reply discussion"),
    )
```

12.1.2 实战演练：带登录验证功能的简易新闻系统

在下面的实例中，演示了使用 Django 中的 auth 模块开发一个简易新闻系统的过程。本实例具有如下所示的功能。

- 前台会员注册：用户可以注册成为系统会员。
- 登录验证：验证会员登录信息是否合法。
- 前台显示新闻信息：包括列表显示新闻和某条新闻的详情信息。
- 后台管理：管理员可以发布、修改或删除新闻信息。

源码路径：**daima\12\yanzheng**

（1）通过如下命令新建一个名为"yanzheng"的工程，在"yanzheng"目录下新建一个名为"blog"的 app 项目。

```
django-admin.py startproject yanzheng
cd yanzheng
python manage.py startapp blog
```

（2）在文件 settings.py 中，将上面创建的 app 项目名"blog"添加到 INSTALLED_APPS 中。

```
INSTALLED_APPS = [
    'django.contrib.admin',
    'django.contrib.auth',
    'django.contrib.contenttypes',
    'django.contrib.sessions',
    'django.contrib.messages',
    'django.contrib.staticfiles',
    'blog',
]
```

（3）在路径导航文件 urls.py 中设置 URL 链接，主要实现代码如下所示。

```
urlpatterns = [
    path(r'admin/', admin.site.urls),
    path(r'', views.index),
    path(r'regist/', views.regist),
    path(r'login/', views.login),
    path(r'logout/', views.logout),
    path(r'article/', views.article),
    path(r'(?P<id>\d+)/', views.detail, name='detail'),
]
```

（4）在视图文件 views.py 中分别实现各个页面的视图。

- index：来到系统主页。
- regist：实现会员注册视图，将表单中的注册数据添加到系统数据库中。

- login：实现登录验证视图，对登录表单中的数据进行验证。
- logout：实现用户退出视图。
- article：获取系统数据库中的表"article"中的信息，将 article 标题列表显示出来。
- detail：获取并显示某条 article 的详细信息。

文件 views.py 的主要实现代码如下所示。

```python
class UserForm(forms.Form):
    username = forms.CharField(label='用户名', max_length=100)
    password = forms.CharField(label='密 码', widget=forms.PasswordInput())

# Django 的 form 的作用：
# 1.生成 html 标签
# 2.用来做用户提交的验证
# 前端：form 表单
# 后台：创建 form 类，当请求到来时，先匹配，匹配出正确信息和错误信息
def index(request):
    return render(request, 'index.html')


def regist(request):
    if request.method == 'POST':
        uf = UserForm(request.POST)   # 包含用户名和密码
        if uf.is_valid():
            # 获取表单数据
            username = uf.cleaned_data['username']  # cleaned_data 类型是字典，里面是提交成功后的信息
            password = uf.cleaned_data['password']
            # 添加到数据库
            # registAdd = User.objects.get_or_create(username=username, password=password)
            registAdd = User.objects.create_user(username=username, password=password)
            # print registAdd
            if registAdd == False:
                return render(request, 'share1.html', {'registAdd': registAdd, 'username': username})

            else:
                # return HttpResponse('ok')
                return render(request, 'share1.html', {'registAdd': registAdd})
                # return render_to_response('share.html',{'registAdd':registAdd},context_instance = RequestContext(request))
    else:
        # 如果不是 post 提交数据，就不传参数创建对象，并将对象返回给前台，直接生成 input 标签，内容为空
        uf = UserForm()
    # return render_to_response('regist.html',{'uf':uf},context_instance = RequestContext(request))
    return render(request, 'regist1.html', {'uf': uf})


def login(request):
    if request.method == 'POST':
        username = request.POST.get('username')
```

```python
        password = request.POST.get('password')
        print(username, password)

        re = auth.authenticate(username=username, password=password)   # 用户认证
        if re is not None:   # 如果数据库里有记录(即与数据库里的数据相匹配或者对应或者符合)
            auth.login(request, re)   # 登录成功
            return redirect('/', {'user': re})   # 跳转--redirect 指从一个旧的 url 转
到一个新的 url
        else:   # 数据库里不存在与之对应的数据
            return render(request, 'login.html', {'login_error': '用户名或密码错
误'})   # 注册失败
    return render(request, 'login.html')


def logout(request):
    auth.logout(request)
    return render(request, 'index.html')


def article(request):
    article_list = Article.objects.all()
    # print article_list
    # print type(article_list)
    # QuerySet 是一个可遍历结构，包含一个或多个元素，每个元素都是一个 Model 实例
    # QuerySet 类似于 Python 中的 list, list 的一些方法 QuerySet 也有，比如切片、遍历
    # 每个 Model 都有一个默认的 manager 实例，名为 objects, QuerySet 有两种来源：通过
manager 的方法得到、通过 QuerySet 的方法得到。mananger 的方法和 QuerySet 的方法大部分同
名同含义,如 filter()、update() 等，但也有些不同,如 manager 有 create()、get_or_create(),
而 QuerySet 有 delete() 等
    return render(request, 'article.html', {'article_list': article_list})


def detail(request, id):
    # print id
    try:
        article = Article.objects.get(id=id)
        # print type(article)
    except Article.DoesNotExist:
        raise Http404
    return render(request, 'detail.html', locals())
```

(5) 在模型文件 models.py 中创建数据库表 Article，因为本实例并没有特意创建会员信息表，而是直接使用了 Django 自带的 user 表，所以在文件 models.py 中没有创建表 user。文件 models.py 的主要实现代码如下所示。

```python
class Article(models.Model):
    title = models.CharField(u'标题', max_length=256)
    content = models.TextField(u'内容')
    pub_date = models.DateTimeField(u'发表时间', auto_now_add=True, editable=True)
    update_time = models.DateTimeField(u'更新时间', auto_now=True, null=True)

    def __unicode__(self):   # 在 Python3 中用 __str__ 代替 __unicode__
        return self.title
```

根据上面的模型,通过如下命令创建数据库。

```
python manage.py makemigrations
python manage.py migrate
```

(6) 在文件 admin.py 中设置在后台显示 ArticleAdmin 模块,只有这样才能在后台显示文章管理模块。文件 admin.py 的主要实现代码如下所示。

```
from blog.models import Article
class ArticleAdmin(admin.ModelAdmin):
    list_display = ('title', 'title','pub_date', 'update_time',)
admin.site.register(Article, ArticleAdmin)
```

(7) 在模板文件 login.html 中实现用户登录表单效果,一定要在<form>标记后面添加{% csrf_token %},否则不会通过 Session 验证。文件 login.html 的主要实现代码如下所示。

```
<form action="/login/" method="POST">{% csrf_token %}
    <h2>请登录</h2>
    <input type="text" name="username" />
    <input type="password" name="password" />
    <button type="submit">登录</button>
    <p style="color: red">{{ login_error }}</p>
</form>
```

在模板文件 regist1.html 中实现用户注册界面效果,一定要在<form>标记后面添加{% csrf_token %},否则不会通过 Session 验证。文件 regist1.html 的主要实现代码如下所示。

```
<form method="POST" enctype="multipart/form-data">
{% csrf_token %}
    {{uf.as_p}}
    <input type="submit" value="OK" />
</form>
{#    <a href="http://127.0.0.1:8000/login">注册</a>#}
{% endblock %}
```

前台新闻列表界面的执行效果如图 12-1 所示。

图 12-1 前台新闻列表界面

新用户注册界面的执行效果如图 12-2 所示。

后台新闻管理界面效果如图 12-3 所示。

图 12-2 新用户注册界面

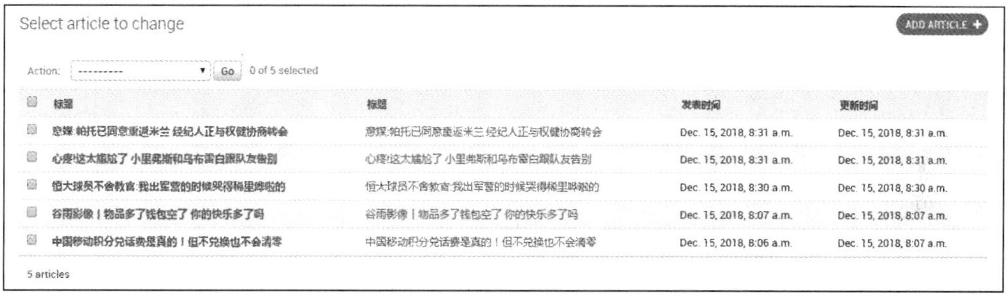

图 12-3 后台新闻管理界面

12.2 使用 django-allauth 实现登录验证系统

框架 django-allauth 是一款管理用户登录与注册的第三方 Django 安装包,利用它能够实现用户注册、用户登录、退出登录、第三方 auth 登录(微信、微博等)、邮箱验证、登录后密码重置、忘记密码邮箱发送密码重置等功能。在本节的内容中,将详细讲解使用 django-allauth 框架的知识。

扫码观看本节视频讲解

12.2.1 django-allauth 框架基础

1. django-allauth 的安装与设置

我们可以使用如下 pip 命令安装 django-allauth。

```
pip install django-allauth
```

安装成功后,在新建 Django 工程时,需要在文件 settings.py 中将和 allauth 相关的 app 加入到 INSTALLED_APP 中。读者需要注意的是,因为 allauth 需要依赖于 django.contrib.sites,所以必须将 django.contrib.sites 加入到 INSTALLED_APP 中,同时设置 SITE_ID。例如下面的设置:

```python
INSTALLED_APPS = [
    'django.contrib.admin',
    'django.contrib.auth',
    'django.contrib.contenttypes',
    'django.contrib.sessions',
    'django.contrib.messages',
    'django.contrib.staticfiles',
    'django.contrib.sites',
    'allauth',
    'allauth.account',
    'allauth.socialaccount',
    'allauth.socialaccount.providers.github',
]
    SITE_ID = 1
```

另外,还需要设置 BACKENDS 并提供用户登录验证的方法和用户登录后跳转的链接。假设我们需要让用户通过 Email 或者用户名登录,则需要进行邮箱验证和通过电子邮箱发送密码重置链接,设置服务器的电子邮箱地址。如果没有邮件服务器,完全可以使用自己的 QQ 或 163 邮箱,具体设置方法和前面 11.2.2 节中介绍的一样。例如下面的设置代码:

```python
# 基本设定
ACCOUNT_AUTHENTICATION_METHOD = 'username_email'
ACCOUNT_EMAIL_REQUIRED = True
LOGIN_REDIRECT_URL = '/accounts/profile/'

AUTHENTICATION_BACKENDS = (
    'django.contrib.auth.backends.ModelBackend',
    'allauth.account.auth_backends.AuthenticationBackend',
)

# 邮箱设定
EMAIL_HOST = 'smtp.qq.com'
EMAIL_PORT = 25
EMAIL_HOST_USER = 'xxxx@qq.com'   # 你的 QQ 账号和授权码
EMAIL_HOST_PASSWORD = 'xxxx'
EMAIL_USE_TLS = True   # 这里必须是 True,否则发送不成功
EMAIL_FROM = 'xxxx3116@qq.com'   # 你的 QQ 账号
DEFAULT_FROM_EMAIL = 'xxxx3116@qq.com'
```

接下来还需要将 allauth 加到 app 的 urls.py 中:

```python
urlpatterns = [
    path('admin/', admin.site.urls),
    path('accounts/', include('allauth.urls')),
]
```

2. django-allauth 常用的设置选项

我们也可以添加其他选项来设置想要使用的功能,例如设置邮件确认过期时间,限制用户使用错误密码登录的持续时间。在 django-allauth 模块中常用的设置选项如下所示。

(1) ACCOUNT_AUTHENTICATION_METHOD(="username" | "email" | "username_email"):设置要使用的登录方法,可以使用用户名或电子邮件地址。

(2) ACCOUNT_EMAIL_CONFIRMATION_EXPIRE_DAYS (=3)：邮件确认的截止日期(天数)。

(3) ACCOUNT_EMAIL_VERIFICATION (="optional")：在用户注册中是否使用邮件验证方法，取值有"强制(mandatory)""可选(optional)"或"否(none)"。

(4) ACCOUNT_EMAIL_CONFIRMATION_COOLDOWN (=180)：邮件发送后的冷却时间(以秒为单位)。

(5) ACCOUNT_LOGIN_ATTEMPTS_LIMIT (=5)：可以登录失败的次数。

(6) ACCOUNT_LOGIN_ATTEMPTS_TIMEOUT (=300)：从上次登录失败后，用户被禁止再次登录的持续时间。

(7) ACCOUNT_LOGIN_ON_EMAIL_CONFIRMATION (=False)：如果设置为 True，用户一旦确认他们的电子邮件地址，就会自动登录。

(8) ACCOUNT_LOGOUT_ON_PASSWORD_CHANGE (=False)：设置在更改或设置密码后是否自动退出。

(9) ACCOUNT_LOGIN_ON_PASSWORD_RESET (=False)：如果设置为 True，用户将在重置密码后自动登录。

(10) ACCOUNT_SESSION_REMEMBER (=None)：控制会话的生命周期，可选项还有 False 和 True。

(11) ACCOUNT_SIGNUP_EMAIL_ENTER_TWICE (=False)：设置用户在注册时是否需要两次输入邮箱进行确认。

(12) ACCOUNT_SIGNUP_PASSWORD_ENTER_TWICE (=True)：设置用户在注册时是否需要输入密码两次进行确认。

(13) ACCOUNT_USERNAME_BLACKLIST (=[])：用户不能使用的用户名列表。

(14) ACCOUNT_UNIQUE_EMAIL (=True)：设置电子邮件地址的唯一性。

(15) ACCOUNT_USERNAME_MIN_LENGTH (=1)：用户名允许的最小长度，是一个整数。

(16) LOGIN_REDIRECT_URL (="/")：设置用户登录后跳转到的链接。

(17) ACCOUNT_LOGOUT_REDIRECT_URL (="/")：设置退出登录后跳转到的链接。

3. URLs 和视图

在 django_allauth 框架中提供了多个可以访问的内置 URLs，具体说明如下所示。

- /accounts/login/(URL 名 account_login)：用户登录页面。
- /accounts/signup/ (URL 名 account_signup)：用户注册页面。
- /accounts/password/reset/(URL 名 account_reset_password)：重置密码页面。
- /accounts/logout/ (URL 名 account_logout)：退出登录页面。
- /accounts/password/set/ (URL 名 account_set_password)：设置密码页面。
- /accounts/password/change/ (URL 名 account_change_password)：修改密码页面(需登录)。

- /accounts/email/(URL 名 account_email)：用户可以添加和移除 email 功能，并验证。
- /accounts/social/connections/(URL 名 socialaccount_connections)：管理第三方账户页面。

4．执行效果

通过上面的介绍，已经简单配置了设置文件 settings.py 和 url.py，接下来通过如下命令即可运行 django-allauth 程序。

```
python manage.py makemigrations
python manage.py migrate
python manage.py runserver
```

首先需要登录 admin 后台，后台界面如图 12-4 所示。在后台需要将 site 域名设置为 127.0.0.1:8000，并任意设置一个名字，再退出登录。

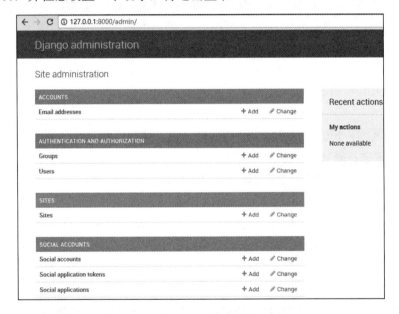

图 12-4　后台界面

接下来就可以通过访问如下链接查看 allauth 的执行效果了。如果已经设置好了服务器邮箱，那么所有涉及邮箱验证和密码重置部分都可以正常使用邮箱实现。

```
http://127.0.0.1:8000/accounts/signup/
http://127.0.0.1:8000/accounts/login/
http://127.0.0.1:8000/accounts/logout/
http://127.0.0.1:8000/accounts/password/reset/
```

12.2.2　实战演练：在 django-allauth 中使用百度账户实现用户登录系统

在现实应用中，很多 Web 都提供了第三方账户登录功能，例如可以使用 QQ 账号、百度账号等登录论坛。在下面的实例代码中，演示了在 django-allauth 中使用百度账户的过程。

源码路径：daima\12\myproject

（1）通过如下命令创建一个名为"myproject"的工程，在工程里面新建一个名为"yanzheng"的 app。

```
django-admin.py startproject myproject
cd myproject
python manage.py startapp yanzheng
```

（2）在配置文件 settings.py 中首先将"yanzheng"添加到 INSTALLED_APPS，然后将 allauth 框架用到的模块添加到 INSTALLED_APPS，最后将第三方账户百度功能添加到 INSTALLED_APPS。

```
INSTALLED_APPS = [
    'django.contrib.admin',
    'django.contrib.auth',
    'django.contrib.contenttypes',
    'django.contrib.sessions',
    'django.contrib.messages',
    'django.contrib.staticfiles',
    'django.contrib.sites',
    'allauth',
    'allauth.account',
    'allauth.socialaccount',
    'allauth.socialaccount.providers.github',
    'yanzheng',
    'allauth.socialaccount.providers.baidu',
]
```

（3）在配置文件 settings.py 中设置服务器邮箱，可以用 QQ 邮箱、126、网易等第三方邮箱账号，例如笔者使用的是 QQ 邮箱。

```
STATIC_URL = '/static/'
# 邮箱设定
EMAIL_HOST = 'smtp.qq.com'
EMAIL_PORT = 25
EMAIL_HOST_USER = '729017304@qq.com'      # 你的 QQ 账号和授权码
EMAIL_HOST_PASSWORD = ''                  #密码
EMAIL_USE_TLS = True                      # 这里必须是 True，否则发送不成功
EMAIL_FROM = '729017304@qq.com' # 你的 QQ 账号
DEFAULT_FROM_EMAIL = '729017304@qq.com'
```

（4）编写 URL 导航文件 myproject/urls.py，主要实现代码如下所示。

```
urlpatterns = [
    path('admin/', admin.site.urls),
    path('accounts/', include('allauth.urls')),
    path('accounts/', include('myaccount.urls')),
]
```

（5）编写模型文件 models.py，在 Django 自带的 User 模型的基础上对其扩展，创建 UserProfile 模型，在里面添加了 org 和 telephone 两个字段。文件 models.py 的主要实现代码

如下所示。

```python
class UserProfile(models.Model):
    user = models.OneToOneField(User, on_delete=models.CASCADE,
related_name='profile')
    org = models.CharField(
        'Organization', max_length=128, blank=True)
    telephone = models.CharField(
        'Telephone', max_length=50, blank=True)
    mod_date = models.DateTimeField('Last modified', auto_now=True)
    class Meta:
        verbose_name = 'User Profile'
    def __str__(self):
        return "{}'s profile".format(self.user.username)

    def account_verified(self):
        if self.user.is_authenticated:
            result = EmailAddress.objects.filter(email=self.user.email)
            if len(result):
                return result[0].verified
        return False
```

(6) 编写 URL 路径导航文件 yanzheng/urls.py，设置通过 URL 来到用户登录成功显示的视图 profile，通过 URL 来到用户修改资料的视图 profile_update。文件 yanzheng/urls.py 的主要实现代码如下所示。

```python
urlpatterns = [
    re_path(r'^profile/$', views.profile, name='profile'),
    re_path(r'^profile/update/$', views.profile_update,
name='profile_update'),
]
```

(7) 编写视图文件 views.py，分别创建用户登录成功视图和用户资料修改视图，主要实现代码如下所示。

```python
@login_required
def profile(request):
    user = request.user
    return render(request, 'profile.html', {'user': user})


@login_required
def profile_update(request):
    user = request.user
    user_profile = get_object_or_404(UserProfile, user=user)
    if request.method == "POST":
        form = ProfileForm(request.POST)

        if form.is_valid():
            user.first_name = form.cleaned_data['first_name']
            user.last_name = form.cleaned_data['last_name']
            user.save()

            user_profile.org = form.cleaned_data['org']
            user_profile.telephone = form.cleaned_data['telephone']
```

```
            user_profile.save()

            return HttpResponseRedirect(reverse('yanzheng:profile'))
    else:
        default_data = {'first_name': user.first_name, 'last_name': user.last_name,
                        'org': user_profile.org, 'telephone': user_profile.telephone, }
        form = ProfileForm(default_data)

    return render(request, 'profile_update.html', {'form': form, 'user': user})
```

(8) 用户登录后可以通过表单修改自己的资料，在表单文件 forms.py 中创建两个表单，一个在更新用户资料时使用，另一个用于重写用户登录表单。文件 forms.py 的主要实现代码如下所示。

```
class ProfileForm(forms.Form):
    first_name = forms.CharField(label='First Name', max_length=50, required=False)
    last_name = forms.CharField(label='Last Name', max_length=50, required=False)
    org = forms.CharField(label='Organization', max_length=50, required=False)
    telephone = forms.CharField(label='Telephone', max_length=50, required=False)

class SignupForm(forms.Form):

    def signup(self, request, user):
        user_profile = UserProfile()
        user_profile.user = user
        user.save()
        user_profile.save()
```

(9) 在配置文件 settings.py 中添加如下代码，告诉 django-allauth 使用自定义的登录表单。

```
ACCOUNT_SIGNUP_FORM_CLASS = 'yanzheng.forms.SignupForm'
```

(10) 编写模板文件 profile.html，用户登录成功后通过此页面显示用户的账户信息。文件 profile.html 的主要实现代码如下所示。

```
{% block content %}
{% if user.is_authenticated %}
<a href="{% url 'yanzheng:profile_update' %}">Update Profile</a> | <a href="{% url 'account_email' %}">Manage Email</a> | <a href="{% url 'account_change_password' %}">Change Password</a> |
<a href="{% url 'account_logout' %}">Logout</a>
{% endif %}
<p>Welcome, {{ user.username }}.</p>


<h2>My Profile</h2>

<ul>
    <li>First Name: {{ user.first_name }} </li>
```

```
        <li>Last Name: {{ user.last_name }} </li>
        <li>Organization: {{ user.profile.org }} </li>
        <li>Telephone: {{ user.profile.telephone }} </li>
</ul>


{% endblock %}
```

(11) 编写模板文件 profile_update.html,用户登录成功后通过此页面修改自己的账户信息。文件 profile_update.html 的主要实现代码如下所示。

```
{% block content %}
{% if user.is_authenticated %}
<a href="{% url 'yanzheng:profile_update' %}">Update Profile</a> | <a href="{% url 'account_email' %}">Manage Email</a> | <a href="{% url 'account_change_password' %}">Change Password</a> |
<a href="{% url 'account_logout' %}">Logout</a>
{% endif %}
<h2>Update My Profile</h2>

<div class="form-wrapper">
  <form method="post" action="" enctype="multipart/form-data">
    {% csrf_token %}
    {% for field in form %}
      <div class="fieldWrapper">
     {{ field.errors }}
     {{ field.label_tag }} {{ field }}
     {% if field.help_text %}
       <p class="help">{{ field.help_text|safe }}</p>
     {% endif %}
       </div>
    {% endfor %}
    <div class="button-wrapper submit">
     <input type="submit" value="Update" />
    </div>
  </form>
</div>

{% endblock %}
```

(12) 登录百度开发者中心 http://developer.baidu.com/,创建一个项目(名字自取,例如命名为"django"),百度会自动为你分配 API Key 和 Secret Key,如图 12-5 所示。

(13) 单击图 12-5 左侧导航中的"安全设置"链接,在弹出的页面中设置回调 URL。这样当百度授权登录完成后,可以跳转回自己的网站。本地回调 URL 的地是 http://127.0.0.1:8000/accounts/baidu/login/callback/,如图 12-6 所示。

(14) 通过如下命令创建一个管理员账户:

```
python manage.py createsuperuser
```

在浏览器中输入 http://127.0.0.1:8000/admin 登录后台管理界面,单击 Social applications 链接,在新界面中单击 ADD SOCIAL APPLICATION 按钮,在弹出的表单界面中输入前面申请到的 API Key 和 Secret Key。其中 Provider 选项的值是"Baidu";Name 选项的值是

"django",和百度开发者中心的名字相同;在 Client id 文本框输入 API key,在 Secret key 文本框输入 Secret key;在 Sites 选项中设置允许的站点,因为我们是本地调试,建议将 http://127.0.0.1:8000 和 http://127.0.0.1 都添加进来。最终界面效果如图 12-7 所示。

图 12-5 项目"django"的 API Key 和 Secret Key

图 12-6 "安全设置"界面

图 12-7 后台设置

(15) 在浏览器中输入 http://127.0.0.1:8000/accounts/login/ 来到登录页面,此时会显示 Baidu 链接,如图 12-8 所示。

单击 Baidu 链接后会弹出百度登录表单页面，可以通过百度账号登录系统，如图 12-9 所示。

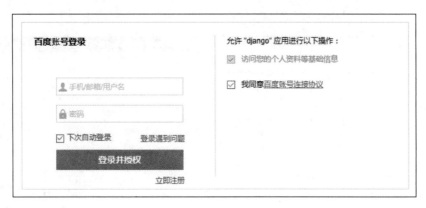

图 12-9　输入百度账号

另外，在用户注册会员时还提供了邮箱验证功能，系统会向用户的邮箱中发送一封验证邮件，如图 12-10 所示。只有单击邮件中的链接后才可以实现用户注册功能。

图 12-10　系统发送的验证邮件

第 13 章

计数器模块

在现实应用中，我们经常需要在 Web 程序中统计并显示一个页面的浏览次数，比如统计某一文件的下载次数，统计某一用户在单位时间内的登录次数等。在本章的内容中，将详细讲解在 Django Web 程序中开发网页计数器的方法，为读者步入本书后面知识的学习打下基础。

13.1 实战演练：一个简单的网页计数器

在下面的实例代码中，将演示使用 Django 框架开发一个简易网页计数器的过程。本实例的策略是，将统计访问本网页的次数保存到服务器本地文件中。

扫码观看本节视频讲解

源码路径：**daima\13\jishu1**

（1）通过如下命令创建一个名为"jishu1"的工程，定位到工程根目录，然后新建一个名为"online"的 app。

```
django-admin.py startproject jishu1
cd jishu1
python manage.py startapp online
```

（2）在文件 urls.py 中设置系统统计页面的 url 是 index/，主要实现代码如下所示。

```
from online.views import myHelloWorld
urlpatterns = [
   path('index/', myHelloWorld),
]
```

（3）统计页面对应的视图文件是 views.py，在里面定义函数 getCount()实现对本地文件的读写工作，定义函数 myHelloWorld()获取并显示函数 getCount()统计的数据。文件 views.py 的主要实现代码如下所示。

```
def getCount():#获取访问次数
    countfile = open('count.dat','r+')#以读写形式打开文件
    counttext = countfile.read()
    try:
        count = int(counttext)+1
    except:
        count = 1
    countfile.seek(0)
    countfile.truncate()#清空文件
    countfile.write(str(count))#重新写入新的访问量
    countfile.flush()
    countfile.close()
    return count

def myHelloWorld(request):
    t = loader.get_template("index.html")#导入模板
    time = getTime()
    count = getCount()
    para = {"title":"我的个人主页",'welcome':"欢迎","content":"这是一个访问站点计数器","count":count,"time":time}
    return HttpResponse(t.render(para))
```

（4）在模板文件 index.html 中统计访问次数和时间，具体实现代码如下所示。

```
<html>
<body>
<title>{{title}}</title>

<h1>{{welcome}}</h1>

<p>网页访问计数器</p>
<li>{{content}}</li>
<li>您是第{{count}}个访问本站的朋友</li>
<li>访问时间:{{time}}</li>
</body>
</html>
```

在浏览器中输入 http://127.0.0.1:8000/index/ 后的执行效果如图 13-1 所示。

图 13-1　执行效果

13.2　实战演练：使用数据库保存统计数据

前面介绍的使用本地文件存储统计次数的方式有很大的缺点，如果有大量的用户访问就会造成对文件资源操作的资源消耗，并且还会引起数据不准确的问题。在接下来的实例中，我们将使用数据库保存统计数据。

源码路径：**daima\13\demo**

扫码观看本节视频讲解

13.2.1　创建 Django 工程

(1) 通过如下命令新建一个名为"demo"的工程，然后定位到工程根目录，新建一个名为"blog"的 app。

```
django-admin.py startproject demo
cd demo
python manage.py startapp blog
```

(2) 在系统设置文件 settings.py 中，将上面创建的"blog"添加到 INSTALLED_APP 中。为了节省篇幅，将不再列出文件 settings.py 的代码。

13.2.2　实现数据库

(1) 编写模型文件 blog/models.py，创建一个保存博客文章的表 Article，在表中创建一个名为 views 的字段，用来记录浏览次数。另外还定义一个称作 viewed 的方法，使 views 在每次访问后增加 1。在 viewed 方法中使用方法 save(update_fields=['views'])只保存 views 字段的更新信息，而不是更新表 Article 的所有信息，这样做的好处是减轻数据库写入的工作量，降低服务器的负担。模型文件 models.py 的主要实现代码如下所示。

```python
class Article(models.Model):

    STATUS_CHOICES = (
        ('d', '草稿'),
        ('p', '发表'),
    )

    title = models.CharField('标题', max_length=200, unique=True)
    slug = models.SlugField('slug', max_length=60)
    body = models.TextField('正文')
    pub_date = models.DateTimeField('发布时间', default= now, null=True)
    create_date = models.DateTimeField('创建时间', auto_now_add=True)
    mod_date = models.DateTimeField('修改时间', auto_now=True)
    status = models.CharField('文章状态', max_length=1, choices=STATUS_CHOICES, default='p')
    views = models.PositiveIntegerField('浏览量', default=0)
    author = models.ForeignKey(User, verbose_name='作者', on_delete=models.CASCADE)

    def __str__(self):
        return self.title

    class Meta:
        ordering = ['-pub_date']
        verbose_name = "文章"
        verbose_name_plural = verbose_name
        get_latest_by = 'create_date'

    def get_absolute_url(self):
        return reverse('blog:article_detail', args=[str(self.id)])

    def viewed(self):
        self.views += 1
        self.save(update_fields=['views'])
```

(2) 使用如下命令创建数据库表。

```
python manage.py makemigrations
python manage.py migrate
```

13.2.3 配置 URL

(1) 首先设置"demo"目录下的导航文件 urls.py,主要实现代码如下所示。

```
urlpatterns = [
    path('admin/', admin.site.urls),
    path('blog/', include('blog.urls')),
]
```

(2) 然后在"blog"目录中新建文件 urls.py,并添加如下所示的代码。

```
urlpatterns = [
```

```
    # 展示文章详情
    re_path(r'^article/(?P<pk>\d+)/$',
            views.ArticleDetailView.as_view(), name='article_detail'),
    # 展示所有文章
    path('', views.ArticleListView.as_view(), name='article_list'),
]
```

13.2.4 实现视图

编写视图文件 views.py，使用了 Django 自带的通用视图来显示文章列表和文章详情，使用 ListView 来显示系统内博客文章的列表，使用 DetailView 来显示系统内某篇博客文章的详细信息。文件 views.py 的主要实现代码如下所示。

```
class ArticleDetailView(DetailView):
    model = Article

    def get_object(self, queryset=None):
        obj = super().get_object(queryset=queryset)
        obj.viewed()
        return obj
class ArticleListView(ListView):
    queryset = Article.objects.filter(status='p').order_by('-pub_date')
    paginate_by = 6
```

说明：

- 假如想要访问/blog/article/2/这篇文章，服务器会根据 URL 的映射关系，调用 ArticleDetailView 来显示文章 id 为 2 的这篇文章。
- ArticleDetailView 通过 URL 传递过来的参数(例如 id=2)获取当前文章对象，并通过模板 blog/article_detail.html 显示。每次通过 get_object 方法获取文章对象后，还调用该对象的 viewed 的方法，使计数增加 1。
- 每当用户重新访问/blog/article/2/或刷新浏览器，计数器都会增加 1。

13.2.5 实现模板

使用模板文件 article_list.html 显示系统博客文章列表，主要实现代码如下所示。

```
<div id="result"></div>

{% if article_list %}
<p>共找到 {{ article_list | length }} 条记录。</p>
  <ul>
   {% for article in article_list %}
  <li><a href="{% url 'blog:article_detail' article.id %}">
{{ article.title }}</a> {{ article.pub_date | date:"Y-m-j" }}</li>
    {% endfor %}
  </ul>
{% endif %}
```

使用模板文件 article_detail.html 显示某篇博客文章的详细信息,在此页面实现访问统计功能,主要实现代码如下所示。

```
<body>
<h3>{{ article.title }}</h3>
<p>日期: {{ article.pub_date | date:"Y-m-j" }}</p>
<p>{{ article.body }}</p>
<p>浏览次数: {{ article.views }}</p>
</body>
```

13.2.6 调试运行

(1) 使用如下命令创建一个后台管理员:

```
python manage.py createsuperuser
```

(2) 使用如下命令运行程序:

```
python manage.py runserver
```

先登录 http://127.0.0.1:8000/admin/ 页面,使用上面创建的管理员登录后台界面,然后添加两篇博客文章。此时在浏览器输入 http://127.0.0.1:8000/blog/后会列表显示刚刚添加的两篇文章,如图 13-2 所示。单击某条文章标题会来到文章详情页面,在此页面会显示这篇文章的详细内容和被访问次数,如图 13-3 所示。

图 13-2 文章列表

图 13-3 在文章详情页面显示统计次数

13.3 实战演练:使用第三方库实现访问计数器

在本书前面的内容中,已经讲解了实现本地文件计数器和数据库计数器的知识。其实在 Django 框架中还有一种更为简单的计数器方案,那便是使用第三方库 django-hitcount。在下面的实例代码中,演示了在 Django Web 中使用 django-hitcount 实现访问统计的过程。

扫码观看本节视频讲解

13.3.1 准备工作

在使用 django-hitcount 之前需要先通过如下命令安装:

```
pip install django-hitcount
```

本项目是框架 django-hitcount 官方自带的演示实例,代码在 https://github.com/thornomad/django-hitcount 托管,读者可以随时关注代码的变化和升级情况。

源码路径:daima\13\example_project

(1) 在 GitHub 下载框架 django-hitcount 的源码,复制里面的演示实例目录"example_project"到本地存储。

(2) 在系统文件 settings.py 的 INSTALLED_APPS 中添加"hitcount"和"blog"。

```
INSTALLED_APPS = (
    'django.contrib.admin',
    'django.contrib.auth',
    'django.contrib.contenttypes',
    'django.contrib.sessions',
    'django.contrib.messages',
    'django.contrib.staticfiles',
    'blog',     # 创建的应用程序的名字
    'hitcount'
)
```

13.3.2 配置 URL

在文件 urls.py 中设置 app 的链接导航,主要实现代码如下所示。

```
urlpatterns = [
    url(r'^$', views.IndexView.as_view(), name="index"),

    url(r'^generic-detail-view-ajax/(?P<pk>\d+)/$',
        views.PostDetailJSONView.as_view(),
        name="ajax"),
    url(r'^hitcount-detail-view/(?P<pk>\d+)/$',
        views.PostDetailView.as_view(),
        name="detail"),
    url(r'^hitcount-detail-view-count-hit/(?P<pk>\d+)/$',
        views.PostCountHitDetailView.as_view(),
        name="detail-with-count"),

    # for our built-in ajax post view
    url(r'hitcount/', include('hitcount.urls', namespace='hitcount')),
]
```

13.3.3 实现数据库

(1) 在模型文件 models.py 中导入框架 django-hitcount 中的 HitCount 和 HitCountMixin 模块,然后创建数据库表 Post。文件 models.py 的主要实现代码如下所示。

```
from hitcount.models import HitCount, HitCountMixin
```

```python
@python_2_unicode_compatible
class Post(models.Model, HitCountMixin):
    title = models.CharField(max_length=200)
    content = models.TextField()
    hit_count_generic = GenericRelation(
        HitCount, object_id_field='object_pk',
        related_query_name='hit_count_generic_relation')

    def __str__(self):
        return "Post title: %s" % self.title
```

(2) 通过如下命令可以根据上述模型文件创建数据库表。

```
python manage.py makemigrations
python manage.py migrate
```

13.3.4 实现视图

在视图文件 views.py 中定义了不同实现的功能类，每个实现对应一个类，并且对应一个 URL 导航路径。文件 views.py 的主要实现代码如下所示。

```python
class PostMixinDetailView(object):
    """
    Mixin to same us some typing. Adds context for us!
    """
    model = Post

    def get_context_data(self, **kwargs):
        context = super(PostMixinDetailView, self).get_context_data(**kwargs)
        context['post_list'] = Post.objects.all()[:6]
        context['post_views'] = ["ajax", "detail", "detail-with-count"]
        return context


class IndexView(PostMixinDetailView, TemplateView):
    template_name = 'blog/index.html'


class PostDetailJSONView(PostMixinDetailView, DetailView):
    template_name = 'blog/post_ajax.html'

    @classmethod
    def as_view(cls, **initkwargs):
        view = super(PostDetailJSONView, cls).as_view(**initkwargs)
        return ensure_csrf_cookie(view)


class PostDetailView(PostMixinDetailView, HitCountDetailView):
    """
    Generic hitcount class based view.
    """
    pass
```

```python
class PostCountHitDetailView(PostMixinDetailView, HitCountDetailView):
    """
    Generic hitcount class based view that will also perform the hitcount logic.
    """
    count_hit = True
```

因为框架 django-hitcount 提供了 3 种统计访问次数的方式，所以在本实例中演示了使用 3 种方式实现访问统计的方法。具体说明如下所示。

- Ajax 统计方式：使用 jQuery 方式实现统计，通过 "Hit counted" 显示当前访问是否被统计，使用 "Hit response" 显示当前访问是否被统计的原因。
- Detail 统计方式：将统计数据的详细信息嵌入到页面。
- Detail-With-Count 统计方式：将实现更加细节化的统计，和 Detail 统计方式相比，增加了使用 Hit counted 和 Hit response 属性的功能。

13.3.5 实现模板

在模板文件 post_ajax.html 中使用 "Ajax 统计方式" 来显示统计信息，对于初学者来说建议使用这种方式。这是因为此种方式使用 jQuery 技术实现，不用修改后台程序代码，只需在模板文件中进行设置即可。文件 post_ajax.html 的主要实现代码如下所示。

```
{% extends "blog/base.html" %}
{% load hitcount_tags %}

{% block content %}

{% get_hit_count_js_variables for post as hitcount %}
{% get_hit_count for post as total_hits %}

<div class="row">

  <div class="col-md-12">
    <h1>{{post.title}}</h1>
  </div>
  <div class="col-md-8">
    <p class="lead">{{ post.content }}</p>
  </div>

  <div class="col-md-4 bg-info">

    <h2>Hitcount Info</h2>

    <dl class="dl-horizontal">
      <dt>Total Hits:</dt>
      <dd>{{ total_hits }}</dd>
      <dt>Ajax URL is:</dt>
      <dd>{{ hitcount.ajax_url }}</dd>
      <dt>The unique PK is:</dt>
      <dd>{{ hitcount.pk }}</dd>
      <dt>Hit counted?</dt>
```

```
      <dd id="hit-counted"></dd>
      <dt>Hit response:</dt>
      <dd id="hit-response"></dd>
    </dl>


  </div>
</div>
{%endblock%}

{% comment %}
If you do not wish to perform any additional JavaScript actions after POST,
you can use this template tag to insert all the JavaScript you need, as in:

{% insert_hit_count_js for post%}

Or you can use with 'debug' for some output:

{% insert_hit_count_js for post debug %}

The code below is used to update the page view so we can test it with selenium.
{% endcomment %}

{% block inline_javascript %}

{% load staticfiles %}
<script src="{% static 'hitcount/jquery.postcsrf.js' %}"></script>

{% get_hit_count_js_variables for post as hitcount %}
<script type="text/javascript">
jQuery(document).ready(function($) {
  $.postCSRF("{{ hitcount.ajax_url }}", { hitcountPK : "{{ hitcount.pk }}" })
    .done(function(data){
      $('<i
/>').text(data.hit_counted).attr('id','hit-counted-value').appendTo('#hit-c
ounted');
      $('#hit-response').text(data.hit_message);
  }).fail(function(data){
      console.log('POST failed');
      console.log(data);
  });
});
</script>

{% endblock %}
```

在模板文件 post_detail.html 中同时实现了 Detail 和 Detail-With-Count 这两种统计方式。两者的区别是，后者比前者多了 Hit counted 和 Hit response 两个统计选项。在模本文件中是否显示这两个属性，取决于视图文件 views.py 中的 count_hit，如果设置为 True 则显示这两个属性信息。在本实例的视图方法 PostCountHitDetailView 中设置了 count_hit = True，所以会在"hitcount-detail-view-count-hit/"链接页面显示这两个属性信息。而在本实例的视图方法 PostDetailView 中只有一个空语句代码行 pass,所以不会在"hitcount-detail-view/"链接参数页面显示这两个属性信息。文件 post_detail.html 的主要实现代码如下所示。

```
{% extends "blog/base.html" %}

{% block content %}
<div class="row">
 <div class="col-md-12">
   <h1>{{object.title}}</h1>
 </div>

 <div class="col-md-8">
   <p class="lead">{{ object.content }}</p>
 </div>

 <div class="col-md-4 bg-info">
   <h2>Hitcount Info</h2>
   <dl class="dl-horizontal">
    <dt>Total Hits:</dt>
    <dd>{{ hitcount.total_hits }}</dd>
    <dt>The unique PK is:</dt>
    <dd>{{ hitcount.pk }}</dd>
    <dt>Hit counted?</dt>
    <dd id="hit-counted">
      <i id="hit-counted-value">{{ hitcount.hit_counted }}</i>
    </dd>
    <dt>Hit response:</dt>
    <dd id="hit-response">{{ hitcount.hit_message }}</dd>
   </dl>
 </div>
</div>
{%endblock%}
```

通过上述两个模板文件的对比可知,虽然 Ajax 方式比较简单,不会涉及后台程序,但是其模板文件的代码比较烦琐,而另外两种方式的模板文件代码就显得更加精简。读者可以根据自己的需求来选择适合自己的访问统计方式。

13.3.6 调试运行

运行程序后,在浏览器中输入 http://127.0.0.1:8000/显示系统主页,如图 13-4 所示。

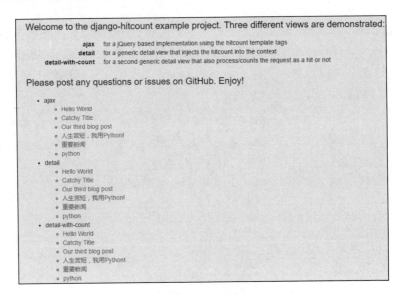

图 13-4　系统主页

使用 Ajax 方式统计某条信息的页面如图 13-5 所示。

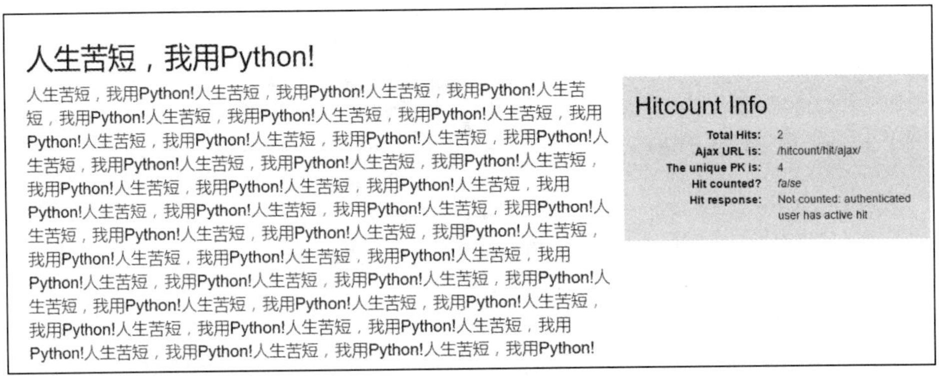

图 13-5　Ajax 统计方式展示

第 14 章

Ajax 模块

　　Ajax 是一种创建交互式网页的网页开发技术，其最大优点是页面内的 JavaScript 脚本可以不用刷新页面，而可以直接和服务器完成数据交互，这样大大提高了用户体验。在本章的内容中，将详细讲解在 Django Web 程序中使用 Ajax 技术的方法，为读者步入本书后面知识的学习打下基础。

14.1 Ajax 技术的原理

Ajax 是一种创建交互式网页的网页开发技术，其最大优点是页面内的 JavaScript 脚本可以不用刷新页面，而可以直接和服务器完成数据交互，这样大大提高了用户体验。在本节的内容中，将讲解 Ajax 技术的原理。

扫码观看本节视频讲解

在传统的 Web 应用模型中，浏览器负责向服务器提出访问请求，并显示服务器返回的处理结果。而在 Ajax 处理模型中，使用了 Ajax 中间引擎来处理上述通信。Ajax 中间引擎实质上是一个 JavaScript 对象或函数，只有当信息必须从服务器上获得的时候才调用它。和传统的处理模型不同，Ajax 不再需要为其他资源提供链接，而只是当需要调度和执行时才执行这些请求。而这些请求都是通过异步传输完成的，而不必等到收到响应之后才执行。图 14-1 和图 14-2 分别列出了传统模型和 Ajax 模型的处理方式。

图 14-1 传统模型处理

图 14-2 Ajax 模型处理

从图 14-1 和图 14-2 所示的处理模式可以看出：Ajax 技术在客户端实现了高效的信息交互，通过 Ajax 引擎可以和用户浏览界面实现数据传输。即当 Ajax 引擎收到服务器响应时，将会触发一些操作，通常是完成数据解析，以及基于所提供的数据对用户界面做一些修改。

14.2 实战演练：无刷新计算器

在下面的实例代码中，演示了使用 Django 框架和 Ajax 开发一个无刷新计算器的过程。

扫码观看本节视频讲解

源码路径：daima\14\ajax1

(1) 通过如下命令新建一个名为"ajax1"的工程，然后在工程里面新建一个名为"tools"的 app。

```
django-admin.py startproject ajax1
cd ajax1
python manage.py startapp tools
```

(2) 将"tools"添加到设置文件 settings.py 的 INSTALLED_APPS 中。

```
INSTALLED_APPS = [
    'django.contrib.admin',
    'django.contrib.auth',
    'django.contrib.contenttypes',
    'django.contrib.sessions',
    'django.contrib.messages',
    'django.contrib.staticfiles',
    'tools',
]
```

(3) 编写视图文件 views.py，计算从表单中获取的 a 和 b 的和，具体实现代码如下所示。

```python
def index(request):
    return render(request, 'index.html')

def add(request):
    a = request.GET['a']
    b = request.GET['b']
    a = int(a)
    b = int(b)
    return HttpResponse(str(a + b))
```

(4) 在 urls.py 文件中实现路径导航，主要实现代码如下所示。

```python
from django.contrib import admin
from django.urls import path
from tools import views

urlpatterns = [
    path(r'index/', views.index),
    path(r'add/', views.add),
]
```

(5) 在模板文件 index.html 中实现计算两个数字和的表单文本框，然后使用 Ajax 技术获取计算结果，并以无刷新效果显示出来。文件 index.html 的主要实现代码如下所示。

```html
<body>
<p>请输入两个数字</p>
<form action="/add/" method="get">
   a: <input type="text" id="a" name="a"> <br>
   b: <input type="text" id="b" name="b"> <br>
   <p>result: <span id='result'></span></p>
   <button type="button" id='sum'>提交</button>
</form>
<script
src="http://apps.bdimg.com/libs/jquery/1.11.1/jquery.min.js"></script>
<script>
   $(document).ready(function(){
     $("#sum").click(function(){
       var a = $("#a").val();
       var b = $("#b").val();

       $.get("/add/",{'a':a,'b':b}, function(ret){
          $('#result').html(ret)
       })
     });
   });
</script>
```

在浏览器中输入 http://localhost:8000/index/ 后，会无刷新计算在表单中输入的两个整数的和，如图 14-3 所示。

图 14-3　执行效果

14.3　Ajax 上传和下载系统

在本章前面的内容中，曾经讲解了通过表单实现文件上传系统的过程。在下面的实例代码中，演示了使用 Django 和 Ajax 开发一个同时实现文件上传和下载功能的过程。

 源码路径：**daima\14\file-upload-download**

扫码观看本节视频讲解

14.3.1　实现文件上传功能

在 Django 框架中，可以使用如下 3 种方式实现文件上传功能。
- 使用表单上传，在视图中编写文件上传代码。

- 使用由模型创建的表单(ModelForm)实现上传,使用方法 form.save()自动存储。
- 使用 Ajax 方式实现无刷新异步上传,在上传页面中无须刷新即可显示新上传的文件。

本实例将实现上述三种上传功能,具体实现流程如下所示。

(1) 通过如下命令创建一个名为"file_project"的工程,然后定位到"file_project"的根目录,在里面新建一个名为"file_upload"的 app。

```
django-admin.py startproject file_project
cd file_project
python manage.py startapp file_upload
```

(2) 将上面创建的 file_upload 这个 app 加入到系统设置文件 settings.py 中。代码如下:

```
INSTALLED_APPS = [
    'django.contrib.admin',
    'django.contrib.auth',
    'django.contrib.contenttypes',
    'django.contrib.sessions',
    'django.contrib.messages',
    'django.contrib.staticfiles',
    'file_upload',
]
```

(3) 设置/media/和/STATIC_URL/文件夹,将上传的文件放在/media/目录中,因为还需要用到 CSS 和 JavaScript 这些静态文件,所以需要设置 STATIC_URL。设置文件 settings.py 中的对应代码如下所示。

```
STATIC_URL = '/static/'
STATICFILES_DIRS = [os.path.join(BASE_DIR, "static"), ]

# specify media root for user uploaded files,
MEDIA_ROOT = os.path.join(BASE_DIR, 'media')
MEDIA_URL = '/media/
```

(4) 规划 URL,路径导航文件 urls.py 的具体实现代码如下所示。

```
from django.contrib import admin
from django.urls import path, include
from django.conf import settings
from django.conf.urls.static import static

urlpatterns = [
    path('admin/', admin.site.urls),
    path('file/', include("file_upload.urls")),
] + static(settings.MEDIA_URL, document_root=settings.MEDIA_ROOT)
```

(5) 创建模型文件 models.py,设置 File 模型包括 file 和 upload_method 两个字段。通过 upload_to 选项指定文件上传后存储的地址,并对上传的文件进行重命名。具体实现代码如下所示。

```
def user_directory_path(instance, filename):
    ext = filename.split('.')[-1]
    filename = '{}.{}'.format(uuid.uuid4().hex[:10], ext)
```

```
    return os.path.join("files", filename)

class File(models.Model):
    file = models.FileField(upload_to=user_directory_path, null=True)
    upload_method = models.CharField(max_length=20, verbose_name="Upload Method")
```

(6) 本项目一共包括 5 个 urls，分别对应如下所示的页面。
- 表单上传页面。
- ModelForm 上传页面。
- Ajax 上传页面。
- 已上传文件列表页面。
- 处理 ajax 请求页面。

在 file_upload 目录下编写文件 urls.py 导航上述 5 个页面，具体实现代码如下所示。

```
urlpatterns = [

    # Upload File Without Using Model Form
    re_path(r'^upload1/$', views.file_upload, name='file_upload'),

    # Upload Files Using Model Form
    re_path(r'^upload2/$', views.model_form_upload, name='model_form_upload'),

    # Upload Files Using Ajax Form
    re_path(r'^upload3/$', views.ajax_form_upload, name='ajax_form_upload'),

    # Handling Ajax requests
    re_path(r'^ajax_upload/$', views.ajax_upload, name='ajax_upload'),

    # View File List
    path('', views.file_list, name='file_list'),

]
```

(7) 编写程序文件 forms.py，分别实现使用普通表单上传和使用 ModelForm 上传方式，具体实现代码如下所示。

```
# 普通表单
class FileUploadForm(forms.Form):
    file = forms.FileField(widget=forms.ClearableFileInput(attrs={'class': 'form-control'}))
    upload_method = forms.CharField(label="Upload Method", max_length=20,
                                    widget=forms.TextInput(attrs={'class': 'form-control'}))

    def clean_file(self):
        file = self.cleaned_data['file']
        ext = file.name.split('.')[-1].lower()
        if ext not in ["jpg", "pdf", "xlsx"]:
            raise forms.ValidationError("Only jpg, pdf and xlsx files are allowed.")
        # return cleaned data is very important.
```

```python
        return file


# Model 方式
class FileUploadModelForm(forms.ModelForm):
    class Meta:
        model = File
        fields = ('file', 'upload_method',)

        widgets = {
            'upload_method': forms.TextInput(attrs={'class': 'form-control'}),
            'file': forms.ClearableFileInput(attrs={'class': 'form-control'}),
        }

    def clean_file(self):
        file = self.cleaned_data['file']
        ext = file.name.split('.')[-1].lower()
        if ext not in ["jpg", "pdf", "xlsx"]:
            raise forms.ValidationError("Only jpg, pdf and xlsx files are allowed.")
        # return cleaned data is very important.
        return file
```

对上述代码的具体说明如下所示。

- 定义 FileUploadForm 实现普通表单上传功能，先定义一个一般表单 FileUploadForm，并通过 clean 方法对用户上传的文件进行验证，如果上传的文件名不以 jpg、pdf 或 xlsx 结尾，将显示表单验证错误信息。在使用方法 clean 验证表单字段时，请不要忘记 return 验证过的数据，即 cleaned_data。只有返回了 cleaned_data，才可以在视图中使用 form.cleaned_data.get('xxx')获取验证过的数据。
- 定义 FileUploadModelForm 实现使用 ModelForm 上传功能，在模型中通过 upload_to 选项自定义用户上传文件存储地址，并对文件进行了重命名。

(8) 编写视图文件 views.py，分别实现使用普通表单上传和使用 ModelForm 上传方式的视图，具体实现代码如下所示。

```python
def file_list(request):
    files = File.objects.all().order_by("-id")
    return render(request, 'file_upload/file_list.html', {'files': files})


# Regular file upload without using ModelForm
def file_upload(request):
    if request.method == "POST":
        form = FileUploadForm(request.POST, request.FILES)
        if form.is_valid():
            # get cleaned data
            upload_method = form.cleaned_data.get("upload_method")
            raw_file = form.cleaned_data.get("file")
            new_file = File()
            new_file.file = handle_uploaded_file(raw_file)
            new_file.upload_method = upload_method
            new_file.save()
```

```python
            return redirect("/file/")
    else:
        form = FileUploadForm()

    return render(request, 'file_upload/upload_form.html', {'form': form,
                        'heading': 'Upload files with Regular Form'})


def handle_uploaded_file(file):
    ext = file.name.split('.')[-1]
    file_name = '{}.{}'.format(uuid.uuid4().hex[:10], ext)
    # file path relative to 'media' folder
    file_path = os.path.join('files', file_name)
    absolute_file_path = os.path.join('media', 'files', file_name)

    directory = os.path.dirname(absolute_file_path)
    if not os.path.exists(directory):
        os.makedirs(directory)

    with open(absolute_file_path, 'wb+') as destination:
        for chunk in file.chunks():
            destination.write(chunk)

    return file_path


# Upload File with ModelForm
def model_form_upload(request):
    if request.method == "POST":
        form = FileUploadModelForm(request.POST, request.FILES)
        if form.is_valid():
            form.save()
            return redirect("/file/")
    else:
        form = FileUploadModelForm()

    return render(request, 'file_upload/upload_form.html', {'form': form,
                        'heading': 'Upload files with ModelForm'})

def ajax_form_upload(request):
    form = FileUploadModelForm()
    return render(request, 'file_upload/ajax_upload_form.html', {'form': form,
                        'heading': 'File Upload with AJAX'})


# handling AJAX requests
def ajax_upload(request):
    if request.method == "POST":
        # 1. Regular save method
        # upload_method = request.POST.get("upload_method")
        # raw_file = request.FILES.get("file")
        # new_file = File()
        # new_file.file = handle_uploaded_file(raw_file)
        # new_file.upload_method = upload_method
        # new_file.save()
```

```
        # 2. Use ModelForm als ok.
        form = FileUploadModelForm(data=request.POST, files=request.FILES)
        if form.is_valid():
            form.save()
            # Obtain the latest file list
            files = File.objects.all().order_by('-id')
            data = []
            for file in files:
                data.append({
                    "url": file.file.url,
                    "size": filesizeformat(file.file.size),
                    "upload_method": file.upload_method,
                    })
            return JsonResponse(data, safe=False)
        else:
            data = {'error_msg': "Only jpg, pdf and xlsx files are allowed."}
            return JsonResponse(data)
    return JsonResponse({'error_msg': 'only POST method accpeted.'})
```

对上述代码的具体说明如下所示。

- 定义方法 file_upload 实现普通文件上传视图，当用户的请求方法为 POST 时，通过 form.cleaned_data.get('file') 获取通过验证的文件，并调用自定义方法 handle_uploaded_file 重命名文件，然后写入文件。如果用户的请求方法不是 POST，则在 upload_form.html 中渲染一个空的 FileUploadForm。另外，还定义了方法 file_list 来显示文件清单。
- 在方法 handle_uploaded_file 中，文件写入地址必须是包含/media/的绝对路径，例如/media/files/xxxx.jpg。而该方法返回的地址是相对于/media/文件夹的地址，例如/files/xxx.jpg。注意，这个地址是相对地址，而不是绝对地址。
- 因为不同操作系统的目录分隔符不同，所以建议使用方法 os.path.join 构建文件的写入绝对路径。在写入文件前需使用 os.path.exists 检查目标文件夹是否存在，如果不存在则先创建文件夹，然后再写入。
- 定义方法 model_form_upload()，功能是获取文件上传视图界面中的上传数据，然后使用 form.save()将这些上传数据保存起来，无须再手动编写代码写入文件。
- 方法 ajax_upload 负责处理 Ajax 请求的视图，该方法首先将 Ajax 发过来的数据与 FileUploadModelForm 结合，然后直接调用方法 form.save 存储，最后以 JSON 格式返回更新过的文件清单。如果用户的上传文件不符合要求则返回错误信息。

14.3.2　实现文件下载功能

（1）通过如下命令定位到 file_project 根目录，然后创建一个名为 "startapp file_upload" 的 app。

```
cd file_project
python manage.py startapp file_upload
```

(2) 将上面创建的"startapp file_upload"app 添加到设置文件,然后新建路径导航文件 urls.py,在 URL 中包含一个文件的相对路径 file_path 作为参数,其对应视图是 file_download 方法。我们现在就开始尝试用不同方法来处理文件下载。具体实现代码如下所示。

```
urlpatterns = [

   re_path(r'^download/(?P<file_path>.*)/$', views.file_download,
name='file_download'),

]
```

(3) 在视图文件 views.py 中实现三种文件下载方式,具体实现代码如下所示。

```
def file_download(request, file_path):
   #第一种下载方法,使用 open()直接打开
   with open(file_path) as f:
      c = f.read()
   return HttpResponse(c)


# 第二种下载方法,使用 HttpResponse 下载,适合 txt 格式的小文件,不适合大的二进制文件
def media_file_download(request, file_path):
   with open(file_path, 'rb') as f:
      try:
         response = HttpResponse(f)
         response['content_type'] = "application/octet-stream"
         response['Content-Disposition'] = 'attachment; filename=' + os.path.basename(file_path)
         return response
      except Exception:
         raise Http404


#第三种下载方式,使用 StreamingHttpResponse 下载,适合流式传输的大型文件,例如 CSV 文件
def stream_http_download(request, file_path):
   try:
      response = StreamingHttpResponse(open(file_path, 'rb'))
      response['content_type'] = "application/octet-stream"
      response['Content-Disposition'] = 'attachment; filename=' + os.path.basename(file_path)
      return response
   except Exception:
      raise Http404


# 第四种下载方式,使用 FileResponse 下载,适合大文件
def file_response_download1(request, file_path):
   try:
      response = FileResponse(open(file_path, 'rb'))
      response['content_type'] = "application/octet-stream"
      response['Content-Disposition'] = 'attachment; filename=' + os.path.basename(file_path)
      return response
   except Exception:
      raise Http404
```

```python
#第五种下载方式，限制文件下载类型，推荐用这种类型
def file_response_download(request, file_path):
    ext = os.path.basename(file_path).split('.')[-1].lower()
    # cannot be used to download py, db and sqlite3 files.
    if ext not in ['py', 'db', 'sqlite3']:
        response = FileResponse(open(file_path, 'rb'))
        response['content_type'] = "application/octet-stream"
        response['Content-Disposition'] = 'attachment; filename=' + os.path.basename(file_path)
        return response
    else:
        raise Http404
```

对上述代码的具体说明如下所示。

- 使用 HttpResonse 方式下载：通过方法 file_download 从 url 获取 file_path，打开文件，然后读取文件，最后通过 HttpResponse 方法输出。但是方法 file_download 存在一个问题，如果下载文件是一个二进制文件，通过 HttpResponse 输出后将会显示为乱码。对于一些二进制文件(图片、pdf)，用户可能更希望直接将它们作为附件进行下载。当把二进制文件下载到本机后，用户就可以用自己喜欢的程序(如 Adobe)打开阅读文件了。所以通过方法 media_file_download 做出了针对二进制文件的改进，为 response 设置 content_type 和 Content_Disposition 参数。
- 使用 SteamingHttpResonse 方式下载：通过方法 stream_http_download 实现。
- 使用 FileResonse 方式下载：编写方法 file_response_download1 实现，FileResponse 方法是 SteamingHttpResponse 的子类，如果为 file_response_download 加上 @login_required 装饰器，那么就可以实现用户需要先登录才能下载某些文件的功能。
- 方法 file_response_download：在上面的 file_response_download1 中，即使加上了 @login_required 装饰器，用户只要获取了文件的链接地址，他们依然可以通过浏览器直接访问那些文件。我们等会儿再谈保护文件的链接地址和文件私有化，因为此时我们还有个更大的问题需要解决。我们定义的下载方法可以下载所有文件，不仅包括.py 文件，还包括不在 media 文件夹里的文件(比如非用户上传的文件)。比如当直接访问 127.0.0.1:8000/file/download/file_project/settings.py/时会发现连同 file_project 目录下的设置文件 settings.py 都被下载了。所以在编写下载方法时，一定要限定哪些文件可以下载，哪些文件不能下载，或者限定用户只能下载 media 目录中的文件。

注 意

上面第一种下载方式 HttpResponse 有一个很大的弊端，其工作原理是先读取文件，载入内存，然后再输出。如果下载文件很大，该方法会占用很多内存。对于下载大文件，Django 更推荐 StreamingHttpResponse 和 FileResponse 方法，这两个方法将下载文件分批(Chunks)写入用户本地磁盘，先不将它们载入服务器内存。

输入下面的命令运行程序,在浏览器中输入"http://localhost:8000/file/"后显示上传文件列表,如图 14-4 所示。

```
python manage.py runserver
```

Filename & URL	Filesize	Upload Method
/media/files/21dabeec32.jpg	297.3 KB	444
/media/files/f6477afdb5.jpg	297.3 KB	111
/media/files/19c8f6862e.jpg	297.3 KB	123

图 14-4　上传文件列表

单击顶部导航中的三个链接 RegularFormUpload、ModelFormUpload 和 AjaxUpload,会弹出三种方式的上传表单界面,并实现对应的文件上传功能。例如单击 ModelFormUpload 链接后的效果如图 14-5 所示。

图 14-5　执行效果

第 15 章

分 页 模 块

  在开发 Web 项目的过程中，经常需要用到分页功能，例如将数据库中的多条新闻信息、商品信息、留言信息等分页显示出来。在本章的内容中，将详细讲解在 Django Web 程序中实现分页功能的知识，为读者步入本书后面知识的学习打下基础。

15.1 类 Paginator 和类 Page

Django 作为著名的 Web 开发框架,提供了分页功能类 Paginator 和 Page。在本节的内容中,将详细讲解使用 Django 的内置类 Paginator 和类 Page 实现分页功能的过程。

扫码观看本节视频讲解

15.1.1 类 Paginator

在 Django 框架中,类 Pagination 实现分页管理功能,数据被分在不同的页面中,并带有"上一页/下一页"标签。类 Pagination 的原型如下所示。

```
class Paginator(object):
    def __init__(self, object_list, per_page, orphans=0,
                 allow_empty_first_page=True):
        self.object_list = object_list
        self.per_page = int(per_page)
        self.orphans = int(orphans)
        self.allow_empty_first_page = allow_empty_first_page
        self._num_pages = self._count = None
```

说明:

- object_list:可以是列表、元组、查询集或其他含有 count() 或 __len__()方法的可切片对象。对于连续的分页,查询集应该有顺序,例如有 order_by()项或默认 ordering 参数。
- per_page:每一页中包含条目数目的最大值,不包括独立成页的那页。
- orphans=0:当使用此参数时说明不希望最后一页只有很少的条目。如果最后一页的条目数少于等于 orphans 的值,则这些条目会被归并到上一页中(此时的上一页变为最后一页)。例如有 23 项条目,per_page=10,orphans=0,则有 3 页,分别为 10、10、3。如果 orphans>=3,则为 2 页,分别为 10、13。
- allow_empty_first_page=True:默认允许第一页为空。

(1) 类方法。

类方法 Paginator.page(number)的功能是根据参数 number 返回一个 Page 对象,number 为 1 的倍数。

(2) 类属性。

- Paginator.count:所有页面对象总数,即统计 object_list 中 item 的数目。当计算 object_list 所含对象的数量时,Paginator 会首先尝试调用 object_list.count()。如果 object_list 没有 count() 方法,Paginator 接着会回退使用 len(object_list)。
- Pagnator.num_pages:页面总数。
- pagiator.page_range:页面范围,从 1 开始,例如[1,2,3,4]。

15.1.2 类 Page

在 Django Web 程序中，也可以使用类 Page 实现分页功能，在下面列出了类 Page 中包含的常用成员。

(1) 类方法。
- Page.has_next()：如果有下一页，则返回 True。
- Page.has_previous()：如果有上一页，返回 True。
- Page.has_other_pages()：如果有上一页或下一页，返回 True。
- Page.next_page_number()：返回下一页的页码。如果下一页不存在，则抛出 InvalidPage 异常。
- Page.previous_page_number()：返回上一页的页码。如果上一页不存在，则抛出 InvalidPage 异常。
- Page.start_index()：返回当前页上的第一个对象，相对于分页列表的所有对象的序号，从 1 开始。比如，将五个对象的列表分为每页两个对象，第二页的 start_index() 会返回 3。
- Page.end_index()：返回当前页上的最后一个对象，相对于分页列表的所有对象的序号，从 1 开始。比如，将五个对象的列表分为每页两个对象，第二页的 end_index() 会返回 4。

(2) 类属型。
- Page.object_list：当前页上所有对象的列表。
- Page.number：当前页的序号，从 1 开始。
- Page.paginator：相关的 Paginator 对象。

15.1.3 实战演练：实现简单的分页

在下面的实例代码中，演示了使用 Django 框架实现简单分页功能的过程。

源码路径：daima\15\fen1

(1) 在模型文件 models.py 中创建数据表，主要实现代码如下所示。

```
class Articles(models.Model):
    title=models.CharField(max_length=32)
    content=models.TextField()

    def __str__(self):
        return self.title
```

(2) 通过如下代码向数据库中添加测试数据。

```
from projectname.wsgi import *
from app01.models import *
```

```
for i in range(1,96):
    title="title"+str(i)
    content="text"+str(i)
    Articles.objects.create(title=title,content=content)
```

数据库中的测试数据如图 15-1 所示。

92	title91	text91
93	title92	text92
94	title93	text93
95	title94	text94
96	title95	text95

图 15-1 测试数据

(3) 视图文件 views.py 的主要实现代码如下所示。

```
def page_demo(request):
    articles=Articles.objects.all()
    paginator_obj=Paginator(articles,5) #每页5条
    # print(paginator_obj.page_range)

    request_page_num=request.GET.get('page',1)
    # print(request_page_num)
    page_obj=paginator_obj.page(request_page_num)

    total_page_number=paginator_obj.num_pages

    return render(request,'page_demo.html',{'page_obj':page_obj,'paginator_obj':paginator_obj})
```

(4) URL 路径导航文件 urls.py 的主要实现代码如下所示。

```
urlpatterns = [
    url(r'page_demo/', views.page_demo),
]
```

(5) 模板文件 page_demo.html 的具体实现代码如下所示。

```
<!DOCTYPE html>
<html lang="en">
<head>
    <meta charset="UTF-8">
    <title>Title</title>
</head>
<body>
{% for article in page_obj %}
    <div>## {{ article.title }} ##</div>
    <div>{{ article.content }}</div>
{% endfor %}

{% for page_num in paginator_obj.page_range %}
    <a href="?page={{ page_num }}">{{ page_num }}</a>
{% endfor %}
```

```
</body>
</html>
```

执行程序后会显示分页效果,例如输入 http://127.0.0.1:8000/page_demo/?page=15 后的效果如图 15-2 所示。

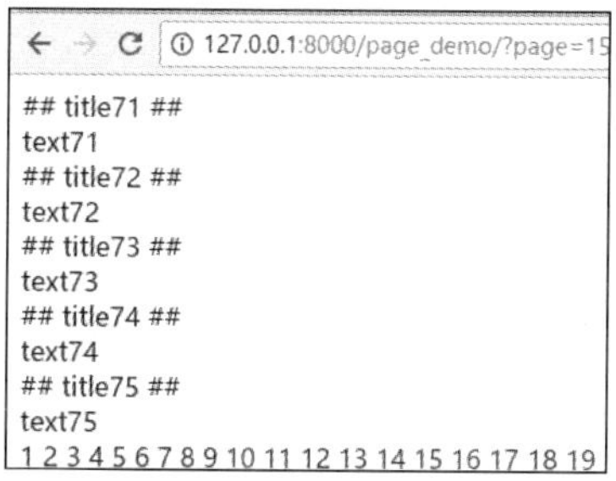

图 15-2 执行效果

15.2 实战演练:自定义的美观的分页程序

本章前面的分页实例样式不够美观,在下面的实例代码中,演示了在 Django Web 程序中自定义编写一个分页功能类,并使用 bootstrap 样式修饰这个分页程序的过程。

扫码观看本节视频讲解

源码路径:**daima\15\Django_customizing_pagination**

(1) 创建一个名为 "Django_customizing_pagination" 的 Django 工程,然后新建一个名为 "app01" 的 app,并将这个 app 添加到设置文件 settings.py 的 INSTALLED_APPS 中。

(2) 编写文件 mypage.py,在里面创建一个封装好的分页类,此类实现完整的分页功能。读者可以直接将此类复制到自己的项目中。文件 mypage.py 的具体实现代码如下所示。

```
class Page():
    def __init__(self, page_num, total_count, url_prefix, per_page=10, max_page=11):
        '''

        :param page_num: 当前页码数
        :param total_count: 数据总数
        :param url_prefix: a标签href的前缀
        :param per_page: 每页显示多少条数据
        :param max_page: 页面上最多显示几个页码
        '''
        self.url_prefix = url_prefix
        self.max_page = max_page
        # 总共需要多少页码来展示
        total_page, m = divmod(total_count, per_page)
        if m:
            total_page += 1
        self.total_page = total_page
```

```python
    try:
        page_num = int(page_num)
        # 如果输入的页码数大于最大的页码数，默认返回最后一页
        if page_num > total_page:
            page_num = total_page
    except Exception as e:
        # 当输入的页码不是数字的时候，默认返回第一页的数据
        page_num = 1
    self.page_num = page_num

    # 定义两个变量，保存数据从哪里取到哪里
    self.data_start = (page_num - 1) * per_page
    self.data_end = page_num * per_page

    # 页面上总共展示多少页码
    if total_page < self.max_page:
        self.max_page = total_page

    half_max_page = self.max_page // 2
    # 页面上展示的页码从哪里开始
    page_start = page_num - half_max_page
    # 页面上展示的页码从哪里结束
    page_end = page_num + half_max_page
    # 如果当前页为1，则强制设置开始页为1
    if page_start <= 1:
        page_start = 1
        page_end = self.max_page
    # 如果当前页加一半，比总页码数还大
    if page_end >= total_page:
        page_end = total_page
        page_start = total_page - self.max_page + 1
    self.page_start = page_start
    self.page_end = page_end

@property
def start(self):
    return self.data_start

@property
def end(self):
    return self.data_end

def page_html(self):
    # 生成分页功能的 HTML 代码
    html_str_list = []
    # 加上第一页
    html_str_list.append('<li><a href="{}?page=1">首页</a></li>'.format(self.url_prefix))

    # 判断，如果是第一页，就没有上一页
    if self.page_num <= 1:
        html_str_list.append('<li class="disabled"><a href="#"><span aria-hidden="true">&laquo;</span></a></li>'.format(self.page_num-1))
    else:
        # 加一个上一页标签
```

```python
        html_str_list.append('<li><a href="{}?page={}"><span aria-hidden="true">&laquo;</span></a></li>'.format(self.url_prefix, self.page_num-1))

        for i in range(self.page_start, self.page_end+1):
            # 如果是当前页就加上一个active样式类
            if i == self.page_num:
                tmp = '<li class="active"><a href="{0}?page={1}">{1}</a></li>'.format(self.url_prefix, i)
            else:
                tmp = '<li><a href="{0}?page={1}">{1}</a></li>'.format(self.url_prefix, i)

            html_str_list.append(tmp)

        # 加一个下一页的按钮
        # 判断，如果是最后一页，就没有下一页
        if self.page_num >= self.total_page:
            html_str_list.append(
                '<li class="disabled"><a href="#"><span aria-hidden="true">&raquo;</span></a></li>')
        else:
            html_str_list.append(
                '<li><a href="{}?page={}"><span aria-hidden="true">&raquo;</span></a></li>'.format(self.url_prefix, self.page_num + 1))
        # 加最后一页
        html_str_list.append('<li><a href="{}?page={}">尾页</a></li>'.format(self.url_prefix, self.total_page))

        page_html = "".join(html_str_list)
        return page_html
```

(3) 在视图文件 views.py 中使用上面的分页功能类，以分页样式显示数据库表 book 中的数据，主要实现代码如下所示。

```python
def books(request):
    page_num = request.GET.get('page')
    print(page_num, type(page_num))

    # 总数据是多少
    total_count = models.Book.objects.all().count()

    from utils.mypage import Page
    page_obj = Page(page_num, total_count, per_page=5, url_prefix="/books/", max_page=5)

    ret = models.Book.objects.all()[page_obj.start:page_obj.end]

    page_html = page_obj.page_html()

    return render(request, 'books.html', {'books': ret, 'page_html': page_html})
```

(4) 在模型文件 models.py 中创建数据表 Book，主要实现代码如下所示。

```python
class Book(models.Model):
    title = models.CharField(max_length=32)
```

```python
    def __str__(self):
        return self.title
    class Meta:
        db_table = 'book'
```

运行如下命令根据上述模型文件生成数据库表。

```
python manage.py makemigrations
python manage.py migrate
```

(5) 在文件 admin.py 中将数据库数据绑定到后台视图，主要实现代码如下所示。

```python
from app01.models import Book
class ArticleAdmin(admin.ModelAdmin):
    list_display = ('title',)
admin.site.register(Book)
```

(6) URL 路径导航文件 urls.py 的主要实现代码如下所示。

```python
urlpatterns = [
    url(r'^admin/', admin.site.urls),
    url(r'^books/$', views.books),
]
```

(7) 在模板文件 books.html 中使用分页显示数据库中的图书信息，主要实现代码如下所示。

```html
<head>
    <meta charset="UTF-8">
    <title>书籍列表</title>
    <link rel="stylesheet" href="/static/bootstrap/css/bootstrap.min.css">
</head>
<body>

<div class="container">
    <table class="table table-bordered">
        <thead>
        <tr>
            <th>序号</th>
            <th>id</th>
            <th>书名</th>
        </tr>
        </thead>
        <tbody>
        {% for book in books %}
            <tr>
                <td>{{ forloop.counter }}</td>
                <td>{{ book.id }}</td>
                <td>{{ book.title }}</td>
            </tr>
        {% endfor %}

        </tbody>
    </table>

    <nav aria-label="Page navigation">
```

```
        <ul class="pagination">
            {{ page_html|safe }}
        </ul>
    </nav>
</div>
</body>
```

登录后台页面 http://127.0.0.1:8000/admin/，然后添加一些图书信息，如图 15-3 所示。

图 15-3　添加图书信息

输入 http://127.0.0.1:8000/books/ 显示前台图书列表页面，图书信息是以分页样式展示的，例如 http://127.0.0.1:8000/books/?page=2 链接显示第二个分页，如图 15-4 所示。

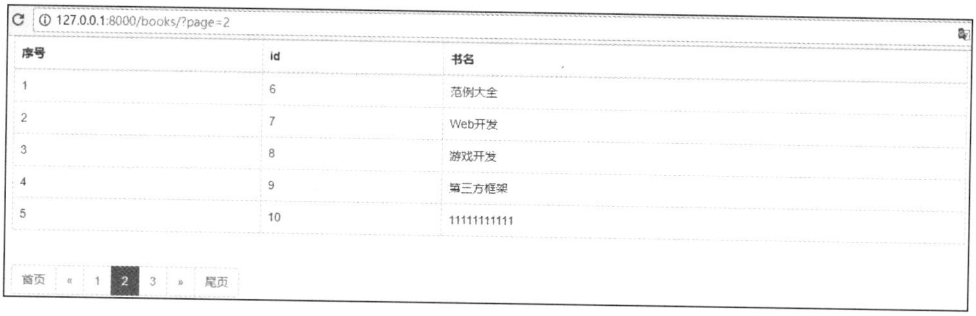

图 15-4　第二个分页

15.3　实战演练：使用分页显示网络信息

在下面的实例代码中，演示了在 Django Web 程序中使用分页程序显示网络信息的过程。

源码路径：**daima\15\django-paginator-api-example**

扫码观看本节视频讲解

15.3.1 创建工程

创建一个名为"mysite"的 Djang 工程,然后新建一个名为"core"的 app,并将这个 app 添加到设置文件 settings.py 的 INSTALLED_APPS 中。代码如下:

```python
INSTALLED_APPS = [
    'django.contrib.admin',
    'django.contrib.auth',
    'django.contrib.contenttypes',
    'django.contrib.sessions',
    'django.contrib.messages',
    'django.contrib.staticfiles',
    'core',
]
```

15.3.2 设计视图

在视图文件 views.py 中使用 requests 获取指定 url 的 JSON 信息,然后通过 paginator 对获取的 JSON 信息进行分页处理。文件 views.py 的主要实现代码如下所示。

```python
def home(request):
    api_requests = requests.get('https://min-api.cryptocompare.com/data/v2/news/?lang=EN')
    api = json.loads(api_requests.content)

    page = request.GET.get('page', 1)

    paginator = Paginator(api['Data'], 6)

    try:
        news_list = paginator.page(page)
    except PageNotAnInteger:
        news_list = paginator.page(1)
    except EmptyPage:
        news_list = paginator.page(paginator.num_pages)

    return render(request,'home.html', {
        'news_list': news_list
    })
```

15.3.3 设计 URL 导航

编写文件 urls.py,代码如下:

```python
from django.contrib import admin
from django.urls import path
from core import views

urlpatterns = [
    path('', views.home),
```

```
    path('admin/', admin.site.urls),
]
```

15.3.4 实现模板文件

在模板文件 home.html 中解析并分页显示获取的网络信息,主要实现代码如下所示。

```
<!doctype html>
<html>
<head>
 <meta charset="utf-8">
 <title>News</title>
</head>
<body>
 {% for news in news_list %}
  <h3>{{ news.title }}</h3>
  <p>{{ news.body }}</p>
 {% endfor %}
 <hr>
 <div>
   {% for i in news_list.paginator.page_range %}
    {% if news_list.number == i %}
      <span>{{ i }}</span>
    {% else %}
      <a href="?page={{ i }}">{{ i }}</a>
    {% endif %}
   {% endfor %}
 </div>
</body>
</html>
```

在浏览器中输入 http://127.0.0.1:8000/ 后会分页显示获取的网络信息,如图 15-5 所示。

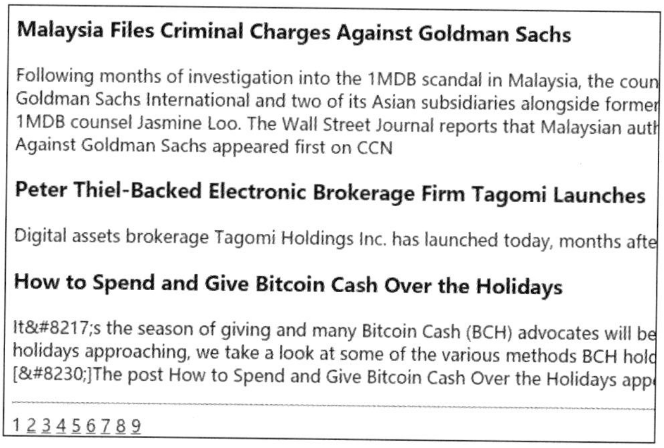

图 15-5 执行效果

第 16 章

富文本编辑器模块

在开发 Web 网站系统的过程中，经常需要使用富文本编辑器。富文本编辑器是一种可内嵌于浏览器，所见即所得的文本编辑器。例如在 BBS 论坛系统中，我们可以使用富文本编辑器发送样式美观的帖子，例如可以设置帖子文本的大小、颜色，可以在帖子中添加图片和视频。在本章的内容中，将详细讲解在 Django 程序中使用富文本编辑器的知识。

16.1 第三方库 django-mdeditor

django-mdeditor 是基于 Editor.md 的一个第三方 Django Markdown 文本编辑器框架。在本节的内容中，将详细讲解在 Django 中使用 django-mdeditor 实现富文本编辑器功能的过程。

16.1.1 django-mdeditor 介绍

扫码观看本节视频讲解

第三方库 django-mdeditor 的功能如下。

(1) 支持 Editor.md 大部分功能。
- 支持标准的 Markdown 文本、CommonMark 文本和 GFM(GitHub Flavored Markdown，Markdown 风格编辑器)文本。
- 支持实时预览、图片上传、格式化代码、搜索替换、皮肤设置、多语言设置等。
- 支持 TOC 目录和表情设置。
- 支持 TeX(文本排版系统)流程图、时序图等图表扩展。

(2) 可以自定义 Editor.md 工具栏。
(3) 提供了 MDTextField 字段用来支持模型字段使用。
(4) 提供了 MDTextFormField 字段用来支持 Form 和 ModelForm。
(5) 提供了 MDEditorWidget 字段用来支持使用自定义 admin 样式。

在使用第三方库 django-mdeditor 之前，需要先通过如下命令进行安装：

```
pip install django-mdeditor
```

16.1.2 实战演练：使用 django-mdeditor 实现富文本编辑器

在下面的实例代码中，演示了在 Django 中使用 django-mdeditor 实现富文本编辑器功能的过程。

> 源码路径：**daima\16\untitled**

(1) 通过如下命令创建一个名为"untitled"的工程，然后在工程目录下新建一个名为"mdeditorexample"的 app。

```
django-admin.py startproject untitled
cd untitled
python manage.py startapp mdeditorexample
```

(2) 在配置文件 settings.py 中加入"mdeditorexample"和"mdeditor"，如果使用的是 Django 3.0+ 版本，还需要添加 X_FRAME_OPTIONS = 'SAMEORIGIN'，然后设置使用 SQLite3 数据库，添加媒体文件的路径，最后在工程的根目录中创建文件夹"uploads/editor"。配置文件 settings.py 的代码如下。

```python
INSTALLED_APPS = [
    'django.contrib.admin',
    'django.contrib.auth',
    'django.contrib.contenttypes',
    'django.contrib.sessions',
    'django.contrib.messages',
    'django.contrib.staticfiles',
    'mdeditor',
    'mdeditorexample',
]

DATABASES = {
    'default': {
        'ENGINE': 'django.db.backends.sqlite3',
        'NAME': BASE_DIR / 'db.sqlite3',
    }
}

STATIC_URL = '/static/'
MEDIA_ROOT = os.path.join(BASE_DIR, 'uploads')
MEDIA_URL = '/media/'
```

(3) 在工程根目录文件 urls.py 中，添加扩展 URL 和媒体文件 URL。代码如下：

```python
from django.contrib import admin
from django.urls import path, include
from django.conf import settings
from django.conf.urls.static import static

urlpatterns = [
    path('admin/', admin.site.urls),
    path('mdeditor/', include('mdeditor.urls'))
]

if settings.DEBUG:
    # static files (images, css, javascript, etc.)
    urlpatterns += static(settings.MEDIA_URL,
document_root=settings.MEDIA_ROOT)
```

(4) 在模型文件 models.py 中创建数据表，主要实现代码如下所示。

```python
from django.db import models
from mdeditor.fields import MDTextField

class ExampleModel (models.Model):
    name = models.CharField (max_length = 10)
    content = MDTextField ()
```

(5) 在后台文件 admin.py 中注册上面的数据库模型，代码如下：

```python
from django.db import models
from mdeditor.fields import MDTextField

class ExampleModel (models.Model):
    name = models.CharField (max_length = 10)
    content = MDTextField ()
```

通过如下命令根据模型文件创建数据库表：

```
python manage.py makemigrations
python manage.py migrate
```

通过如下命令创建一个后台管理员账号：

```
python manage.py createsuperuser
```

最后通过如下命令运行这个 Django 工程：

```
python manage.py runserver
```

在浏览器中输入 http://127.0.0.1:8000/admin/ 进入后台，成功登录后的效果如图 16-1 所示。

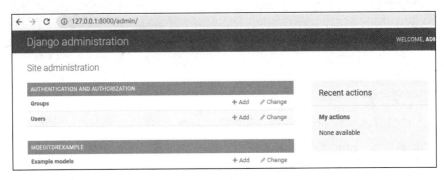

图 16-1　后台主页效果

单击 Example models 选项后面的 Add 链接，在弹出的添加信息页面中会显示 django-mdeditor 富文本编辑器，效果如图 16-2 所示。

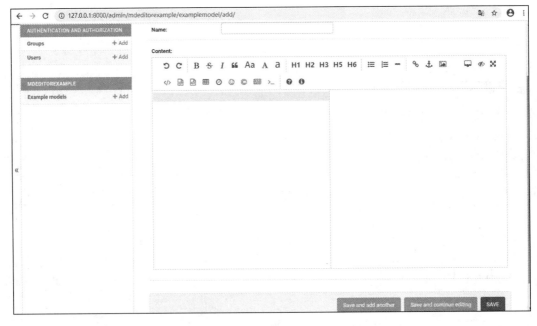

图 16-2　发布信息页面的富文本编辑器

在编辑器中可以上传文件，上传的图片会被保存在"uploads/editor"目录中。

在配置文件 settings.py 中可以设置在编辑器中显示的元素，例如通过下面的代码设置使用指定的编辑器元素，其中没有使用传统格式中的"撤销"和"重做"图标，如图 16-3 所示。

```
MDEDITOR_CONFIGS = {
    'default': {
        'width': '90% ',  # Custom edit box width
        'heigth': 500,  # Custom edit box height
        'toolbar': [
                    "bold", "del", "italic", "quote", "ucwords", "uppercase", "lowercase", "|",
                    "h1", "h2", "h3", "h5", "h6", "|",
                    "list-ul", "list-ol", "hr", "|",
                    "link", "reference-link", "image", "code", "preformatted-text", "code-block", "table", "datetime", "emoji",
                    "html-entities", "pagebreak", "goto-line", "|",
                    "help", "info",
                    "||", "preview", "watch", "fullscreen"],  # custom edit box toolbar
        'upload_image_formats': ["jpg", "jpeg", "gif", "png", "bmp", "webp"],  # image upload format type
        'image_folder': 'editor',  # image save the folder name
        'theme': 'default',  # edit box theme, dark / default
        'preview_theme': 'default',  # Preview area theme, dark / default
        'editor_theme': 'default',  # edit area theme, pastel-on-dark / default
        'toolbar_autofixed': True,  # Whether the toolbar capitals
        'search_replace': True,  # Whether to open the search for replacement
        'emoji': True,  # whether to open the expression function
        'tex': True,  # whether to open the tex chart function
        'flow_chart': True,  # whether to open the flow chart function
        'sequence': True,  # Whether to open the sequence diagram function
        'watch': True,  # Live preview
        'lineWrapping': False,  # lineWrapping
        'lineNumbers': False,  # lineNumbers
        'language': 'zh'  # zh / en / es
    }
}
```

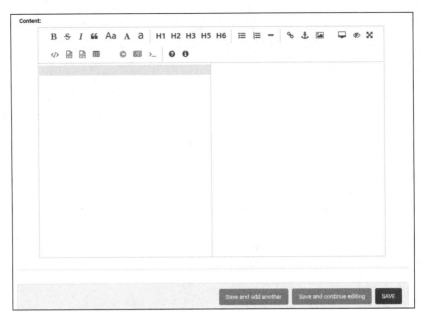

图 16-3　在编辑器中没有"撤销"和"重做"图标

16.2　第三方库 django-ckeditor

　　CKEditor 是全球最优秀的网页在线文字编辑器之一，因其惊人的性能与可扩展性而广泛地被运用于各大网站。第三方库 django-ckeditor 集成了 CKEditor 的功能，可以在 Django 工程中使用 django-ckeditor 实现富文本编辑器功能。

16.2.1　django-ckeditor 介绍

扫码观看本节视频讲解

　　CKEditor 是一款功能强大的开源在线文本编辑器。其所见即所得的特点，使你在编辑时所看到的内容和格式，能够与发布后看到的效果完全一致。CKEditor 完全是基于 JavaScript 脚本语言开发的，因此不必在客户端进行任何安装，并且兼容各大主流浏览器。CKEditor 的前身是 FCKEditor。目前，有很多公司都在使用 CKEditor 作为 Web 编辑的解决方案。

　　如果想在 Django 工程中使用 CKEditor，需要先通过如下命令安装 django-ckeditor。

```
pip install django-ckeditor
```

　　另外还需要安装库 Pillow，这是 Python 的一个图像处理库，django-ckeditor 需要依赖库 Pillow。我们可以通过如下命令安装 Pillow。

```
pip install pillow
```

16.2.2 实战演练：在博客系统中使用 django-ckeditor 富文本编辑器

在下面的实例代码中，演示了在 Django 中使用 django-ckeditor 实现富文本编辑器功能的过程。

> 源码路径：**daima\16\django-ckeditor-tutorial**

(1) 通过如下命令创建一个名为"mysite"的工程，然后在工程目录下新建一个名为"blog"的 app。

```
django-admin.py startproject mysite
cd mysite
python manage.py startapp blog
```

(2) 在配置文件 settings.py 中加入"blog.apps.BlogConfig"和"ckeditor"，然后设置使用 SQLite3 数据库，添加 media 文件的路径，最后在工程的根目录中创建文件夹 "media/uploads"。配置文件 settings.py 的代码如下。

```
INSTALLED_APPS = [
    'blog.apps.BlogConfig',
    'ckeditor',
    'ckeditor_uploader',
    'django.contrib.admin',
    'django.contrib.auth',
    'django.contrib.contenttypes',
    'django.contrib.sessions',
    'django.contrib.messages',
    'django.contrib.staticfiles',
]

WSGI_APPLICATION = 'mysite.wsgi.application'
DATABASES = {
    'default': {
        'ENGINE': 'django.db.backends.sqlite3',
        'NAME': os.path.join(BASE_DIR, 'db.sqlite3'),
    }
}

CKEDITOR_UPLOAD_PATH = 'uploads/'
MEDIA_URL = '/media/'
MEDIA_ROOT = 'media/'
```

(3) 在工程根目录文件 urls.py 中，添加扩展 URL 和媒体文件 URL。代码如下：

```
from django.contrib import admin
from django.urls import include, path
from django.conf.urls.static import static
from django.conf import settings

urlpatterns = [
    path('blog/', include('blog.urls')),
    path('ckeditor/', include(
```

```
            'ckeditor_uploader.urls')),
    path('admin/', admin.site.urls)
] + static(settings.MEDIA_URL,
           document_root=settings.MEDIA_ROOT)
```

(4) 在模型文件 models.py 中导入类 RichTextUploadingField 创建数据表，主要实现代码如下所示。

```
from django.db import models
from ckeditor_uploader.fields import RichTextUploadingField

class Post(models.Model):
    body = RichTextUploadingField(blank=True,
                         config_name='special')
```

(5) 在视图文件 views.py 中创建类 PostDetailView 和 PostCreateView。其中类 PostDetailView 用于实现博客详情视图，指向模板文件 blog/post_detail.html；类 PostCreateView 用于实现发布新博客视图。代码如下：

```
from django.views.generic import DetailView
from django.views.generic.edit import CreateView
from django.urls import reverse
from .models import Post

class PostDetailView(DetailView):
    model = Post
    template_name = 'blog/post_detail.html'

class PostCreateView(CreateView):
    model = Post
    fields = ['body']
    def get_success_url(self):
        return reverse('blog:detail',
                   args=[self.object.pk])
```

(6) 在后台文件 admin.py 中注册上面的数据库模型，使博客信息在后台显示。代码如下：

```
from django.contrib import admin
from .models import Post

admin.site.register(Post)
```

通过如下命令根据模型文件创建数据库表：

```
python manage.py makemigrations
python manage.py migrate
```

通过如下命令创建一个后台管理员账号：

```
python manage.py createsuperuser
```

最后通过如下命令运行这个 Django 工程：

```
python manage.py runserver
```

在浏览器中输入 http://127.0.0.1:8000/admin/ 进入后台，成功登录后的效果如图 16-4 所示。

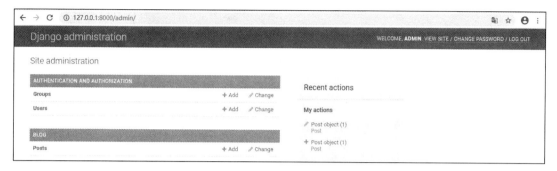

图 16-4　后台主页效果

单击 Posts 选项后的 Add 链接，在弹出的添加信息页面中会显示 django-ckeditor 富文本编辑器，效果如图 16-5 所示。

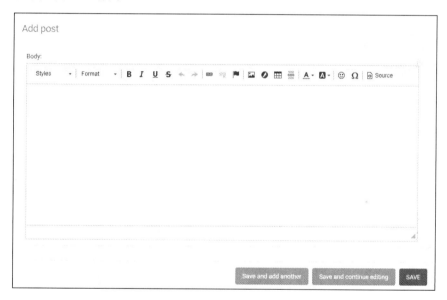

图 16-5　发布信息页面的富文本编辑器

在编辑器中可以上传文件，上传的图片被保存在"media/uploads"目录中。

在配置文件 settings.py 中可以设置在编辑器中显示的元素，例如通过下面的代码设置使用了指定的编辑器元素，其中只设置显示 4 个选项：Bold、Link、Unlink 和 Image，如图 16-6 所示。

```
CKEDITOR_CONFIGS = {
    'default':
        {'toolbar': 'Custom',
         'toolbar_Custom': [
            ['Bold', 'Link', 'Unlink', 'Image'],
        ],
        },
    'special':
        {'toolbar': 'Special', 'height': 500,
```

```
        'toolbar_Special':
          [
            ['Bold', 'Image'],
            ['CodeSnippet'],
          ], 'extraPlugins': 'codesnippet',
      }
}
```

图 16-6　定制的编辑器

第 17 章
综合实战：民宿信息可视化

在本书前面的内容中，已经详细讲解了使用 Django 开发 Web 程序的知识，也通过实例对这些知识点的用法进行了演示。在本章内容中，将讲解一个使用 Django 开发大型综合实例的实现过程。向读者详细讲解爬虫抓取民宿信息的方法，并讲解使用 Django 可视化展示这些民宿信息的过程。

很多工作累了想出来放松放松的游客，不住酒店，而是想通过民宿这个平台，感受一下当地的风土人情，体验一下不同的生活方式，为此市场需要打造"民宿+当地文化"的个性民宿。基于越来越多的人们喜欢住在民宿的市场需求下，分析民宿市场的发展和市场定位变得愈发重要。在本章的内容中，将通过一个综合实例的实现过程，详细讲解爬虫抓取民宿信息的方法，并讲解可视化分析这些民宿信息的过程。

17.1 系统背景介绍

绿水青山就是金山银山。近年来，随着国家建设美丽乡村政策的实施，各地纷纷加大对特色小镇的建设力度，相继出台对民宿的补贴扶持方案。之所以选择从事民宿行业，大部分人是因为自己喜欢旅行，也有"隐于野"的诗意情结，他们或是放弃了稳定的工作，或是逃离了大都市的生活，希望能通过民宿传递自己的生活理念。

扫码观看本节视频讲解

在新的消费观念下，越来越多的客人已经厌倦了千篇一律的酒店住宿形态。在以后的旅游过程中，游客更希望体验多样化的住宿形态，深入当地的特色文化中。大众化景点路线会越来越被轻视，个性化住宿、个性化旅游线路的选择会带给客栈民宿更多的发展机遇。

作为一种新兴的非标准住宿业态，民宿对传统标准酒店住宿业起到明显的补充作用。目前美团民宿交易额约占美团酒店交易额的 4.8%，且整体呈现上升趋势。从各省份民宿交易额看，广东省民宿交易份额占据首位，交易额占全国市场的 11.6%，交易额排前 10 位的省市依次为广东省、北京市、四川省、江苏省、山东省、陕西省、重庆市、上海市、浙江省、湖北省，上述十省市交易额占全国民宿市场交易额比例超过 65%。

数据显示，2017 年民宿预订以女性消费者为主，占比 55.7%。从民宿产品用户年龄层分布来看，40 岁以下人群占整体消费者比例达到 86.2%，可见国内民宿产品受众偏向年轻化。其中，"90 后"是民宿消费的主力军，"90 后"消费者的订单量占比约 58.9%，"80 后"占比约 27.3%。从消费品类偏好看，用户在住民宿期间，同时消费餐饮品类的比例约占 30.8%，同时消费非餐饮品类的比例约占 28.2%。这说明民宿消费对其他品类的消费具有一定的带动作用。

在民宿市场大发展的前提下，可视化分析民宿市场的发展现状对商家来说具有重要的意义。另外，对于消费者来说，也可以通过可视化系统及时了解民宿行情，帮助自己取得更加物美价廉的服务。

17.2 爬虫抓取信息

本项目将使用 Scrapy 作为爬虫框架，使用代理 IP 爬取业内知名民宿网中的数据信息，然后将爬取的信息保存到 MySQL 数据库中。最

扫码观看本节视频讲解

后使用 Django 可视化展示数据库中保存的民宿数据信息。在本节的内容中，将首先讲解爬虫功能的具体实现过程。

17.2.1 系统配置

在 Django 模块中设置整个项目的配置信息，在文件 settings.py 中设置数据库和缓存等配置信息，主要代码如下所示。

```
# 配置 MySQL 数据库
DATABASES = {
    'default': {
        'ENGINE': 'django.db.backends.mysql',
        'NAME': "scrapy_django",
        'USER': 'root',
        'PASSWORD': '66688888',
        'OPTIONS': {
                'charset':'utf8mb4',
                #"init_command":SET foreigh_key_chenks=0;"
            },      #值编码方式，避免 emoji 无法存储
        'HOST': "127.0.0.1",           #IP 地址
        'PORT': '3306'                 #端口
    }
}
CACHES = {  # redis 做缓存
    'default': {
        'BACKEND': 'django_redis.cache.RedisCache',
        "LOCATION": "redis://127.0.0.1:6379/3",  # 本机 django 的 redis 缓存路径
        # 'LOCATION':"redis://127.0.0.1:6378/3",
        'OPTIONS':{
            "CLIENT_CLASS":"django_redis.client.DefaultClient",
        }
    }
}
```

17.2.2 Item 处理

Scrapy 为我们提供了 Item 类，这些 Item 类可以让我们自己来指定字段。比如在某个 Scrapy 爬虫项目中定义了一个 Item 类，在这个 Item 类里面包含了 title、release_date、url 等，然后通过各种爬取方法抓取这些字段的信息，再通过 Item 类进行实例化。这样做的好处是抓取的信息条理分明，不容易出错，因为我们在一个地方统一定义过字段，而且这个字段具有唯一性。在本项目实例文件 items.py 中设置了 4 个 ORM 对象，这 4 个对象和本项目数据库字段是一一对应的。文件 items.py 的具体实现代码如下所示。

```
class HouseItem(DjangoItem):
    django_model = House
    jsonString = scrapy.Field()     # 这里需要增加临时字段，以便把多个其他对象的属性一次性传过来
```

```python
class HostItem(DjangoItem):
    django_model = Host

class LabelsItem(DjangoItem):
    django_model = Facility

class FacilityItem(DjangoItem):
    django_model = Labels

class CityItem(DjangoItem):
    django_model = City

class urlItem(scrapy.Item):  # master 专用 item
    # define the fields for your item here like:
    url = scrapy.Field()
```

17.2.3 具体爬虫

编写文件 hotel.py，功能是实现具体的网络爬虫功能，具体实现流程如下所示。

（1）创建类 HotwordspiderSpider，设置爬虫项目的名字是"hotel"，然后分别设置爬虫的并发请求数、延时、最大的并发请求数量、保存数据的方式和使用的代理等信息。

（2）设置爬虫网页 HTTP 请求协议的请求报文(Request Headers)。

（3）编写函数 regexMaxNum()获取页数，使用正则表达式返回匹配到的最大的数字就是页数。函数 regexMaxNum()的实现代码如下：

```python
def regexMaxNum(self,reg,text):
    temp = re.findall(reg,text)
    return max([int(num) for num in temp if num != ""])
```

（4）编写函数 start_requests()，功能是设置爬虫启动时要爬取的城市列表。

（5）编写函数 getRSXFPrice()，功能是提取爬虫数据中的价格信息。

（6）编写函数 detail()，功能是获取每个民宿的详细信息，包括面积、标签、标题、地址、房型、位置、城市、留言数量等信息。

（7）编写函数 detail()展示某个民宿的详细信息，其中抓取民宿价格的操作比较难以实现，因为民宿网对价格进行了数据加密，所以需要专门的逻辑来破解这个反扒机制。

（8）在函数 detail()中提取留言回复信息。

（9）在函数 detail()中提取促销和普通标签的信息。

（10）在函数 detail()中提取评价信息和回复率信息。

（11）在函数 detail()中提取好评平均分信息。

17.2.4 破解反扒字体加密

在网站中的价格信息是加密的，为了获取每个民宿的价格信息，需要对".woff"格式的加密字体进行破解。编写文件 parseTool.py，功能是破解".woff"格式的价格信息，主要实现代码如下所示。

```python
# 获得 j-gallery 这段的字符串
def getFontUrl(UserJson):
    j_gallery_text = UserJson              # 准备过滤
    UserJson = j_gallery_text.replace("'",'"').replace("\\","")    # 过滤斜杠
    test = re.findall('(?<=cssPath\\"\\:\\").*?(?=\\}\\,)',UserJson)[0]

    print()
    wofflist = re.findall('(?<=\\(\\").*?(?=\\)\\;)',test)    # j-gallery 字段处理
#  print(wofflist)
    print()
    font_url = ''
    for woffurl in wofflist:
        if woffurl.find("woff")!=-1:
            tempwoff = re.findall('(?<=\\").*?(?=\\")',woffurl)
#  print(tempwoff)
            for j in tempwoff:
                if j.find("woff")!=-1:
                    print("https:"+j)
                    font_url = "https:"+ j

    # 提取字体成功
#  print(font_url)
    return font_url

def download_font(img_url,imgName,path=None):
    headers = {'User-Agent':"Mozilla/5.0 (Windows NT 6.1; WOW64) AppleWebKit/537.1 (KHTML, like Gecko) Chrome/22.0.1207.1 Safari/537.1",
               }  ##浏览器请求头(大部分网站没有这个请求头会报错,请务必加上)
    try:
        img = requests.get(img_url, headers=headers)
        dPath = os.path.join("woff",imgName)    # 传递 imgName
        # print(dPath)
        print("字体的文件名 "+dPath)
        f = open(dPath, 'ab')
        f.write(img.content)
        f.close()
        print("下载成功")
        return dPath
    except Exception as e:
        print(e)


# 从字体文件中获得字形数据用来备用待对比
def getGlyphCoordinates(filename):
    """
    获取字体轮廓坐标,手动修改 key 值为对应数字
    """
    font = TTFont("woff/"+f'{filename}')    # 自动带上了 woff 文件夹
    # font.saveXML("bd10f635.xml")
    glyfList = list(font['glyf'].keys())
    data = dict()
    for key in glyfList:
```

```python
            # 剔除非数字的字体
            if key[0:3] == 'uni':
                data[key] = list(font['glyf'][key].coordinates)
    return data


def getFontData(font_url):
    # 获取字体信息
    filename = os.path.basename(font_url)
    font_data = None
    if os.path.exists("woff/"+filename):
        # 直接读取
        font_data = getGlyphCoordinates(filename)  # 读取时自带 woff 文件夹
    else:
        # 先下载再读取
        download_font(font_url, filename, path=None)
        font_data = getGlyphCoordinates(filename)
    if font_data == None:
        print("字题文件读取出错,请检查")
    else:
        # print(font_data)
        return font_data

# 自动分割大写形式的价格
def splitABC(price_unicode):
    raw_price = price_unicode.split("&amp;")
    temp_price_unicode = []
    for x in raw_price:
        if x != "":
            temp_price_unicode.append(x.upper().replace("#X", "").replace(";", ""))
    return temp_price_unicode  #提取出简化大写的价格,例如 400 是原价,折扣价是 280

def getBothSplit(UserJson):
    UserJson = UserJson.replace("\\", "").replace("'", '"')
    result_price = []
    result_discountprice = []
    try:
        price_unicode = re.findall('(?<=price\\"\\:\\").*?(?=\\"\\,)', UserJson)[0]
        result_price = splitABC(price_unicode)
    except Exception as e:
        print("没有找到价格")
        print(e)

    try:  # 可能没有找到表示价格的数字,那就会有乱码符号
        discountprice_unicode = re.findall('(?<=discountPrice\\"\\:\\").*?(?=\\"\\,)', UserJson)[0]
        result_discountprice = splitABC(discountprice_unicode)
    except Exception as e:
        print("没有找到折扣价")
        print(e)
    if result_discountprice == [] and result_price != []:
        result_discountprice = result_price  # 如果折扣价为 0,那么就等于原价
    return result_price, result_discountprice  # 没有折扣时返回的价格编码
```

```
def pickdict(dict):    # 序列化这个字典
    with open(os.path.join(os.path.abspath('.'),"label_dict.pickle"), "wb") as f:
        pickle.dump(dict, f)
```

17.2.5 下载器中间件

下载器中间件是在引擎及下载器之间的特定钩子(specific hook)，用于处理 Downloader 下载器传递给引擎的 response(也包括引擎传递给下载器的 Request)。其提供了一个简便的机制，通过插入自定义代码来扩展 Scrapy 功能。在本项目中的下载器中间件文件 middlewares.py 中，主要实现了在线代理 IP 功能。具体实现流程如下所示。

(1) 准备好基础工作，先创建类 EnvironmentIP 和 EnvironmentFlag，对应实现代码如下所示。

```
class EnvironmentIP:                    # 设置一个全局变量，单例模式
    _env = None
    def __init__(self):
        self.IP = 0                     # 存储 IP 属性

    @classmethod
    def get_instance(cls):
        """
        返回单例 Environment 对象
        """
        if EnvironmentIP._env is None:
            cls._env == cls()
        return cls._env

    def set_flag(self, IP):  # 里面放的是数字
        self.IP = IP

    def get_flag(self):
        return self.IP

envVarIP = EnvironmentIP()  # 是否切换使用的代理

class EnvironmentFlag:  # 设置一个全局变量，单例模式
    _env = None
    def __init__(self):
        self.flag = False    # 默认不使用代理

    @classmethod
    def get_instance(cls):
        """
        返回单例 Environment 对象
        """
        if EnvironmentFlag._env is None:
            cls._env == cls()
        return cls._env
```

```python
    def set_flag(self, flag):
        self.flag = flag

    def get_flag(self):
        return self.flag

envVarFlag = EnvironmentFlag()  # 这个变量用于设置是否切换使用代理

class Environment:  # 设置一个全局变量，单例模式
    _env = None
    def __init__(self):
        self.countTime = datetime.datetime.now()

    @classmethod
    def get_instance(cls):
        """
        返回单例 Environment 对象
        """
        if Environment._env is None:
            cls._env == cls()
        return cls._env

    def set_countTime(self, time):
        self.countTime = time

    def get_countTime(self):
        return self.countTime

envVar = Environment()  # 初始化一个默认的 Environment 实例
```

(2) 定义类 RandomUserAgent 实现随机生成 IP 功能，通过函数 process_request()和 process_response()及时获取响应信息，这样可以判断这个 IP 是否可用。对应实现代码如下所示。

```python
class RandomUserAgent(object):  # ua 中间件
    # def __init__(self):

    @classmethod
    def from_crawler(cls, crawler):
        s = cls()
        crawler.signals.connect(s.spider_opened, signal=signals.spider_opened)
        return s

    def process_request(self, request, spider):
        ua = UserAgent()
        print(ua.random)
        request.headers['User-Agent'] = ua.random
        return None

    def process_response(self, request, response, spider):
        # 使用从 downloader 获取的响应
        print(f"请求的状态码是 {response.status}")
        print("调试 ing")
```

```python
        print(request.url)
        HTML = response.body.decode("utf-8")
        # print(HTML)
        print(HTML[:200])
        try:
            # print('进来中间件调试')
            if HTML.find("code")!=-1:
                if re.findall('(?<=code\\"\\:).*?(?=\\,)',HTML)[0]=='406':
                    # 转义保留双引号
                    print("正在重新请求(网络不好)")
                    return request
        except Exception as e:
            print(request.url)
            print(e)

        try:
            temp = json.loads(HTML)
            if temp['code'] == 406:  #
                print("正在重新请求(网络不好)状态码406")
                request.meta["code"] = 406
                return request      # 重新发给调度器，重新请求
        except Exception as e:
            print(e)
        return response

    def process_exception(self, request, exception, spider):
        pass

    def spider_opened(self, spider):
        spider.logger.info('Spider opened: %s' % spider.name)
```

(3) 定义类 proxyMiddleware 实现在线代理 IP 功能，创建 redis 代理连接池，用列表 remote_iplist 中的 IP 轮询访问，并打印输出对应的响应信息。对应实现代码如下所示。

```python
class proxyMiddleware(object):  # 代理中间件
    # 这里是使用代理ip
    # MYTIME = 0  # 类变量用来设定切换代理的频率

    def __init__(self):
        # self.count = 0
        from redis import StrictRedis, ConnectionPool
        # 使用默认方式连接到数据库
        pool = ConnectionPool(host='localhost', port=6378, db=0,password='Zz123zxc')
        self.redis = StrictRedis(connection_pool=pool)

    @classmethod
    def from_crawler(cls, crawler):
        # 这个方法被Scrapy用来创建我们的爬虫
        s = cls()
        crawler.signals.connect(s.spider_opened, signal=signals.spider_opened)
        return s
```

```python
    def get_proxy_address(self):
        proxyTempList = list(self.redis.hgetall("useful_proxy"))
        # proxyTempList = list(redis.hgetall("useful_proxy"))
        return str(random.choice(list(proxyTempList)), encoding="utf-8")

    def process_request(self, request, spider):
        # 这里是用来代理的
        remote_iplist = ['125.105.70.77:4376', '58.241.203.162:4386',
'117.5.181.109:4358', '14.134.186.95:4372', '125.111.150.25:4305',
'122.246.173.161:4375']

        print()
        print("proxyMiddleware")
        now = datetime.datetime.now()
        print("flag")
        print("time")
        print(f"现在时间{now}")
        print(f"变量内时间{envVar.get_countTime()}")
        print("变量状态{True}才使用代理")
        print(envVarFlag.get_flag())
        print("相减后的结果")
        print((now-envVar.get_countTime()).seconds / 40)
        if envVarFlag.get_flag() ==True:
            if (now-envVar.get_countTime()).seconds / 20 >= 1:
                envVarFlag.set_flag(not envVarFlag.get_flag())  # 切换为使用代理
                envVar.set_countTime(now)
            print("使用代理池中的ip")
            proxy_address = None
            try:
                proxy_address = self.get_proxy_address()
                if proxy_address is not None:
                    print(f'代理IP -- {proxy_address}')
                    request.meta['proxy'] = f"http://{proxy_address}"  # 如果出现
了302错误，有可能是因为代理的类型不对
                else:
                    print("代理池中没有代理ip存在")
            except Exception as e:
                print("检查到代理池里面已经没有ip了,使用本地")
        else:  # 不使用代理,这里轮流使用本地ip和外面的ip
            if (now-envVar.get_countTime()).seconds / 40 >= 1:  # 传进来的是切换状
态的时间
                envVarFlag.set_flag(not envVarFlag.get_flag())  # 切换为使用代理
                envVar.set_countTime(now)

            if envVarIP.get_flag() <= len(remote_iplist)-1:    ## 直接本地IP
                remoteip = remote_iplist[envVarIP.get_flag()]
                print(f'使用远程ip -- {remoteip}')
                request.meta['proxy'] = f"http://{remoteip}"
                envVarIP.set_flag(envVarIP.get_flag()+1)
            else:
                envVarIP.set_flag(0)   # 把这个ip设置成0,这个是使用本地的ip
                print("使用到本地ip")
            pass
```

17.2.6 保存爬虫信息

编写实例文件 pipelines.py，功能是将爬取的民宿房源信息保存到本地数据库中。具体实现流程如下所示。

(1) 编写类 urlItemPipeline，保存房源的 URL 信息，对应实现代码如下所示。

```python
class urlItemPipeline(object):                    # master专用管道
    def __init__(self):
        self.redis_url = "redis://Zz123zxc:@localhost:6379/"  # master端是本地redis的
        self.r = redis.Redis.from_url(self.redis_url,decode_response=True)

    def process_item(self, item, spider):
        if isinstance(item, urlItem):
            print("urlItem item")
            try :
                # item.save()
                self.r.lpush("Meituan:start_urls",item['url'])
            except Exception as e:
                print(e)
        return item
```

(2) 编写类 cityItemPipeline，保存房源的城市信息，对应实现代码如下所示。

```python
class cityItemPipeline(object):
    def process_item(self, item, spider):
        if isinstance(item, CityItem):
            print("CityItem item")
            try :
                item.save()
            except Exception as e:
                print(e)
        return item
```

(3) 编写类 houseItemPipeline，保存房源的详细信息，主要包括 Labels、Facility 和 Host 等信息。对应实现代码如下所示。

```python
class houseItemPipeline(object):
    def __init__(self):
        pass

    def process_item(self, item, spider):
        if isinstance(item, HouseItem):
            print("HouseItem item")
            house = None
            try:
                house = item.save()     # 保存信息
                print("house 保存成功")

            except Exception as e:
                print("house 保存失败，后面的跳过保存")
                print(e)
```

```python
            print(item)
            return item

        jsonString = item.get("jsonString")
        labelsList = jsonString['Labels']
        facilityList = jsonString['Facility']
        hostInfos = jsonString['Host']

        # house = House.objects.filter(**{'house_id':
item.get("house_id"),"house_date":item.get("house_date")}).first()
        # 查询一次就可以，多条件查询，查询两个联合的主键
        #查询房源的具体信息,多条件查询
        for onetype in labelsList:  # 1.查询 labelsList 中的每一个标签
            for one in labelsList[onetype]:  # one 表示一个标签
                try:  # 验证查询标签是否已经存在
                    label = Labels.objects.filter(**{'label_name':
one[0],"label_desc":one[1]})  # 找到的话就直接加入另一个 Meiju 对象中
                    # print("长度")
                    if len(label) == 0:
                        # print("需要创建后添加")
                        l = Labels()
                        l.label_name = one[0]
                        l.label_desc = one[1]
                        # print("检查 label")
                        # print(f"onetype:{onetype}")
                        # print(one)
                        if onetype == "1":  # 优惠标签
                            l.label_type = 1  # 标签类型
                        else:
                            l.label_type = 0
                        l.save()
                        house.house_labels.add(l)  # 添加标签
                    else:
                        # print("找到有直接添加")
                        # print(label)
                        house.house_labels.add(label.first())
                        # print("添加成功")
                except Exception as e:
                    print(e)
                    print("label 已存在，跳过插入")
                    # print(e)


        # 写入 Facility
        # print(facilityList)
        # 遍历 facilityList 列表
        for facility in facilityList:
            if 'metaValue' in facility:
                print()  # 执行先检查后添加
                try:  # 先验证是否存在，然后把已经有的添加进来
                    fac = Facility.objects.filter(**{'facility_name':
facility['value']})  # 找到后添加到另一个 Meiju 对象中
                    # print("长度")
                    if len(fac) == 0:
```

```python
                    # print("需要创建后添加")
                    l = Facility()
                    l.facility_name = facility['value']
                    l.save()
                    house.house_facility.add(l)
                else:
                    # print("找到有直接添加")
                    house.house_facility.add(fac.first())  # 添加到第一个位置
                    # print("添加成功")
            except Exception as e:
                print("facility已经存在，跳过插入")
                # print(e)

    print("下面开始host信息的添加")
    '''{'hostId': '36438164',
        'host_RoomNum': '51',
        'host_commentNum': '991',
        'host_name': '店家1',
        'host_replayRate': '100'}'''
    print(hostInfos)

    try:    # 先查询是否已经有这条信息，然后把已经有的添加进来
        hosts = Host.objects.filter(**{
            'host_id':hostInfos['hostId'],
            'host_updateDate':house.house_date})   # 找到的话就直接加入另一个
Meiju对象中
        print("长度")
        print("输出查找到的结果")
        print(hosts)
        try:
            # print("需要创建后添加")
            l = Host()
            l.host_name = hostInfos['host_name']
            l.host_id = hostInfos['hostId']
            l.host_RoomNum = hostInfos['host_RoomNum']
            l.host_commentNum = hostInfos['host_commentNum']
            l.host_replayRate = hostInfos['host_replayRate']
            l.save()    # 保存不成功会自然进行处理
            # fac = Facility.objects.filter(**{'label_name':
facility['value']})   # 找到的话就直接添加到另一个Meiju对象中
            # print("__label_")
            # print(l)
            house.house_host.add(l)
        except Exception as e:
            print(e)
        # else:
            # print("不为空找到了")
            # print("__label_")
            # print(fac)
            house.house_host.add(hosts.first())  # 添加到第一个位置
            print("已有的情况下添加成功")
    except Exception as e:
        print("已有这个host跳过插入")
        print(e)
```

```
        return item    # 返回 item 信息
    return item
```

通过如下命令运行爬虫程序：

```
scrapy crawl hotel
```

爬虫的数据被保存在 MySQL 数据库中，如图 17-1 所示。

图 17-1　数据库中的爬虫数据

17.3　数据可视化

本项目使用 Django 框架实现可视化功能，提取在 MySQL 数据库中保存的民宿数据信息，然后使用 Echarts 实现数据可视化功能。在本节的内容中，将详细讲解实现数据可视化功能的具体过程。

17.3.1　数据库设计

扫码观看本节视频讲解

编写文件 models.py 实现数据库模型设计功能，在此文件中每个类和 MySQL 数据库中的表一一对应，每个变量和数据库表中的字段一一对应。代码如下：

```
class City(models.Model):
    city_nm = models.CharField(max_length=50,unique=True)     # 城市名字
    city_pynm = models.CharField(max_length=50,unique=True)   # 减少冗余的代价是
时间代价
    city_statas = models.BooleanField(default=False)
    # 这个是让爬虫选择是否进行爬取的城市(缩小爬取范围才可以全部爬下来)
```

```python
    def __str__(self):
        return self.city_nm + " " + self.city_pynm


class Labels(models.Model):
    TYPE_CHOICE = (
        (0, "普通标签"),
        (1, "优惠标签"),
    )
    label_type = models.IntegerField(choices=TYPE_CHOICE)  # 类型1为营销,0为默认标签
    label_name = models.CharField(max_length=171,unique=True)
    label_desc = models.CharField(max_length=171,unique=True)  # 减少冗余的代价是时间代价

    def __str__(self):
        return str(self.label_type)+" "+self.label_name+" "+self.label_desc

    class Meta:
        # 设置约束
        unique_together = ('label_name',"label_desc")


class Host(models.Model):
    '''自己会自动创建一个id的'''
    host_name = models.CharField(max_length=171)  # 房东名字
    host_id = models.IntegerField()  # 房东id
    host_replayRate = models.IntegerField(default=0)  # 回复率
    host_commentNum = models.IntegerField(default=0)  # 评价总数,会变也要存,这样可以看变化率
    host_RoomNum = models.IntegerField(default=0)  # 不同时间段的房子含有数量
    host_updateDate = models.DateField(default=timezone.now)  # 自动创建时间,不可修改

    def __str__(self):
        return str(self.host_id)+ " "+self.host_name

    class Meta:
        # 新房子在一天最多显示一个价格数据
        unique_together = ('host_id',"host_updateDate")

class Facility(models.Model):
    ''' 设施类型,本系统设置了85个设施
        todo django 建议不用外键约束,因为这样可以更高效
    '''
    facility_name = models.CharField(max_length=50,unique=True)  # 设施名字

    def __str__(self):
        return self.facility_name


# 表示酒店公寓的类model
class House(models.Model):
```

```python
'''
house_id 为主键
'''

    house_img = models.CharField(max_length=171,default="static/media/default.jpg")
# 设置房子的默认预览图

    house_id = models.IntegerField(default=0)    # 房子编号
    house_cityName = models.CharField(max_length=50,default="未知城市")
    house_title = models.CharField(max_length=171)   # 标题
    house_url = models.CharField(max_length=171)  # 地址信息
    house_date = models.DateField(default=timezone.now)   # 爬取时间
    house_firstOnSale = models.DateTimeField(default=datetime.datetime(1770, 1, 1, 1, 1, 1, 499454))  # 发布时间
    # 用户评价
    house_favcount = models.IntegerField(default=0)  # 房子页面的点赞数
    house_commentNum = models.IntegerField(default=0)   # 评分人数(也是评论人数)

    house_descScore = models.FloatField(default=0)   # 房子四个分数
    house_talkScore = models.FloatField(default=0)
    house_hygieneScore = models.FloatField(default=0)
    house_positionScore = models.FloatField(default=0)
    house_avarageScore = models.FloatField(default=0)   # 总的平均分, 5.0满分, 0分当成未评价

    # 房子的具体内容信息
    house_type = models.CharField(max_length=50,default="未分类")   # 整套/单间/合住
    house_area = models.IntegerField(default=0)    # 房子的面积单位 m²
    house_kitchen = models.IntegerField(default=0)   # 厨房数量, 0 就是没有
    house_living_room = models.IntegerField(default=0)  # 客厅数量
    house_toilet = models.IntegerField(default=0)   # 卫生间数量
    house_bedroom = models.IntegerField(default=0)  # 卧室数量
    house_capacity = models.IntegerField()   # 可以容纳的人数
    house_bed = models.IntegerField(default=1)   # 床的数量

    # 房子的价格信息
    house_oriprice = models.DecimalField(max_digits=16,decimal_places=2)  # 刚发布价格
    house_discountprice = models.DecimalField(max_digits=16,decimal_places=2,default=0.00)
    # 折扣价格,如果没有discountPrice,那么表示原价就是折扣价

    # 房源位置
    house_location_text = models.CharField(max_length=171) # 因为使用utf8mb4格式,char最长为171,四个字节为一个字符
    house_location_lat = models.DecimalField(max_digits=16,decimal_places=6) # 纬度,小数点后6位
    house_location_lng = models.DecimalField(max_digits=16,decimal_places=6)  # 经度

    #房源设施
    house_facility = models.ManyToManyField(Facility)   # 房屋内的设施信息

    #房东信息
```

```python
    house_host = models.ManyToManyField(Host)  # 多对多

    #普通标签和优惠标签
    house_labels = models.ManyToManyField(Labels)

    earliestCheckinTime = models.TimeField(default="00:00")  # 可以最早入住的时间

    def __str__(self):
        return str(self.house_id)+":" +f"{str(self.house_cityName)}" + ":"+ f"{str(self.house_title[0:15])}..." + ":"+str(self.house_oriprice) + "¥/晚"

    class Meta:
        # 约束
        unique_together = ('house_id',"house_date")  # 一个房子一天最多一个价格
# class TestUser(models.Model):
#     account = models.IntegerField(default=0)
#     password = models.CharField(max_length=171)

class Favourite(models.Model):  # 收藏夹
    user = models.OneToOneField(User,unique=True,on_delete=models.CASCADE)
    #   # fav_house = models.CharField(max_length=50,unique=True)    # 城市名字

    fav_city = models.ForeignKey(City, on_delete=models.CASCADE)    # 设置系统偏好城市，默认是广州
    fav_houses = models.ManyToManyField(House)

    def __str__(self):
        return str(self.user.username) + ":" + str(self.fav_city)
```

17.3.2 视图显示

在本项目中，数据可视化功能是通过 View 视图文件和模板文件实现的。本项目的 View 视图文件是 drawviews.py，具体实现流程如下所示。

（1）编写函数 bar_base() 获取系统内的爬虫数量，分别显示房源总数和城市数量。代码如下：

```python
def bar_base() -> Bar:    # 返回给前端用来显示图的 json 设置,按城市分组来统计数量
    nowdate = time.strftime('%Y-%m-%d', time.localtime(time.time()))
    count_total_city = House.objects.filter(house_date=nowdate).values("house_cityName").annotate(
        count=Count("house_cityName")).order_by("-count")
    # for i in count_total_city:
    #     print(i['house_cityName']," ",str(i['count']))
    c = (
        Bar(init_opts=opts.InitOpts(theme=ThemeType.WONDERLAND))
            .add_xaxis([city['house_cityName'] for city in count_total_city])
            .add_yaxis("房源数量", [city['count'] for city in count_total_city])
            .set_global_opts(title_opts=opts.TitleOpts(title="今天城市房源数量",subtitle="如图"),
```

```
xaxis_opts=opts.AxisOpts(axislabel_opts=opts.LabelOpts(rotate=-90)),
                )
        .set_global_opts(
            datazoom_opts={'max_': 2, 'orient': "horizontal", 'range_start': 10,
'range_end': 20, 'type_': "inside"})
        .dump_options_with_quotes()
    )
    return c
```

(2) 编写类 PieView 统计数据库中的房型数据并绘制饼图。代码如下：

```
class PieView(APIView):
    def get(self, request, *args, **kwargs):
        result = fetchall_sql(
            "select house_type,count(house_type) from (select distinct 
house_id ,house_type from hotelapp_house  group by house_id,house_type ) hello 
group by house_type")
        c = (
            Pie()
            .add("", [z for z in zip([i[0] for i in result], [i[1] for i in 
result])])
            # .add("",[list(z) for z in zip([x['house_type'] for x in 
house_type_count],[x['count'] for x in house_type_count])])
            .set_global_opts(title_opts=opts.TitleOpts(title="总房屋类型"))
            .set_series_opts(label_opts=opts.LabelOpts(
                formatter="{b}: {c} | {d}%",
            ))
            .dump_options_with_quotes()
        )
        return JsonResponse(json.loads(c))
```

(3) 编写类 getMonthPostTime、getMonthPostTime2 和 timeLineView，获取数据库中保存的发布时间信息，并绘制发布时间折线图和最近 7 天的折线图。

(4) 编写类 drawMap 绘制房源分布热力图。代码如下：

```
class drawMap(APIView):  # 要加 apiview  # 美团房源数量热力图
    def get(self, request, *args, **kwargs):
        from pyecharts import options as opts
        from pyecharts.charts import Map
        from pyecharts.faker import Faker

        result = cache.get('house_city', None)  # 使用缓存实现共享
        if result is None:  # 如果在缓存中无相关数据，则在数据库查询数据
            print("使用缓存统计有房源的城市")
            result = fetchall_sql(
                """select house_cityName,count(house_cityName) as count from 
(SELECT distinct(house_id),house_cityName FROM  hotelapp_house) hello group by 
house_cityName""")
            cache.set('house_city', result, 3600 * 12)  # 设置缓存

        else:
            pass
        c = (
            Geo()
```

```
        .add_schema(maptype="china")
        .add(
            "房源",
            [z for z in zip([i[0] for i in result], [i[1] for i in result])],
            type_=ChartType.HEATMAP,
        )
        .set_series_opts(label_opts=opts.LabelOpts(is_show=False))
        .set_global_opts(
            visualmap_opts=opts.VisualMapOpts(),
            title_opts=opts.TitleOpts(title="美团民宿房源热力图"),
        )
        .dump_options_with_quotes()
    )
    return JsonResponse(json.loads(c))  # 返回JSON结果
```

为了节省本书篇幅，本项目介绍到此为止，有关本项目的具体实现流程，请参考本书的配套源码。执行本项目可视化模块后的效果如图 17-2 所示。

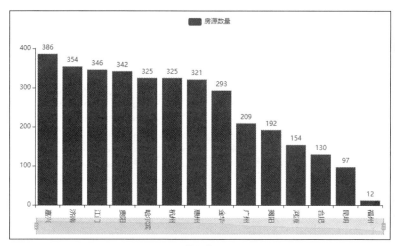

(a) 数据概览

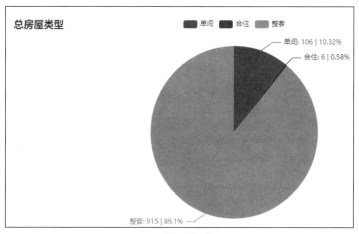

(b) 房屋类型

图 17-2 可视化执行效果

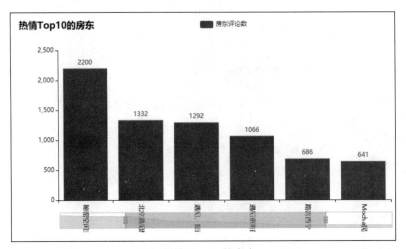

(c) 热情 Top10 的房东

(d) 房屋设施分析

(e) 搜索房源

图 17-2 可视化执行效果(续)